Edgar March Crookshank

A Textbook of Bacteriology

Third Edition

Edgar March Crookshank

A Textbook of Bacteriology
Third Edition

ISBN/EAN: 9783744689465

Printed in Europe, USA, Canada, Australia, Japan

Cover: Foto ©berggeist007 / pixelio.de

More available books at **www.hansebooks.com**

TEXT-BOOK

OF

BACTERIOLOGY

BY

CARL FRAENKEL, M.D.

Professor of Hygiene, University of Königsberg.

THIRD EDITION

TRANSLATED AND EDITED BY

J. H. LINSLEY, M.D.

Professor of Pathology and Bacteriology, Medical Department of the University of Vermont; Demonstrator of Pathology and Bacteriology, New York Post-Graduate Medical School and Hospital; Pathologist to the New York Post-Graduate Hospital and the New York Infant Asylum.

NEW YORK

WILLIAM WOOD AND COMPANY

1891

ELECTROTYPED AND PRINTED BY
THE PUBLISHERS' PRINTING COMPANY
30 & 32 WEST 13TH STREET
NEW YORK

PREFACE.

At the present time there is no work that deserves the name of a text-book on bacteriology published in the English language on this side of the Atlantic.

The want of such a book has long been felt.

While a student in Professor Koch's laboratory at the Hygienic Institute in Berlin the past summer, the third edition of "Fraenkel's Grundriss der Bakterienkunde" was recommended to me as a text-book.

It is now published in six different languages.

In addition to this testimony to the merit of the work, I heard many favorable comments on it by teachers in several different laboratories on the Continent which I visited during the summer of 1890.

A few changes have been made in the book and the presentation of the matter changed from the lecture style.

I append some extracts taken from the preface to the first German edition.

The book "does not claim complete or exhaustive discussion of the subject and presents no statements from literature. . . . Only such facts and observations have been given as were examined by myself. . . . I have at all times been assisted by Dr. Koch's weighty advice, and am thus fortunate enough to know that my views are in complete harmony with those of the master of recent bacteriology. Being convinced that this circumstance will enhance the value of the work, I avail myself of this opportunity to express my heartfelt gratitude to my revered instructor and chief."

Jo. H. Linsley, M.D.,
39 Gramercy Park.

April, 1891.

CONTENTS.

CHAPTER I.

CHAPTER II.

CHAPTER III.

CHAPTER IV.

CHAPTER V.

NON-PATHOGENIC BACTERIA.

CHAPTER VI.

PATHOGENIC BACTERIA.

CHAPTER VII.

APPENDIX.

TEXT-BOOK OF BACTERIOLOGY.

CHAPTER I.

Classification, Morphology, Biology, Structure, Formation of Groups, Flagella, Powers of Resistance of Spores, Multiplication of Bacteria, Sporulation, Varieties of Spore-formation, Ptomaines, Development of Pigment, Formation of Gases. Production of Odors.

I. CLASSIFICATION, MORPHOLOGY, BIOLOGY.

SINCE Anthony Van Leeuwenhoek (1683), with his single lenses, first saw bacteria in mucus of the mouth, and illustrated his discovery with excellent drawings, progress in the field of bacteriology has been exceedingly slow until the middle of the present century.

Even Ehrenberg, who studied the bacteria more closely, knew little else to do with them except to classify them. In fact, he considered them to be the lowest members of the animal kingdom, and thought he saw in them little bladders, which represented stomachs and eggs.

Ferdinand Cohn, shortly before 1860, showed clearly that they belonged to the vegetable kingdom, remarking that the individuals grow and divide like plant cells; that they agree with these in structure, and particularly in the manner of fruit-forming, and that they pass through a series of intermediate steps into a higher class—that of the colorless algæ.

Cohn also made an attempt to systematize them, but on quite a different basis from that of Ehrenberg.

Seeing clearly that the necessary arrangement of their various relations to each other in proper classification was still in its infancy, he described the outward form only, hoping thereby to bring something like order into this world of apparent chaos. He therefore distinguished globular, cylinder or rod-shaped, and screw-shaped bacteria, as well as intermediate varieties. Such a classification can, however, only be superficial; for instance, one would, with such a system, have to place the blind-worm with the snakes and the whale with the fishes. But Cohn knew well that he had only done a preliminary work, and that it belonged to later investigators to show whether or not his form-classes and form-species would agree with proper natural-history classes and species. He was, however, misunderstood, and people objected to what he called "the constancy of forms." The following will show the views on this subject held by observers at that time: "Within the last ten years," says Nageli, "I have examined thousands of dividing yeast forms, and I should be unable to maintain (if I except sarcinæ) that there is a need even of a division into two specific forms." And again, there is that often-quoted expression, in which he draws the logical conclusions of his investigation: "If my view is correct, the same species in the course of generations take different morphological and physiological forms, which, in the course of years, bring about sometimes the sourness of milk, or the acid formation of sauerkraut, or the fermentation of wine, or the putrefaction of albuminous substances, or the reddening of foods containing starch; sometimes producing typhus fever, cholera, or intermittent fever."

It is a matter of course that if the above were correct, scientific investigation of bacteria would be an impossibility. Yet at the present day no one who has seriously occupied himself with these matters, doubts that there is a long series of species clearly differing one from the other under all circumstances, both in their physiological and morphological character.

We know that certain species of bacteria, at all times and under whatever conditions they are placed, in their development show the same expressions of life which in essentials always remain in harmony, thereby distinguishing them from other bacteria. We also know bacteria which, under the most varied conditions, always possess the same form, and thus are always distinguishable from others. Practice alone enables us to recognize these fine differences in form. Not only the ultimate varieties, but also variations in the same species, and even within the same cell differences are perceptible on examination.

Let us now consider the different conditions to which we must submit the bacteria in preparing them for examination. If we examine anthrax bacilli from a young gelatin culture with a high power ($\frac{1}{12}$), we shall see a number of separate motionless and colorless bacteria, which appear as long rods, apparently uniform.

If bacilli from the same source be treated with coloring agents (gentian-violet), the rods will appear thicker and shorter. The coloring matter has entered them, has formed a layer upon them, has —so to speak—formed a mantle around them, and hence the apparent enlargement. If we examine a properly-stained preparation of anthrax bacilli in the process of sporulation, a very different appearance from the above is developed, namely, roundish blue grains, lying at regular distances from each other, and also a single row of them surrounded by a light border (like a halo). These anthrax spores before being stained are first strongly heated, then stained, and afterward treated with a solution of iodine. Different methods of preparation may cause them to present dissimilar appearances under the microscope. If we examine unstained anthrax bacilli from a bouillon culture (with system 7, ocular 2) long fibres will be seen which are perfectly uniform, and in which even on close examination, no division into separate pieces can be observed.

If a preparation be treated with gentian-violet, we immediately notice that the threads consist of a row of cells of equal length, closely connected, and whose points of division had previously escaped observation.

In fact, the investigation as to the true shape has caused the greatest difficulty to the investigator.

At this point let us consider some appearances which have raised doubt in regard to the constancy of their form.

If we employ a fluid preparation which has been made from the liver tissue of a Guinea-pig whose death has resulted from inoculation with anthrax and stain it with anilin-red (fuchsin), large thick rods will be seen spread in rich profusion over the whole microscopic field of vision. If, on the other hand, we examine a section from the liver of the same animal similarly treated with fuchsin, we shall find the bacilli present in as great numbers as before, but they will appear much more slender and insignificant, scarcely to be compared with the fine-looking specimens of the previous preparation. This result is attributed to the alcohol in which the organ had previously been hardened, for it contracts the tissues and alters the appearance of the bacteria.

Frequently, however, variations in form are due to other circumstances. The bacterium passes through a course of development like plants of a higher order. It is young, it grows, reaches the highest development, and then generates cells of the same kind. It cannot, therefore, surprise us if young bacteria which have just proceeded from others appear smaller, and if old bacteria which are just about to divide appear larger than the average normal cells.

The circumstances of their nourishment exercise a very important influence on the appearance of the bacteria. The degree of perfection of their nutrition affects the vigor of their development.

If we examine the same variety of bacilli which have been developed at a low temperature on the surface of a slice of boiled potato, they can scarcely be recognized. It is, therefore, evident that this kind of nourishment is unsuitable for their development, as we find all sorts of irregularly-developed cells and amorphous bodies massed together, with here and there a long cylinder, which reminds one of what the anthrax cell should be. We also see numerous distinctly rounded forms of bacteria, which might be mistaken for cocci.

Do such cells, therefore, belong in the circle of development of the anthrax bacilli? Certainly not. For if we bring them into favorable conditions of nourishment, it will be seen either that they are no more capable of propagation—that we, therefore, have to do with dead substances—or, on the other hand, that they immediately give birth to the described typical forms of growth, the long rods of regular shape. These forms were, indeed, nothing but the expression of degeneration in the bacteria in question. They are degenerative forms, or, as Nageli has called them, "involution forms," malformations which are not to be mistaken for the healthy bacilli.

While the species of bacteria which we have just considered adhere, with a most remarkable tenacity, to that peculiar form (which they keep under all circumstances), there are others which seem to take a peculiar delight in developing strange forms, which degenerate more frequently than others, quitting their normal external forms, and in preparations exhibiting features calculated to puzzle for the moment even the most practised observer. Now, these changes are not a necessary, inevitable step in the changes which the individual must assume during its growth, but simply a sign that under the influence of unfavorable circumstances a degeneration of the bacterial protoplasm has taken place.

This is just what is meant by the expression "constancy of form;" that a species of bacteria, under varied conditions of life, may change its external appearance, its form, more or less; but under all circumstances a clearly describable form exists, in which this species finds the expression for the maximum of its development, the climax of its well-being. It is of course possible that future researches may point out an exception or exceptions to the above rule. Among the lowest vegetable organisms which stand nearest bacteria, some have been known to botanists which certainly possess the capability of going through a tolerably wide metamorphosis. These are chiefly water-pieces—the crenothrix, cladothrix, and beggiatoa, which, under some circumstances, appear as long threads, then again as larger or smaller rods, or even as clearly circular cells, and lastly also as spiral-shaped bodies. This species has been thought by some to belong to the bacteria, and has been cited to defend the views regarding the apparent changes of form already mentioned.

Now, the above-named organisms most certainly do not belong to the bacteria, although they may be very closely allied to them.

There are genuine bacteria which possess chlorophyl (leaf-green), and which therefore present, when they gather without distinction, all the colorless forms of vegetable life. Crenothrix, cladothrix, and beggiatoa are distinguished by an indubitable point-growth; that is, they strive to prolong themselves by progressive lengthening, and shoot from a narrow beginning into a point, which widens out more and more—a proceeding of which we find no indication whatever among the bacteria. Add to this the peculiar branching of cladothrix, and lastly the pleomorphism, and we have convincing proofs of the difference between these organisms and the bacteria proper.

We may therefore maintain that, thus far at least, a many-formed species of bacteria has not been observed, and the rule, "One can distinguish by the growth and form clearly recognizable genera and species of bacteria, which do not run into each other," remains as the general result of our investigations on this question. The question of classification might be discussed indefinitely, but of what consequence is it? The bacteria interest medical men principally from an etiological standpoint, because in them we find the exciters of a number of virulent diseases. All the rest, the nomenclature and purely theoretical study, we may safely leave to the proper persons, the botanists, in whose hunting-ground we have already ventured far enough. What we know of the bacteria in general, in a word is as follows: The bacteria are the lowest members of the vegetable kingdom, closely related to the lower algæ. They divide themselves in a series of species, well defined by growth and form, which do not run into each other. Of the forms in which the bacteria appear we know the globular bacteria, or micrococci; the rod-shaped bacteria, or bacilli; and the screw-like bacteria, or spirilla.

II. STRUCTURE—FORMATION OF GROUPS—FLAGELLA.

I have already mentioned that we must regard the bacteria as cells, for they grow and divide and form fruit as such.

The strongest argument in favor of their cellular nature is the fact that they possess a central mass composed of albumin, or protoplasm, which is inclosed in a well-defined limiting membrane, or cell wall. But in regard to the existence of a nucleus there seems to be some doubt.

Until lately we were inclined to deny this in toto. It was shown in the examination of unaltered bacteria that nothing could be seen of a nucleus, even with the best optical aids; but under the influence of those coloring agents known as nuclei stains, the whole individual bacterium became evenly and equally colored, and the reaction which is otherwise seen in the nucleus alone extended here to the entire cell contents.

But every one of experience in these matters will see that this does not form a rule without exceptions. Occasionally, under still unknown conditions, especially, however, after a short exposure to active coloring agents, one sees how a part only of the bacterial body greedily sucks up the coloring matter, while the surrounding parts remain paler and show clearly in contrast with the centre. Such pictures, which cannot be explained as artificial productions or faults of preparation, do indeed lead us to suppose that we have a nucleus before us, with its protoplasmic body, and that the presence of the former generally escapes observation only because our method of observation is imperfect and unable to show such fine differences.

This supposition becomes the more probable, as uncolored bacteria have been recently said to show something similar.

We must of course be very careful to avoid self-deception. I by no means wish to say that the cases mentioned proceed from an error. I believe, on the contrary, that these observations are correct, yet a convincing proof for my opinion I cannot give, and it will require further and deeper investigation to bring the question to a final settlement.

The contents are generally seen as a homogeneous, translucent, dull mass, without traces of any particular network.

But now and then we do see a sort of granulation, and such molecular thickenings of the protoplasm may for a moment deceive us into the belief that we see a structure.

Some few bacteria possess chlorophyl in their cell contents; others present a peculiar reaction, reminding one of the similar behavior of granulose under the action of an aqueous solution of iodine: when treated with the same, they take a deep indigo-blue stain.

The membrane consists, perhaps, of a mass similar to cellulose belonging to the members of the carbo-hydrate group. It (the membrane) is hard to recognize under the microscope without further preparation, but if we employ measures which cause a contraction of the protoplasmic contents (iodine, for example), then the membrane becomes clearly visible. It is either stiff or elastic, and thereby determines the behavior of the whole cell; that is to say, whether it can show contortions and bendings, or whether it must remain in an unaltered stiff condition.

Of great importance is the fact that the membrane, in its outer layers, possesses a pronounced inclination to bulge.

By absorbing water it passes into a jelly-like state, and then surrounds the cell with a gelatinous covering, often very thick.

This is well seen in a preparation of Friedländer's so-called capsule-cocci as they are found in pneumonia.

This capsule is, in fact, nothing more than such a gelatinous sheath, which becomes visible as distinguished from the proper bacteria cell by its much smaller capacity for absorbing color.

Still more remarkable is the conduct of the membrane when the bacteria divide. It prevents the immediate separation of the newly-formed members; keeps up their connection, and thereby causes the origin of bacterial groups from their simplest to their most highly-developed forms.

When two cocci still cleave together after their division—diplococci; or joining one to the other in rows—streptococci; arranging themselves in clearly-defined groups—the staphylococci; when the rod bacteria remain chained together in long threads, and when on the surface of nutritive fluids containing bacteria the separate cells unite into dense masses or films of mould; all this is but a consequence of membrane-bulging.

The union of cells of the same species have been called zoöglea, and from their behavior it has been attempted to gain criteria for the distinction of certain species.

The zoöglea develop best in liquid media. This can be seen by examining a pure culture of the Bacillus subtilis (the hay bacillus) in beef bouillon in one of Erlenmeyer's little flasks. On the surface can be seen the even, strong, slightly-furrowed grayish-white covering.

If the plug of cotton-wool be removed from the mouth of the flask and some of the scum taken out with a platinum needle, we observe that it still coheres, and even in distilled water will dissolve but little. The same thing takes place when bacteria are changed from an originally solid field of nourishment to a fluid one.

But the formation of firmly-united groups is by no means confined to liquid substrata. This may be seen on the surface of a slice of potato on which is growing a culture of the so-called potato bacillus. It appears as a curiously-creased grayish-yellow covering.

If a needle be dipped into the covering and slowly drawn out again, an elongating thread follows it which can be drawn out to about 30 cm., and which consists of nothing but bacteria firmly adhering together.

The disposition of the membrane to form a jelly, and so to unite in groups and form such coverings, is very different in the different species of bacteria; in general it is particularly strong with the mobile rod bacilli.

It is still doubtful whether the membrane is the place of deposit for the pigment in the micro-organisms which form coloring matter, and it is not yet definitely determined whether the formation of the coloring matter takes place in the interior of the cells or first in the substratum. A number of close observations would seem to favor the latter possibility. In the Micrococcus prodigiosus, for instance, the coloring matter lies separated in grains outside the bacteria, and other colored products of change of matter, such as a frequently-occurring greenish coloring matter, remain in solution outside the cells, and communicate themselves by diffusion to the neighboring objects.

A large number of rod and screw-shaped bacteria possess the faculty of spontaneous movement—that is, they have the power to leave a place and make an independent change of location.

For a long time we have been acquainted with the special organs by which the movement is effected, and the existence of cilia, or whipping threads (flagella), was clearly shown and proved beyond all doubt by R. Koch, with unstained bacteria, years ago. A particularly clever and careful observation was requisite to see these things. The flagella are very delicate formations, with about the same refractive power as the usual media employed as food-solution, etc. As they, moreover, continue throughout their activity in very brisk movement, there is no possibility of seeing the flagella in bacteria as they usually come under examination—that is to say, in a state of motion. All assertions to the contrary are based on self-deception.

Koch succeeded best in rendering the flagella visible by the following method:

He took a drop of some putrid vegetable infusion—algæ, aquatic plants, or dead leaves—and placed it on a cover-glass. Before it was quite dry he laid the cover-glass on the slide and examined it with the highest magnifying power. Then he saw here and there, wherever a trace of dampness still remained, a stranded bacillus, or spirillum, and at the end of it the flagella, which, having no room for action, lay exposed, and could even be shown in a photograph.

Thus the existence of these extremely fine processes was indisputably proved, and we naturally regarded their presence as extremely probable also in other moving micro-organisms in which no such processes had hitherto been discovered.

This supposition has been fully confirmed since Löffler has succeeded, by means of a peculiar coloring process (the particulars of which will hereafter be described), in showing the existence of the long-sought-for flagella in a large number of important species of bacteria, including some pathogenic ones, as for instance the cholera bacillus.

The same experiments revealed some other facts previously unknown. For example, the thick, clumsy-looking, screw-like bacteria which often occur in stagnant water, known as Spirillum undula, possess at each end not merely one flagellum, but a whole bundle of the finest threads, all curved in the same manner.

In the spirilla of decaying blood the same thing may be observed, while in the cholera bacillus we see a micro-organism which, in contrast to most of the others, presents at one end only a flagellum with wave-like curves. The bacilli in this case have, at either pole, a single cilium, rolled up like a whip-lash or a pig's tail, which illustrates a very singular fact made known by R. Pfeiffer. He discovered, with the aid of Löffler's staining method, that the flagella of typhus and some other bacilli do not proceed from the end, but from the side of the micro-organism, and always in a considerable number, so that such a bacillus has a peculiar appearance, reminding one, with its long projections, of a centipede or a spider.

Among the globular bacteria, too, Ali-Cohen and Mendoza have recently found two motile species, and Löffler has been able to prove the existence of flagella in the Micrococcus agilis of the first-named investigator. In a preparation of this bacterium we may see, with a little care, that the small globular cells which generally cohere in crowded heaps and packets do really show a perceptible locomotion.

These observations, it is true, give us as yet but isolated facts. The possibility that we may still find more motile representatives of the micrococci must of course be granted; yet, on the other hand, it must be noted as a striking fact that we find no traces of locomotion in any of the more common and important species.

The trembling and wavering which can often be perceived in examining unstained preparations of micrococci is purely molecular. It is the Brownian movement. A somewhat longer observation suffices to convince us that what we see is a mere motion at a place, and not motion from a place.

III. POWERS OF RESISTANCE OF SPORES—MULTIPLICATION OF BACTERIA—SPORULATION—VARIETIES OF SPORE-FORMATION.

The bacteria multiply by fission; the cell stretches out in the direction of its length, the membrane pushes a partition wall into the interior, and thus brings about the division into two new germs. These, again, can very speedily divide, and the multiplying powers of the bacteria are, in fact, immeasurable.

If the direction in which the consecutive divisions take place is one and the same, and if the cells after such division remain in connection, we see—as we have already seen—those simple forms of connection known in the case of cocci as streptococci, in the bacilli as threads (apparent threads), leptothrix, etc. These terms are nothing but the expression for the occurrence of growth proceeding in a continued line. Two particularly good examples of these two classes are seen in uncolored preparations of the erysipelas micrococci (in long bead-like chains) and in anthrax bacilli, grown out into threads, which traverse the entire microscopic field of vision.

A remarkable difference between the two will be apparent.

In the cocci, notwithstanding the close connection in their linear arrangement, we will be able to recognize the separate members. We can see everywhere the points of division, slight contractions, something like articulations, but in the anthrax threads we can perceive a point of division here and there in consequence of the difference of refractive power, and we must adopt some other means to demonstrate that such threads consist of many little rods placed end to end.

The fission of bacteria generally proceeds in one direction, occasionally also in two places at right angles to each other, and, lastly, in all three directions of space. The last two phenomena have as yet been observed with certainty only in the micrococci; in one case we have the so-called tetrad forms; in the other those peculiar groups known as sarcinæ are formed. Doubtless from clinical studies the Sarcina ventriculi will be remembered, which may serve to illustrate the cubic, or bale-like, packets formed by these micrococci.

The bacteria are able to propagate—which does not mean to multiply immediately—by another means than that of division. In quite a number of bacilli, and also in some few spirilli, genuine fructification, or the formation of spores in the interior of the cell, has been observed.

The fact, as such, has been seen in many bacteria, but it has as yet only been thoroughly investigated and followed through its details in three different rod species: in the Bacillus subtilis of F. Cohn, in the Bacillus anthracis of Koch, and in the Bacillus megaterium of De Bary.

What has been elicited in brief is as follows.

At the commencement of the spore-formation, the protoplasmic contents of the bacterial cell concentrate at certain points, which appear to the eye as darker portions of different refractive power. These coalesce, while the rest of the cell contents becomes clear and light-colored. The fully-formed spore then appears as a strongly-refracting, bright-gleaming body of definite (generally egg-shaped) form, with a regular dark outline, a thick spore membrane surrounded by the rest of the spore-bearing cell, which is as clear as water.

This latter soon perishes entirely, the membrane dissolves and disappears, the spore is free, and the process is at an end.

This is seen in threads of the Bacillus anthracis after completion of the spore-formation. Cell after cell is seen, each bearing in the middle the bright-gleaming spore, so that the whole resembles a well-arranged string of beads. In addition to these a few free spores may be seen, which, as a rule, show a brisk molecular motion.

One cell forms only one spore, under all circumstances. It may have its seat in the middle, as just described, or in other cases at the end of the rod. The latter often undergoes no change of shape in the sporulation; but it may swell and widen at the place where the spore afterward forms, and the sporulation takes place in the cell thus modified. In the cases of spores which grow in the middle of the bacteria cells, peculiar spindle or shuttle-shaped forms, with a stout body and short pointed ends, the so-called clostridiæ develop, while sporulation from the end of a cell leads, under similar circumstances, to the development of the "drummer bacteria" and "headed bacteria," in which the spore is located at one end of the poles of the cylinder, which presents a club-like swelling.

Of what the contents of the spore really consist is still unknown.

That a very early and thorough separation within the bacterial cell, between that part of the protoplasm which goes to form the spore and the remainder of its substance, occurs, may be perceived from the coloring. Several investigators have called attention to the fact that in the interior of many micro-organisms about to form spores, certain grains and globules are found which, under the influence of suitable staining processes, are easily distinguished from the rest of the body.

The view has even been maintained that these sporogenic grains are connected not only with the origin of the spores, but also with the presence of nuclei in the cell protoplasm.

How far these statements and their application may be correct cannot, as yet, be definitely stated. It is a fact, however, that the fully-formed spores are very differently sensitive to staining than the rest of the cell, and may be very clearly distinguished from it by the color alone.

An important part of the spore is its membrane, the spore-skin. This is an extremely tough and strong covering, inclosing the spore on all sides and surrounding it with an almost impenetrable mantle.

If we bring spores into a fresh nourishing solution, they germinate sooner or later and grow into rods. This process has been carefully watched (by Prazmowski and De Bary) and many important things discovered. A spore about to germinate first stretches somewhat in the direction of its length, its contents lose part of their bright gleaming, and the dark outline—the tough membrane—seems to swell. The more the spore lengthens, the more it resembles a short cylinder. At last the membrane bursts, and a young bacillus is set at liberty. The empty membrane soon dissolves and disappears.

The formation of spores has been seen with certainty only in several bacilli and some few spirilla, while in the micrococci certain indications of a similar process have been observed, but no decided results have as yet been obtained.

Under what particular conditions a bacterium proceeds to sporulate is by no means certain. Formerly it was believed that the micro-organisms began sporulation as soon as the necessary conditions for their growth were withdrawn, either by the nourishing solutions being exhausted or by accumulations of their own excretions rendering their growth and propagation more difficult.

It has been claimed that a bacillus which finds itself, and therefore its species, threatened with extinction, tries to avoid this by being changed into the form of a spore capable of resisting most external influences and waiting for better days.

Beautiful as is this thought, the facts do not bear it out, since we find bacteria forming spores when their nourishment is more than sufficient.

It is, on the other hand, very probable that the bacteria, contrary to the higher plants, proceed to sporulation chiefly or only when the cells have obtained their highest development, when they are in the best conditions as regards nourishment and growth.

The anthrax bacillus, for example, produces no spores at a temperature of 20° C., but produces them most rapidly and certainly at the temperature of the incubator (35°–37° C.), which seems especially suitable. With insufficient oxygen the anthrax bacilli also produce no spores, and this is so noticeable that in the upper strata of fluids in which they sink to the bottom and cannot regain the surface on account of their immobility, the spore-formation is greatly retarded, and within the body of a dead or living animal it entirely ceases.

On the other hand, by measures which in some way or other act prejudicially on the bacterial protoplasm, we can artificially rob it temporarily, or permanently, of its capacity for spore-formation.

K. B. Lehmann and Behring discovered the "non-spore-producing" anthrax, which cannot, under any circumstances, be brought back to spore-production.

Sporulation is, in many respects, a remarkable phenomenon. I have stated that it has been proposed as a basis for a scientific arrangement of the bacteria.

Besides the manner of sporulation above described, in which the spore develops in the interior of the fruit-bearing cell as a special formation, observers have believed there was another way in which such reproductive organs originated.

It has been found that whole cells separate from their connection and form the beginning of new groups. No very striking change of form takes place in these cells; they only appear to increase somewhat in size and refractive power, and also possess a firmer, darker envelope.

This kind of sporulation, in which the entire cells are said to undertake the functions of permanent cells, has been designated as arthrosporic, in contra-distinction to the ordinary endosporic fructification, where special new-formed organs arise in the interior of vegetative cells. From this difference the criteria for a classification have been derived.

As already stated, observations are neither numerous nor certain enough for us at the present time to found upon them the principle of a finally correct classification. When such an experienced man as Prazmowski recently expressed doubts as to the existence of an arthrosporic sporulation in the bacteria in general, and even in the micrococci (in which it was thought to have been most clearly seen), and when he regards endogenic spores as the only possible ones, it is best to proceed cautiously.

From a biological point of view sporulation is also important, since the spore as a "persistent form" can do much more for the preservation of the species than the transitory "growth forms" of bacteria.

The extraordinary capacity of resistance to external influences must be regarded as the most remarkable property of the spore. This is no doubt chiefly owing to the stout, firm membrane which surrounds it, and which possesses an almost unlimited power of resistance. The continued or temporary influence of dryness and moisture, heat and cold, is well borne by the spore, as are also many chemical agencies which it withstands without damage and without injury, though they destroy all other life. One must, in fact, regard them as the most enduring organic formations of our world.

The impunity with which spores can bear high temperature is practically very important. While we can easily succeed in destroying sporeless bacteria, we find great difficulty in destroying those capable of sporulation. A dry heat of not less than 140° C. destroys all life in the spores with certainty after an exposure of several hours, and a boiling heat also requires several minutes to produce the same result.

It is true this does not apply equally to all spores. The resistance of the permanent forms is very different in the different species, and is subject to considerable variations even in the same species. We are acquainted with bacteria which yield to slight injurious influences, and others which are almost miraculously unimpressionable.

Globig has observed a particular sort of potato bacillus the spores of which he had to expose for more than four hours to the action of a jet of steam of 100° C. before he could finally destroy them.

Fortunately such cases are great exceptions, and in general the figures given will be found correct. It was a long time before we arrived at these facts by experiment and utilized them, thus avoiding the mistakes which formerly made our knowledge of the bacteria unsafe in a high degree.

IV. CONDITIONS NECESSARY FOR THE GROWTH OF BACTERIA.

The bacteria originate only in germs of their own species. Easily comprehensible and natural as this assertion may seem to us at the present day, it has taken an immense amount of time and trouble to establish it as a universally-acknowledged fact.

It is not very long ago since the spontaneous generation of the bacteria was seriously defended, and this view was not fairly abandoned till Pasteur made his victorious campaign against generatio æquivoca.

It is true he found the ground well prepared. Already, in the previous century, the wonderful experiments of the learned Abbé Spallanzani had yielded decided proofs against the possibility of spontaneous generation. Cautious and clever investigators, such as F. Schulze, Schwann, Schroder, and von Dursch, had still further perfected these experiments and established their accuracy. But the doctrine of spontaneous generation raised its head again and again, and when it seemed to be fairly destroyed, it reappeared with its old assertions and pretensions.

Thus it required the influential personality of a Pasteur, his energetic labors and his precise investigations to rid us once for all of these traditional errors, and to settle the question so thoroughly that it only now and then, quite stealthily and timidly, reaches us like some voice out of ancient times.

There were two facts in particular which led to and maintained this belief in the self-originating powers of the bacteria. First, one had no knowledge and no conception of the extraordinary powers of resistance possessed by the lasting forms of the bacteria.

We exposed fluid to a boiling heat for a quarter of an hour, and supposed all life in it to be thereby destroyed. But we almost always neglected to convince ourselves that the necessary temperature really penetrated through all parts of the medium in question.

It has been found as a rule that this is not the case unless particular precautions are employed, and that the equable penetration of the boiling temperature into fluids is by no means so easily attained as was formerly supposed. But wherever the heat had not properly penetrated spores could, under certain circumstances, find a place of refuge, where they escaped destruction. One was not a little astonished to see new bacterial vegetation reappear in such liquids, even when the greatest care had been taken to protect them against exterior pollution. Such growths were supposed to have developed "of themselves" by "spontaneous generation," or they were referred to as "atoms of nitrogen" and other innocent molecules, while, in fact, some spores that had escaped scalding, were the cause of this mysterious process. Further, it was thought necessary to suppose a generatio æquivoca, because we had no idea of the wide distribution, the omnipresence of the bacteria. There is, indeed, scarcely anything that is free from these minute, invisible, living forms. The great masses which surround us, the air, the earth, the water, are as much stocked with them as are the objects of daily use; the majority of our vegetables, our clothing and our dwellings, our intestinal canal and the surface of our skin, swarm with micro-organisms, and we only know one field which is unimpregnable to them, and that is the uninjured, unaltered, healthy organs and juices of the bodies of men and animals.

This unlimited extent of bacterial life will seem comprehensible by considering the extremely modest requirements which the lowest representatives of the vegetable kingdom generally make for their development.

The smallest quantities of organic substance serve for their nourishment; wherever they find these and no particular hindrances to their development, there they grow and multiply.

It is true they are distinguished from the majority of other plants by their requiring preformed carbonic combinations of organic nature; that is to say, by their inability to obtain the carbon which they need from carbonic acid pure and simple. They want —with the exception of a few species described by Tieghcim—the necessary chlorophyl, without the presence of which pure carbonic acid cannot be taken and appropriated for use.

The plants without leaf-green have been placed together in a group and called "fungi," the bacteria, in consequence of their manner of increasing by division, being called the "splitting fungi," (schizomycetes), as distinguished from the rest, which are called the "sprouting fungi" and the "mould fungi."

But since there exist, as we have just seen, some bacteria which possess chlorophyl, and since the name "fungi" seems calculated to produce confusion, it is better to renounce it altogether and employ the name "bacteria" for the lowest splitting plants.

The amount of carbonic acid which the bacteria require in their nourishment, together with the carbonic combinations, may be supplied directly by the organic substance, or by inorganic bodies, such as nitric acid or combinations of ammonia.

The bacteria, or at least the great majority of them, also require that their food should show an alkaline, or at least a neutral, reaction. On an acid medium most of the bacteria hardly grow at all, whereas the mould fungi, for example, develop successfully upon it.

An alkaline solution of organic substance is what the bacteria require as a nourishing soil for their development. Nature bountifully supplies them with this. Remains of organic matter are everywhere to be found, and therefore the conditions for the growth of the micro-organisms are almost universally present.

Thus a large number of them develop on dead portions of organic origin, remnants of organic life, on remains of dead plants, decaying corpses in the soil and in the water.

A comparatively small number, however, are more dainty; they grow only in the living bodies of the higher organisms, at whose expense and generally to whose detriment they live as genuine parasites, and without which they cannot exist at all.

These are called the strictly parasitical bacteria, and the first mentioned, which do not thrive in living organisms, are called the strictly saprophytic bacteria. The boundary is, indeed, not impassable. There exists a whole class of bacteria which can find the conditions for life as well outside other living beings—can live saprophytically—as they can by penetrating into foreign organisms and there live as parasites; these are the semi-parasitical or semi-saprophytic bacteria.

Besides the conditions already mentioned, there are some other factors which are of importance in the life of the bacteria.

A certain degree of warmth is quite necessary for their plentiful development. Warmth is a mighty spring in the clock-work of all organic life, and its influence on the bacteria is unmistakable.

The temperature required by micro-organisms for their prosperity is, it is true, very different for the different species. In general, however, we may say that the limits between which the bacteria can exist are from 40° C. down to about 10° C., yet but few of them can exist throughout this entire range. By far the greater number are restricted to a much narrower compass, within which, again, there is another still narrower range, an optimism of temperature, at which they thrive most quickly and luxuriantly.

There is a particularly noticeable difference here between the strictly saprophytic and the strictly parasitical bacteria. The former find the best conditions for their growth in the average temperature of our summer months or the medium temperature of rooms—i.e., about 24° C. This limit some of them can hardly pass; that is, they die in a higher temperature, and are therefore incapable of living parasitically in the warm-blooded creatures and there causing disease.

The parasitic species, on the other hand, develop most favorably and quickly at the temperature of incubation—i.e., at about 35° to 40° C. Some micro-organisms which cling very tenaciously to their parasitic habits, which only with difficulty and by artificial means can be induced to grow on our culture media outside of living bodies, obstinately refuse to bate anything of their due temperature. The tubercle bacillus, for instance, only thrives at a constant temperature of 37° C., and a slight departure therefrom produces defective development.

If the temperature sinks below or rises above the proper degree the bacteria fall into a state of insensibility and inactivity, from which they recover as soon as they are again placed under favorable circumstances. If the temperature departs still further from the favorable mean it acts directly prejudicially, or even destructively, on the majority of the micro-organisms known to us. Heat of 60° C. kills quickly and to a certainty many very important bacteria, such as the typhus, cholera, glanders (malleus), and tubercle bacilli. Others have a greater power of resistance, and particularly in a dry state some, such as the pus cocci, keep their vital power even at 80° C.

In like manner the influence of cold is manifested. If fluids containing bacteria be allowed to freeze, we observe that the greater part of the micro-organisms perish. The bacteria are specially sensitive, as shown by the investigations of Prudden, to repeated freezings and thawings. Yet here also we see differences in the different species, some of which display a high power of resistance, while others are soon destroyed.

Still more important is the fact that even within the same species the separate individuals are by no means similar in their powers of resistance: some yield more quickly, others very much more slowly, to the influences of a prejudicial temperature. As the same difference has been remarked with respect to other external influences also, it may be regarded as certain that among the bacteria, as also among higher organisms, there are strong and weak, young and old, healthy and diseased, individuals, all of which must not be considered subject alike to the same influence.

It is well to note here some further striking observations which in part go to contradict the general rules above given, the correctness of which, however, is beyond all doubt.

On one side, Fischer and Forster have made us acquainted with certain bacteria occurring in sea-water and in the earth which even at the freezing point (0° C.) cannot only exist, but are able to grow without interruption.

On the other hand, bacteria have been discovered by Globig and Miquel which continue to develop and multiply at 60° C., and even at 70° C. Nay, there are even some which require this high temperature, and are unable to thrive at a lower temperature than about 60° C. It is hard to conceive how and where in a state of nature such organisms can find the necessary conditions for their growth and reproduction.

After heat, oxygen is a very important factor in the life of bacteria. By far the greater part of the micro-organisms yet known cannot thrive in the absence of oxygen. Some, indeed, are so sensitive on this point that even a slight diminution of oxygen in their surroundings exercises an unfavorable influence on their vital powers.

These are called aërobic bacteria. River water, the air, and upper portions of the soil are rich in representatives of this class, and the majority of the pigment-forming kinds seem to require oxygen for the development of coloring matter.

Others are not so dependent on the presence of oxygen; they grow, indeed, in an atmosphere rich in oxygen, and often better than in one poor in it, but even the entire want of this gas does not suffice to stop their development completely. These are semi-aërobic, and to them belong most of the pathogenic species. In the interior of living bodies, except in the lungs, there is no oxygen for them; that introduced by hæmoglobin being quickly absorbed by the cells and appropriated by the tissues.

All micro-organisms that are to exist as parasites must, therefore, be able to live without oxygen, at least under certain circumstances.

We may premise this as necessary, even though our experiments and observations may seem to show the contrary. In fact, some of the most important pathogenic bacteria, for instance, can scarcely be grown by our artificial cultures with exclusion of oxygen. But it is to be understood that we do not argue from this as to their natural conditions.

The ability of the cell-protoplasm to exist without oxygen is, in many micro-organisms, dependent also on the other conditions of nourishment, and is therefore not an invariable property.

If these bacteria find in their neighborhood an opportunity for decomposing higher combinations, they can thus obtain their need of oxygen, taking it from the molecule. In such conditions they manage to live without free oxygen, which they cannot do where the opportunity of obtaining it by creating decomposition fails them.

Besides those micro-organisms which can thrive without oxygen, there are also others which can only thrive in the absence of oxygen. These strange bodies, which differ from all other known organic beings, were first discovered by Pasteur. Recently special attention has been directed to these strictly anaërobic bacteria, and it has been found that their occurrence is by no means so rare as might be supposed.

It seems, on the contrary, that these species have an important duty to perform in the household of Nature. The atmospheric conditions necessary for their welfare are produced by the simultaneous presence of the aërobic bacteria, which consume the oxygen and, therefore, create a zone without oxygen.

Thus we find anaërobic germs widely spread, and they are capable of developing in various places, from whence their lasting forms—their spores—are communicated to the water, earth, etc., in which they are almost always to be found. Our knowledge on this interesting subject is still very imperfect. In making our experiments, it is a difficult matter to exclude all oxygen, and thus obtain a prime condition for the growth of anaërobic bacteria, for which oxygen is positively a poison.

A slight trace of it is sufficient not only to prevent them from multiplying, but to even kill their germs, with the exception of spores, in a short time.

Lastly, must be mentioned the influence of light. While the ordinary daylight is unimportant with regard to the development of bacteria, sunlight has, according to the joint testimony of a number of investigators (among whom may be mentioned Duclaux and Arloing), a decidedly prejudicial, destructive effect upon the most tenacious forms of micro-organisms. Anthrax spores, when exposed to the direct rays of the sun for a few hours, lose their power of development, and even before this they show unmistakable signs of a certain debility and diminution of vital power.

V. PTOMAINES—DEVELOPMENT OF PIGMENT—FORMATION OF GASES—PRODUCTION OF ODORS.

When the bacteria have all that is necessary to their development—an alkaline food solution, a suitable temperature, and favorable atmospheric conditions—they increase and flourish luxuriantly. Yet this has a definite limit.

The vital action of the bacteria produces substances some of which have a tendency to check their further development.

It has been noticed, for example, that, earlier or later, the processes by which the micro-organisms bring about the formation of butyric acid, lactic acid, etc., cease, in consequence of the accumulation of the acids themselves. It has been further remarked that the micro-organisms living in the human intestines decompose their contents more and more, till they at length produce substances which, when chemically analyzed, display antiseptic qualities which could not but be deleterious to bacterial life. Such observations have led to a great number of thorough investigations in this direction, from which it has been discovered that the powers of the bacteria in producing definite chemical substances are great and varied. This can be demonstrated by adding a solution of litmus to some test-tubes containing nutrient gelatin, and then planting them with different micro-organisms.

It will be seen by comparison with a tube containing litmus alone, without any bacteria, that the coloring matter will be decidedly altered in nearly all of them by the growth of the bacteria.

In this way, and also by other more accurate proceedings, it can be shown that some micro-organisms produce very considerable quantities of acid, others of alkali. Ammonia and the higher bases, such as trimethylamin, further, also, sulphuretted hydrogen, etc., have been detected as immediate products of the vital action of bacteria. We are particularly indebted to the successful researches of Brieger for the knowledge of some alkaloidal substances of very complicated combination (but chemically well defined) which are regularly produced by certain bacteria. This gives us an important element for judging the pathogenic or disease-producing importance of these same micro-organisms.

We know that many bacteria, especially those of a saprophytic nature, and some parasitic ones also, have a very considerable power of reduction: they change, for instance, nitric into nitrous combinations, or even into ammonia. Others are said to have an opposite action and to produce oxidation.

In what has just been said, the action of the bacteria has been presented in certain details, but it is rendered more tangible and interesting if we consider it from a more general point of view.

It has already been stated that we study the bacteria principally and almost entirely because it has been proved that they are the cause of a whole series of phenomena which are of deep and wide-spread importance in the economy of nature. The micro-organisms are, in the first place, the exciters of fermentation.

It was Pasteur who, in opposition to the prevailing opinion of his time, first maintained the doctrine of the vital principle of fermentation, and proved that this important and for us, in some respects, indispensable process is brought about by the action of certain micro-organisms. Many of them, especially those chiefly concerned in alcoholic fermentation, do not exactly belong to the bacteria; but there are genuine exciters of fermentation among the bacteria, as, for instance, those which cause the formation of lactic and butyric acid.

It is further to be noted that the known varieties of fermentative action which lead to different final results generally owe their origin to different species of micro-organisms.

Still more important, perhaps, is the service rendered by bacteria in the economy of nature, in consequence of their being the only agents for causing the putrefaction of organic substances. It is understood that by putrefaction is meant the stinking decomposition of matter containing albumin. Pasteur referred this process mainly to the activity of anaërobic bacteria. He thought that the exclusion of oxygen or its removal by aërobic micro-organisms simultaneously present was an essential condition of putrefaction. Yet this view has not been fully confirmed. We know aërobic bacteria also which can cause real putrefaction, and it is probable that very many different bacteria are able to do the same.

Putrefaction is not a specific process that might be caused by the action of one particular species of bacteria, but the general name for a number of separate phenomena which produce the same results, being all reduction processes, and all resulting in the formation of ammonia, sulphuretted hydrogen, etc.

On this account we may class putrefaction among those changes that we observe in the soil and call nitrifaction and nitration. These also depend on the action of bacteria, and show a decomposition of organic matter into its simplest component parts, and these in turn causing the production of higher combinations from their union.

In this manner alone is the soil enabled to serve as nourishment to the higher orders of plants, and the importance of the micro-organisms for the development of vegetation is therefore extremely great.

Nearly related to putrefaction is a process which becomes familiar in the course of breeding experiments.

As is already known, the artificial food-solution which we usually employ is changed by the addition of gelatin (an extract of calf's bone and other substances, rich in chondrin and gluten) into a mass capable of solidification. Numbers of bacteria, motile and non-motile bacilli and micrococci, possess the capacity of decomposing the gelatin; to digest it, or, as we may say, to peptonize it. It thereby loses its solid consistence and becomes fluid.

This is so remarkable a fact that it has even been employed to make a division, for practical use, of the bacteria into two classes: the "liquefying" and the "non-liquefying."

Generally, however, it is not the bacteria themselves which directly cause the liquefaction of the food-medium, but, in the great majority of cases, the products of their metabolism. The liquefying species produce a sort of ferment which we can separate from the micro-organisms either by killing them or by means of filtration, and which then causes this change in the gelatin already mentioned.

We have the bacteria, then, as the buriers of the organic world; it is their task to put out of the way the material that has become useless, and thus to make room for new living forms.

Yet, as we already know, they by no means restrict their attacks to dead and useless things; they also penetrate into living organisms, grow and multiply in them, and thus, as they can only do so at the cost and to the injury of the individual attacked, they become the source of quite a number of the most varied pathological phenomena. The diseases known to be caused by some particular species of bacteria are becoming more and more numerous, and as their origin is external they are called infectious diseases.

In a general way we distinguish the bacteria which possess these "infectious" qualities as pathogenic, and those which are harmless and cannot produce these effects in other organisms as non-pathogenic.

Besides these three chief fields of bacterial activity—the excitation of fermentation, of putrefaction, and of infection—there are others of minor importance.

That which most strikes the eye is the formation of coloring matter by a number of species, most of them harmless. All kinds of colors may be observed: white, black, blue, green, brown, red, orange, etc., some of them of the brightest hue. How the formation of coloring matter is accomplished is not yet known with certainty. Probably the majority of micro-organisms do not generate the pigment directly, but only the basis of it—a chromogenic body.

If this is liberated in any way, for instance by its passing through the membrane of the cell by diffusion, or by the death and decomposition of the micro-organisms, it has an opportunity to combine with certain ingredients of the culture medium, or to gain access to the oxygen of the air, and then, but not till then, does the color appear. This explains why the pigment is often observed only on the surface of our cultures, and why the tint is, as a rule, dependent on the nature of the substratum.

There is another remarkable property of some few bacteria, especially investigated by Fischer, which "phosphoresce" or shine in the dark. They do so, under some circumstances, with such a degree of brilliancy that, by the light of a few gelatin cultures, the time may be read from the face of a watch—nay, it has even been possible to photograph such cultures by the light which they themselves emitted.

What processes cause this "phosphorescence" is not yet known, but the researches of Lehmann and Tollhausen have made it at least probable that it depends on the direct intra-cellular vital action of the bacterial protoplasm, which finds expression in this peculiar manner. The molecular changes within the cell, which in other cases causes the formation of heat, of carbonic acid, etc., are here accompanied by a development of light.

The undoubted influence of changes of nourishment on the occurrence of "phosphorescence" is to be explained by the protoplasm answering such changes, sometimes with a higher, sometimes with a lower, degree of specific activity.

Lastly, several of the bacteria known to us possess the power of developing gas in the medium which surrounds them. With the employment of solid foods this is often clearly perceptible, since the bubbles of gas which are formed cannot then escape. It is the anaërobic species more especially that show a decided tendency to the generation of gas. No exact investigations have as yet defined the nature of the gases thus formed.

In connection with the formation of gas may be mentioned the occurrence of smells—sometimes very penetrating—with certain micro-organisms. It is known to everybody that substances of very offensive smell arise in the process of putrefaction. One of the pigment-forming bacteria, the Micrococcus prodigiosus, when grown on potato develops a decided odor of trimethylamine; others, for instance, the cholera-bacillus, produce in cultivation a peculiar aroma, and nearly all of the few known anaërobes have the common property of emitting a truly abominable stench in ordinary cultivation.

We have now completed our general observations on the bacteria. We have determined their position in the domain of nature, discussed the attempts to systematize them, and made acquaintance with their chief morphological and physiological peculiarities.

Manifold and interesting as our attainments already are, it cannot have escaped the reader that we have not yet passed beyond the rudiments of an exact knowledge of the bacteria.

Questions of extreme importance still await their solution; wide fields of research, far from being sufficiently examined, still lie almost untouched by our investigators. Everywhere we meet with doubts and uncertainties.

This need not be wondered at. We owe to Koch and his introduction of the solid-transparent foods the important facts already elicited and the interest now taken in the study of bacteria, although it is not yet ten years since his methods of investigation came into use.

That in so short a time it has not been possible to accomplish all that was required, is almost self-evident; but we may cherish the certain hope that the near future will bring us further useful discoveries, and that many further additions to our stock of knowledge will yet be made by the application of new methods of investigation, next to be treated of.

CHAPTER II.

Methods of Investigation; the Microscope; Microscopical Examination of Bacteria; Stains and Staining; Preparation of Cover-Glass Specimens for Staining.

METHODS OF INVESTIGATION.

In order to learn the peculiarities of the bacteria, to become acquainted with their nature and habits, the only means at command for a long time was to examine them microscopically in the state in which they occur in nature under ordinary conditions.

This was a very imperfect procedure, the more so as the microscope itself was inadequate for our purposes. Of these smallest living forms little could be seen with the aid of the strongest obtainable magnifying power, and this little was so peculiar that we had great difficulty in comprehending it. It was not till a later period that we learned to render the bacteria more accessible to observation by special preparation and staining processes.

Microscopic examination, both of unstained and of stained micro-organisms, is still an indispensable and essential part of bacterial investigation, and in the decision of the great majority of the questions that arise it plays an important part.

Yet we are no longer dependent on the microscope alone for our knowledge of bacterial life, and the rapid progress made in bacteriology in the last few years dates from the moment when we began to detach the bacteria from the accidents and casualties of their natural existence, to breed them artificially under favorable conditions, and to study their behavior under given circumstances.

Microscopic examination and the breeding process work hand in hand in the most satisfactory manner, and open to us, in some cases, a really deep insight into the mysterious and various manifestations of the lowest representatives of organic life.

Yet, that with all these improved means of research we have by no means reached the desired goal, is shown by the simple fact that even the best-known species of micro-organisms still present an abundance of unexplained phenomena.

I. THE MICROSCOPE.

Before going into the procedures at our disposal for preparing the bacteria for microscopic examination, it will be well to direct attention, first of all, to the microscope itself.

The improvement of the microscope essentially contributed to perfect our methods of procedure.

In fact, it has been necessary to make a special study of the handling of the microscope, and it has been requisite to provide it with special auxiliary apparatus in order to make it suitable for the purposes of bacteriology.

The merit of having insisted on this point belongs to Koch. He showed that the manner of using the microscope as was done for histological researches did not suffice for the requirements of bacteriology, and by recommending the adoption of homogeneous immersion and showing the right employment of Abbe's illuminating apparatus in examining colored objects, he made a revolution in the microscopic investigation of the bacteria from which a new era dates.

It is readily understood that for observing such extremely small creatures as the bacteria it is necessary to use a very strong magnifying power, and that we require all possible perfection in the lenses to be employed.

What is demanded, then, in a good, faultless system in these respects?

Three things: First, it must sufficiently and very considerably magnify the object under examination; secondly, it must give a correct image, sharp and well defined; lastly and chiefly, however, the microscope must enable us to analyze the object into its simplest component parts, to distinguish the finest combinations of its lines, the arrangements of its substance, and this distinguishing or resolving power contributes far more to the value of a lens than its mere magnifying power. To ascertain the value of a system the question should not be, How many times does it magnify? but, How does it define?

On what do these three chief points in a microscopic system depend?

The size of the image is in proportion to the focal distance of the lens. For the compound lenses which are now almost exclusively used we calculate from the focal distances of the component parts an average or "equivalent" focal distance for the whole, an imaginary number which only gives the number of times which the system would magnify if it had the given focal distance.

The sharpness of the image is dependent on the exact spherical and chromatic correction of the object-glass.

Recently a great advantage has been gained since Abbe in Jena, in conjunction with C. Zeiss, has succeeded, by means of new glass pastes, in constructing lenses distinguished by a number of superior qualities.

The object-glasses previously in use had always the essential fault of not evenly uniting the different-colored rays—i.e., the chromatic correction was not equal in all parts of the lens. This fault prevented the formation of an evenly correct and clear image to a sensible degree; the lenses could not be employed up to the full limit which one had a right to expect, because the so-called over-enlargement by eye-pieces was already limited by the difficulties of correction increasing with every increased magnifying power. All this is swept away by the new system.

The chromatic variation is almost entirely avoided, and there is no longer any hindrance to increasing the magnifying power by means of eye-pieces. In fact, the advantages of these "apochromatic" object-glasses are extraordinary. They yield an image of remarkable sharpness and clearness.

Now as to the last point: The distinguishing power of a system depends on many factors, among which the most important is the size of the angle of aperture of the lens in question.

Imagine a diameter drawn through the front plane of the lens so that it connects the two most distant points in the circumference of the lens, and imagine, further, that the object is a single point; then connect the ends of the lens circumference with this object, and we have in the angle formed at the object-point the angle of aperture of the system. In other words, this angle is formed by the axial point of the object-plane and the two marginal rays—the last two rays which, proceeding from the object, could pass through the lens. It is self-evident that the size of the angle of aperture has a decisive influence on the amount of light passing through an objective. Abbe of Jena, to whom our knowledge of these abstruse relations is almost exclusively due, has, by a series of extremely ingenious and peculiar experiments, ascertained that the penetrative power of a system also stands in direct relation to the size of its angle of aperture.

He even succeeded in giving to this relation a definite formula, in expressing it by a definite number, and in proving that "the penetrating power of an object-glass equals the sine of half its angle of aperture."

It is true that this is not the only factor to be considered. As is well known, rays of light in passing from one optical medium to another suffer a refraction and reflection which causes much loss of light. If, therefore, rays proceed from the object and through the thin glass cover into the air between it and the objective, they lose as much illuminative power here as on their leaving the air to enter the glass medium of the lens.

This fact is also of essential importance for the definition, and with reference to it Abbe has drawn up the final rule: the definition of every microscopic object-glass depends on its "numerical aperture." This, however, is equal to the sine of half the angle of aperture multiplied by the index of refraction of the stratum which separates the object and the first lens of the object-glass.

It follows from this that in the ordinary dry systems, such as were almost exclusively in use for a long time, in which the index of refraction of that intermediate stratum, the air, is an invariable amount, the numerical aperture—i.e., the definition—cannot be increased beyond a certain limit, which is only determined by the size of the angle of aperture.

The introduction of immersion in water by Amici was, therefore, a very great improvement, since it brought between objective and object a medium of considerably stronger refractive power—i.e., water.

But a still greater progress was seen in the systems with homogeneous immersion as first introduced by Stephenson and afterwards improved by Abbe.

These two investigators also completed their theoretical elaboration and calculation, thus contributing essentially to their general adoption.

Between the object and the objective was placed, instead of the stratum of air, a medium of nearly equal refractive power with glass—namely, oil, and especially a certain kind of cedar oil.

After this it was possible to give to the field of vision an unusually large quantity of light; for the losses which otherwise take place at the separating surfaces by different media were now obviated. The effects of immersion in oil may be illustrated by a simple experiment. Into an empty test-tube place a moderately thick glass rod. There will be no difficulty in perceiving it, for the differences of refraction between the surrounding air and the glass allow the latter to be seen clearly enough.

Now pour some water into the glass: water stands nearer to glass in point of refractive power, and it will be more difficult to distinguish the glass rod. If, however, the glass be filled with cedar oil instead of water, the rod, as far as it lies in the oil, will immedi-

visible, their outlines are clear and sharp; in short, the whole picture is now adapted to the purpose of bacterial investigation.

The reason is that under ordinary conditions the lines and shadows which compose the structure picture are liable to darken and cover up stained objects of small dimensions, which therefore do not emerge from their obscurity until the structural part of the picture disappears. Herein lies the value of Abbe's illuminating apparatus: it is capable of giving decided prominence to the colored portions of a stained preparation, particularly to the nuclei and the bacteria.

This may be elucidated still further by another example. The effect of Abbe's apparatus may be considerably diminished (and even quite destroyed) by diminishing the base of the cone of rays with the aid of diaphragms, or "stops," and thereby decreasing its angle of aperture. The smaller the opening of the diaphragm, the more it shuts off the action of the condenser, and with a very small opening one works, so to say, without the Abbe.

Examine a preparation of micrococci treated with fuchsin first without any diaphragm—i.e., with the full working of the Abbe and the isolation of the color picture. A collection of micrococci will appear as small, uniform, strongly-colored grains. The close observation of them will not be difficult.

If a small stop be employed—that is, the Abbe shut off—the structure of the tissue immediately becomes visible, and the bacteria which before were so distinct are now indistinct and thrown into the shade, so that with all possible efforts they cannot be found again. Nevertheless the place where they ought to be visible is known. How much more unfavorable, then, would be the circumstances if the search had to be made without even this aid.

All colored bacteria preparations in which the effect of the colored picture alone is advantageous must be examined with the undiminished action of the Abbe—i.e., without any stops—but all unstained objects in which only the structure picture is desired must be examined with the smallest possible stop or opening.

By the words "smallest possible" stop is understood a stop which leaves a sufficient illumination of the field; as a rule, the higher the power employed the larger must the stop be.

Special note should be taken of this. Beginners often err from inattention to the rules just given, and they are seen, faithful to their histological custom, either examining stained preparations with the diaphragm, or uncolored objects, hanging drops, etc., under the full effect of the Abbe, without any stop. Under both circumstances, little can be seen of what it is desired to recognize.

By this we do not mean that in all cases and under all circumstances the stained preparations are to be only examined without stop. On the contrary, such an examination is only advisable for the beginning of an investigation, when we wish to know whether our object contains any micro-organisms at all, and if so, what forms, what appearance, and what arrangement they present. This ascertained, the next questions are: What relation do the bacteria bear to the tissues and how are they distributed in them?

These questions can only be answered with the aid of a suitable stopping. The structure picture must be allowed to reappear so far as is compatible with the sufficient preservation of the delicate color picture and without hiding the bacteria. It must be the aim of every careful observer to hit this happy medium between the extremes. The selection of a diaphragm and the consequent illumination of the preparation is greatly facilitated by an instrument first applied in England, but with which most microscopes are now provided, called the iris diaphragm.

The alteration which must be given to the pencil of rays proceeding from the angle of aperture of the condenser is here obtained with certainty by simply moving an arm which regulates the size of the opening, and the distance to be moved must be determined by the appearance of the microscopic picture.

II. MICROSCOPICAL EXAMINATION OF BACTERIA.

In the present state of science and its auxiliary machinery bacteria can be examined unstained or stained.

The former is the simpler process, and though it only yields satisfactory results to a certain limited extent, yet it is an extremely essential and altogether indispensable part of our investigations.

Never be committed to anything approaching an expression of certainty with regard to a species of bacteria before examining it in its uncolored state—i.e., not till it has been studied under circumstances which, at least approximately, correspond with its natural condition. For in examining stained objects we always have to deal with a dead organism and with altered circumstances which only permit us to form a conditional opinion of the state of things which existed during the life of that organism.

In some small flasks of beef-bouillon, which forms an excellent nourishment for bacteria, a cloudy turbidity of the liquid indicates that there are plenty of bacteria in it. A growth of certain bacteria on the surface of slices of boiled potato appears as a whitish-gray moist covering. These examples show the occurrence of bacteria on liquid and solid media, and the rules for the treatment and examination of both are as follows: To examine the micro-organisms in the bouillon in their uncolored state, the simplest way is to take a platinum wire, bent at the point, remove any foreign matter that may adhere to it by heating it to redness in the flame, wait a few moments until it has sufficiently cooled, then dip it into the flask and endeavor to draw up a little of the liquid. The portion thus obtained is spread on a thin cover-glass, and this process is repeated (meanwhile heating the wire each time) till a sufficient quantity of the liquid is placed on the cover-glass. Then turn the latter over and lay it on the slide, so that the fluid to be examined lies between the cover-glass and the slide. The quantity of fluid should be sufficient to form an even capillary layer, without air bubbles or dry spots, yet it should not extend beyond the edges of the thin glass, still less overflow the surface of it. Proceed in a similar manner with the bacteria that have grown on a solid nourishing substance, remembering, however, that for the purpose of examining them they must first be put into a liquid medium. Put some distilled water on a cover-glass as before, and bring into this a small quantity of bacteria which are taken from the surface of the potato with a platinum wire that has been previously heated. All the further steps are identical with those already described.

Even with liquids containing bacteria, it is often well to place some distilled water on the cover-glass, in order to dilute them before examination. The number of micro-organisms in nourishing solutions is often so extremely great that a certain degree of dilution is necessary before they can be examined successfully. When such preparations are brought under the microscope, it is best to at once employ the strongest magnifying power.

Put a drop of immersion oil on the cover-glass, screw down the lens with the coarse adjustment, till it enters the oil, and then proceed to the fine adjustment. Of course the examination must be made with the diaphragm, since the specimen is unstained, and a "structure picture" is desired; an opening about the size of a pea, with tolerably good light, will be found the most suitable.

In examining such a preparation the bacteria will be recognized without difficulty. They move across the field, some knocking against and rolling over each other, others swimming slowly, or even lying quite still for a few moments. But this kind of examination has, nevertheless, its very great disadvantages.

By the pressure of the cover-glass on the slide, inequalities are continually being caused in the layer between them; at the free edges a brisk evaporation is constantly taking place and causing all sorts of currents in the liquid, so that the whole preparation is in continual motion. As this motion is sometimes quick and irregular, it sweeps away the bacteria from under the eye of the observer; they are driven about in rapid currents, and it becomes impossible to decide one of the most important questions with regard to many micro-organisms — whether they possess the faculty of voluntary movement. Besides this, the rapidity of the drift movements is generally so great that it is extremely difficult to satisfactorily observe the finer peculiarities of form in these wandering bacteria. Lastly, too, a prolonged examination of such preparations is rendered impossible by the evaporation, which soon ends in the complete drying up of the liquid.

These are the very grave objections to this method of procedure, and which prevent its general use. As a rule, it is only employed tentatively—for instance, to see whether a fluid contains bacteria or not; further, also, when we wish to take a speedy glance, in order to know whether we have bacilli or micrococci to deal with, etc. In all cases, however, where more exactness is required, this most simple of all proceedings is abandoned, since, fortunately, means have been found to avoid these inconveniences.

The end of a platinum wire is bent into a loop. Special attention is called to the manner of making such a loop. It is no mere caprice that calls forth this advice. Experience will prove that this simple tool is of great importance for many technical manipulations, and particularly in the cultivation of bacteria it is almost constantly in use, while the success of many experiments depends entirely on the suitability of the loop employed. The great point is to see that the loop forms a *closed* circle, so that it is able to raise a full drop. It should, further, be smoothly rounded and neither too large nor too small; the size of a capital O in Roman print will be found to best answer general requirements. If such a loop —after heating in a flame—be plunged into the liquid, a drop will be found hanging to it when it is withdrawn.

Carefully and slowly touch the centre of a cover-glass with it, the drop leaves the loop and passes to the glass, on which it quietly remains. A successful "drop" should be as shallow as possible, should have smooth, even edges, and be about the size of a pea at the most. Now take a hollowed slide with a shallow cavity and brush around the edge some soft vaselin, or some other air-tight unguent, turn the slide over so that the hollow may be on the under side, and press it down upon the cover-glass, which, of course, sticks to the vaselin. Take up the slide with the cover-glass, the latter will lie above the cavity of the former, and in the hollow "hangs" the drop, protected from evaporation and isolated from all that surrounds it by the vaselin.

To examine bacteria which have grown on a solid medium, first with the platinum loop place on the cover-glass a drop of distilled water, and then "inoculate" it with a small number of micro-organisms by means of a platinum needle. The fewer bacteria introduced into the hanging drop, the better the preparation will be.

Beginners almost always make the mistake that in well-meant zeal they try to get as many germs as possible, and then they "cannot see the woods because of all the trees." The fewer micro-organisms present in the field the more exactly can their details be seen.

For this reason it is also advisable, in the case of liquids which contain bacteria, to put a few into a drop of distilled water.

A microscopic examination of such an object is not without its difficulties. If the immersion plan be used, and in order to get the clearest possible picture, the smaller stop is employed, the field will prove rather dark, and some difficulty will be met in merely finding the drop. Seek and seek, push the slide backward and forward, and at last the lens is brought down too far, the cover-glass is smashed, and thus the examination is brought to a premature end. It is, therefore, better to first place the preparation under a low power, bring the edge of the drop into the field, and then use the high power with immersion. Do not, however, make the mistake of unnecessarily adding to the difficulty by using strong eye-pieces. The edge will soon be found, appearing as a wavy line, clearly distinguished from its surroundings, and generally bordered by very small bubbles which have settled on the glass.

It is desirable for several reasons to devote special attention to the edge. Here the liquid is thinnest, and the conditions for a leisurely examination are most favorable; in the middle of the drop it is almost impossible to see through its whole depth with the lens, and non-motile bacteria, which by their own weight sink to the bottom, escape observation. Further, the motile bacteria at the edge are restricted in their very rapid movement from place to place, and we can better observe their peculiarities of form. And lastly, the great majority of motile micro-organisms, yielding to their need of oxygen, proceed to the edge of the drop, and generally remain in its vicinity.

The great advantages of examination in the hollowed slide will now be apparent.

It shows the bacteria in the nearest approach to natural conditions, and enables us to take a glance at them in "real life."

The form of the bacteria, it is true, can be recognized only to a certain extent, and all the peculiarities of form cannot be seen without the application of special means.

But the outlines are seen sharply and clearly, we observe the turbid, equable contents of the separate cells, now and then also a slight granulation, or, in the interior of the germs, those bright, gleaming, egg-shaped bodies, or spores of the bacteria, may be detected.

The ability of voluntary movement is seen very beautifully and clearly with the hollow slide in several of the bacteria; in fact, this kind of examination is the only one that can yield unobjectionable and really reliable information regarding this important function of the micro-organisms.

It is true that slight changes of place made by micro-organisms which have not the power of self-movement may be noticed. But on closer observation, it will be seen that this is not true locomotion, but only a motion at and about a point, the so-called Brown's or molecular motion. The globular bacteria which, as a rule, are non-motile, in particular, almost always show this peculiar dancing up-and-down movement. But it is very different from the decided, almost self-conscious manner in which many of the rod and screw-shaped bacteria move. Differences of gait may even be distinguished. The typhus bacilli taken from a potato culture, in serpent-like windings glide nimbly across the field of vision; the hay-bacilli waddle along, bending from side to side as they go; while the Bacillus megaterium crawls along with his peculiar amœboid movement. Quite different is the scene when examining the blue-milk bacilli, the green-pus bacilli, or even the cholera bacilli. There all appears in a confused jumble, like "a swarm of dancing gnats," and the eye of the observer can scarcely distinguish the individuals in the moving mass.

The special motile organs of the bacteria, the flagella, it is true, cannot be seen by this method of observation; as already noted, it requires other conditions to make them visible. Another particularly valuable property of the hollowed slide consists in the outside air being kept out, so that no considerable amount of evaporation can take place from the surface of the liquid. Therefore the hanging drops are indispensable for all examinations of long duration; they can be kept for days, and even weeks, and that at pretty high temperatures, without drying up; and (as will be described later) this quality also enables us to utilize them in our breeding processes.

Beyond this point we cannot advance with unstained bacteria.

In tissues especially, it is all but impossible to find and to distinguish them, and therefore we must proceed to the observation of stained micro-organisms.

III. STAINS AND STAINING.

By the examination of unstained bacteria in the hanging drop, the peculiarities of form are, for the most part, imperfectly recognized, and many distinctive marks altogether unseen; the preparations are not very durable, and therefore of but little use for comparative investigation. This is very different, however, as soon as the distinguishing power of color is utilized.

In the first place, color is an invaluable means for distinguishing bacteria with certainty from their non-bacterian surroundings, for a certain class of staining matters and the great mass of the bacteria stand in special mutual relation to each other, which can be utilized at pleasure. Thus the staining of many micro-organisms was the first thing that betrayed their existence, and the true sources of the most important pathological conditions now recognized as bacterial diseases have been discovered only by their aid.

Scarcely a quarter of a century has elapsed since the first faint attempts were made to imbue animal and vegetable tissues with coloring matters. Yet these beginnings for some time attracted but little attention, and it was not till about the year 1875 that the new process came into pretty general use. Then came the anilin colors and their application for the staining of bacteria. Rapid progress was now made; everybody began to stain, the isolated and the double staining were introduced, and at the present day the art of staining has already reached a high state of perfection. The names of Weigert, Koch, and Ehrlich are closely connected with these advances in the technics of investigation.

The anilin colors, which are of special importance for bacteriology, are obtained from a secondary product which arises in the manufacture of the gas which lights our streets—from coal tar. This is a somewhat intricately compounded substance, and the number of dyes obtained from it is considerable.

Most of them have figured in the service of science for a longer or shorter period, and have been adopted on the recommendation of an investigator who was partial to them. Yet for ordinary purposes only a limited number of them have remained in use.

In an undissolved state most of them are fine, smooth powders, while some occur in the form of small crystalline scales, with an iridescent gleam.

First we have a violet dye, the *gentian-violet*, which is distinguished by particularly strong staining action. It is not a pure chemical substance, but probably owes its peculiar powers to foreign admixtures. Another somewhat similar violet is the *methyl-violet;* and there is a blue dye, *methyl-blue.* For red, we have *fuchsin*, or rubin; for brown, the *Bismarck-brown* or vesuvin. All these anilin colors have the common property of containing a proper staining ingredient of a basic character; they are the basic anilin dyes. To them belongs by far the most important place in bacterial investigations, yet some of the acid anilin dyes are also used, especially eosin and acid fuchsin.

In addition to these bodies, we also employ a staining agent which comes from the vegetable kingdom. This is *hæmatoxylon* prepared from Campeachy wood (logwood), and lastly an animal product, *carmine*, obtained from the cardinal insect. There exist also, as before mentioned, a great number of other stains, more or less similar to the above, which were formerly in use, but of far less importance than those just named.

In the great majority of cases, the colors mentioned will be found fully sufficient, and there is not for the present any pressing need to enlarge the list.

In what does the peculiar action of these stains consist, and what is it that makes them indispensable for our purposes? Examine a section from the liver of a healthy rabbit. Transfer it from alcohol to a saucer containing a diluted solution of an anilin violet color. After it has remained a short time—about two minutes—take it out, remove the superfluous staining matter by washing in distilled water acidulated with a few drops of acetic acid, place the section on a slide, lay a thin cover-glass over it, and examine it with a medium power.

If, in order to get the color picture separated as much as possible from the structure-picture—i.e., to see the peculiar action of the stains—use no diaphragm, but with the full illuminating power of the Abbe; the texture of the tissue will be found, on the whole, less distinct than formerly with the uncolored object and without the Abbe.

The outlines of the cells are not sharp, the smaller vessels are almost imperceptible, the connective tissue shows only slight traces of stain, and the more delicate parts of the structure are quite unrecognizable. But something strikes the eye at once, and that is the strong distinctive staining of the cell nuclei, which, at the first glance, enables us to distinguish them from surrounding structures. Hardly anything but the nuclei is seen, and while it was somewhat difficult to see these portions of the cells clearly in the uncolored preparation, they now surpass everything else in distinctness, so that it is by the presence of the nuclei that attention is drawn to the cells to which they belong.

Now take another section, this time from the liver of a rabbit which has died from inoculation with anthrax. Treat it in the same way as the other; stain it for the same length of time in the same anilin solution; remove superfluous color with acidulated water, and put it under the microscope with a high power, and, naturally, without diaphragm. The same picture as before! the same indistinct, faint staining of the connective tissue, as well as the other tissues; the same prominence of the nuclei. Along with them, widely dispersed over the whole preparation, anthrax bacilli are seen, which in the distinctness of their coloring perhaps even surpass the nuclei.

The most striking peculiarity of the anilin dyes is that they stain the nuclei and the bacteria more strongly than the surrounding parts, thus giving them an isolated coloring, and it is precisely because all other parts of the object remain so much less distinct than the nuclei and the cells (the bodies of the cells are thrust into the background) that this system of staining is so valuable for us. It seems to point out the bacteria, which we should otherwise have to seek for among many other things, and possibly not find them after all.

It is particularly the basic anilin stains which, by differentiating equally the cell nuclei and the bacteria, are so well adapted to our purposes. Most of the acid anilin stains, as also carmine and hæmatoxylon, act differently; they are chiefly "nuclei stains," and leave the bacteria almost entirely unaffected.

If one of these sections which has just been colored with anilin violet be placed in a solution of simple carmine or picro-carmine and then treated as before, the nuclei will appear distinct under the microscope, the protoplasmic bodies of the cells are also clearly brought out, but of the bacteria there is nothing to be seen. They have not absorbed the stain at all.

Particular attention is called to this distinction between pure nucleus staining and the staining of nuclei and bacteria together. Means have been found to happily utilize this distinction for purposes of investigation.

On what does this difference of behavior in bodies comparatively nearly related depend?

We may say that in producing the stains various processes probably take place.

There are physical processes, depending on the laws of imbibition and diffusion, and on those peculiar phenomena which we call surface attraction. But without doubt the principal part in the development of the coloring and its specific forms must be assigned to processes of a purely chemical nature. Although there may not be such relations as can be expressed by formulas and give origin to definite new combinations, yet the fine differences which are observed in employing the various staining matters, their greater or less coloring power, and in particular the marked preference of special kinds of bacteria for special colors, certainly depend on special chemical affinities, on a certain relationship between certain stains yielding and certain other colors absorbing substances.

We have, therefore, a right to regard the staining as the expression of a micro-chemical reaction, and if it is not always such in a strict sense, it has the same value for us, since it enables us to distinguish from each other by special criteria substances difficult to separate.

In order to be used they must be dissolved. Now, the anilin colors, as also carmine and hæmatoxylon, dissolve equally well in alcohol and in water. We use the first principally and make a saturated alcoholic solution of the anilin colors, by putting into a bottle containing alcohol as much of the dry pigment as the alcohol can dissolve. It is better to put in a still greater quantity, shaking it up well, filtering it after a few days, and using the concentrated solution thus obtained. One point, however, requires particular attention. Anilin colors are seldom if ever sold in complete purity. They are generally mixed with other substances, particularly dextrin and soda, not as an intentional adulteration, but only to meet certain technical requirements.

These admixtures often dissolve slowly and imperfectly in the alcohol, and remain at the bottom of the bottle as a thick sediment, which, when we employ the solution, may cause deposits and other unpleasant consequences.

The methyl-blue especially often contains quantities of foreign substance, and is therefore sometimes dissolved in distilled water instead of in alcohol.

Bismarck-brown, too, keeps better in an aqueous solution, or in a mixture of equal parts of glycerin and water, than in alcohol.

The concentrated aqueous or alcoholic solutions are the source from which we take the staining matter we require, and which is kept on hand for that purpose. An attempt to use these solutions in their concentrated form would produce effects greatly too rapid and too intense. They overstain the preparation.

On the contrary, it is our principle (and this deserves special attention) *to stain with very weak solutions, and allow a longer time for their action.* The peculiar differentiating qualities of the coloring matters then show themselves to the best advantage, enabling us to perceive differences which would be entirely buried under the quantity of color given by strong solutions.

The best way of preparing the diluted solutions is by taking a glass bottle of about 10 cm. in height, which has a pipette in its perforated cork, filling it two-thirds full of tap-water and slowly adding as much of the saturated solution as leaves the liquid in the thick part of the neck of the bottle just transparent. A little less is better than too much. The same result may be obtained by taking a saucer of tap-water and dropping the concentrated solution into it until the liquid begins to lose its transparency.

It is true, solutions thus prepared have the bad quality of decomposing after a time. The coloring matter is deposited more and more, and the solution soon begins to lose its utility. It is therefore well, for delicate staining operations, to prepare the solutions anew every two or three weeks, or even oftener. But before commencing to stain it is well to consider whether we can employ indifferently all the stains hitherto mentioned, or whether certain colors are preferable for certain purposes. In fact, they have their special properties.

Gentian-violet is, as already mentioned, a very intense stain. But this intensity is apt to become a disadvantage, for too long an exposure overstains the object, the distinctions of texture are lost, the whole preparation becomes indistinct and worthless for examination. Cautiously applied, however, gentian-violet is a particularly useful stain, and the more so as its effects are extremely enduring and do not fade.

Methyl-violet colors less intensely than gentian-violet, and is not so apt to overstain an object, but it is less durable.

Methyl-blue has far less coloring power than the above-mentioned dyes. It requires a very long time to produce a perfect staining, seldom overcharges with color, and yields tolerably durable preparations.

Fuchsin is a very beautiful, strong, lively color, which is particularly pleasing to the eye. It is nct very apt to overstain and gives preparations a high degree of durability, so that it is, in many respects, preferable to the others.

Bismarck-brown stains slowly and is no great aid to discrimination; therefore, it would probably not have remained long in use had it not formerly been indispensable for the purposes of microphotography.

The ordinary photographic plates are only sensitive to rays belonging to the blue part of the spectrum; all the others are more or less ignored by the plate. Now, these blue rays are absorbed by a solution of Bismarck-brown. If a preparation treated with vesuvin be placed before a plate and illuminated, the portions of the object which are stained brown, especially therefore the bacteria and the cell nuclei, will stop the really efficacious rays. At these points the plate receives no light, and the bacteria are seen on the negative as transparent points or lines, according to their shapes.

These phenomena were, however, only to be seen in the exact manner here described, with the old collodion plates; the dry plates now universally employed possess a higher degree of sensitiveness to color, and go beyond the blue portion of the spectrum. Recently plates have been made which possess this quality in a far higher degree, and with the introduction of the orthochromatic or isochromatic plates into micro-photography, the use of Bismarck-brown has become superfluous.

In fact, all the varieties of bacteria hitherto discovered may be stained with solutions of the basic anilin colors, of course some more or less quickly and more or less satisfactorily. Yet it is possible to strengthen the staining power of these dyes by special methods, and thereby heighten their effect upon the bacteria.

It is known that, in the process of staining, some substances are able to play a peculiar mediating part between the coloring substance and the object to be stained. They are called mordants, and have been in use for a long time, especially in the dyeing of cottons. As the most important of them may be mentioned certain metallic salts, in particular some combinations of lead, iron, and chrome, and then also alum, tartar-emetic, and tannin. Several of them have also been found useful for histological purposes—for example, acetate of alumina, which in combination with carmine has attained great importance as alum-carmine. It is less adapted for bacteriological investigation, being a pure nucleus stain and leaving the bacteria almost uninfluenced. Yet in special cases, as will be explained later, this particular stain has been utilized to advantage.

This is still more the case with another carmine combination, the so-called picro-carmine, a compound of carmine and picric acid. This compound color stains not only the nucleus, but also the body of the cell and the connective tissue, in a peculiar manner, and is therefore in a high degree suitable for differentiaing the constituent parts of tissues. The bacteria, however, as we have seen, are not stained by picro-carmine.

On the other hand, the combination of coloring matters belonging to the basic anilin group with a particular kind of mordant is extremely useful for bacterial researches, by disclosing the finest peculiarities of form. The staining of the flagella by Löffler's method is effected by treating the preparations with a compound mordant fluid. This consists of two parts—say 10 cm. of a 20-per-cent solution of tannin, a few drops of saturated aqueous solution of ferro-sulphate, and one part—say 4 or 5 cm.—of an infusion of logwood, taking one portion of wood to 8 of water. This gives an inky fluid; and in fact, Löffler was led to the discovery of his method by employing black ink as recommended by Neuhaus.

There are two other substances which act as mordants, though they do not belong to that group, namely, anilin oil and phenol, with their allies.

Anilin oil (the mother of most of the anilin colors) is an oily body of peculiar smell, obtained from tar. It is not a genuine oil, but rather, by its chemical composition, a mere derivative of the combination with the aromatic series of benzol, and it took its name only from its outward behavior. It is employed in aqueous solution, and in connection with it gentian-violet and fuchsin yield extremely valuable stains which we often have to employ. There are numerous directions given for uniting such an anilin coloring solution with a saturated aqueous oil solution.

The latter is thus obtained: take 5 cm. of anilin oil and shake it well for some minutes with 100 cm. of distilled water, and pass it through a moistened filter. The filtered fluid must be as clear as water, must show no more oil-drops, and must not become turbid when shaken.

Yet this anilin water, as it is often called, is not stable. It is apt to decompose, and even the addition of alcohol, as has been recommended, only remedies this defect to a certain degree. It is therefore much better to prepare a small quantity of fresh anilin water for every new requirement. In this way one is surest to avoid deposits and faults of staining. Pour anilin oil into an empty test-tube to the depth of about 2 cm., add distilled water, shake up for a few moments, filter the emulsion, pour the clear results of this filtration (which have a strong smell of anilin oil) into a glass dish, and then add as much of a concentrated alcoholic solution of fuchsin or gentian-violet, or other coloring matter, as is required to produce saturation with coloring matter. This is easily recognizable on the surface by the appearance of a peculiar iridescent film with metallic colors, which consists of undissolved coloring matter and shows the state of saturation. In many cases, however, it is better to stop short of saturation, and stain for a longer time with a thinner solution. Phenol, or oxybenzol, is nearly related to anilin, or amido-benzol, and is employed in like manner. Thus in Ziehl's solution we have a combination of aqueous carbolic acid and fuchsin.

Ziehl's Solution.

Aqua destil., ...	100 grams.
Acid carbolic cryst., ...	5 "
Alcohol, ...	10 "
Fuchsin, ...	1 gram.

This can be made very easily by taking a 5 per cent aqueous solution of carbolic acid and adding concentrated alcoholic solution of fuchsin till saturation is reached. In the same manner we prepare a mixture of methyl-blue and phenol, as recommended by H. Kühne. Put 1 to 2 parts of coloring matter into 10 to 15 parts of pure alcohol—that is, make a concentrated solution—and add this in suitable quantity to about 100 cm. of 5 per cent carbolic acid solution.

We have still to mention the combination of anilin stains with potash. Methyl-blue in particular, which, though it has but slight coloring power, nevertheless yields beautiful and distinct pictures, gains by this mixture a considerably extended field of usefulness and an almost unlimited power of staining all kinds of bacteria. Two solutions of methyl-blue and potash are particularly useful; the one called Koch's solution, or the weak solution, consists of:

Concentrated alcoholic sol. of methyl-blue, .	1 gram.
Distilled water, ...	200 grams.
10-per-cent solution of liquor potassæ, . .	0.2 gram.

The other—Löffler's, or the strong solution—consists of:

Concent. alcoholic sol. methyl-blue, . . .	30 grams.
Liquor potassæ 1:10,000 (0.01 per cent), . .	100 "

The latter especially is a really excellent stain, which hardly ever proves refractory.

Carbonate of ammonium in 0.5 to 1% solution, as recommended by H. Kühne, acts in the same manner as the potash. It is, however, not applied together with the coloring matter, but serves as a sort of preliminary mordant. The preparations are placed in it for a few moments before being transferred to the staining solution. It is a combination of two such proceedings which Löffler has recently recommended for the staining of bacterial preparations in general.

An anilin-water gentian-violet solution or an anilin-water fuchsin solution is mixed with 1% hydrate of sodium, in the proportion of 1 : 100 (1 cm. of the staining solution). The coloring matter is thereby brought to the verge of separation, and in this state develops a high staining power.

We have by no means exhausted all the combinations of color nor all the directions for staining with which our investigators have enriched this branch of research, but it would lead too far should we attempt to go more into detail.

Only one more means to increase the action of our dyes will be mentioned, and that is the warming of the solutions while they are being used. The coloring substance when warmed penetrates very much more quickly and energetically into the objects, the tissues as well as the bacteria take a stronger and more distinct staining, the stain lasts better and becomes less soluble.

This means cannot, of course, be employed in all cases. Cover-glass preparations bear the heating well, and with them the process can be employed without damage so far as to cause the formation of bubbles and even boiling of the fluid. This can best be managed by taking the cover-glass with forceps, letting a few drops of the staining solution fall on it, and holding the preparation, with the fluid, immediately over a flame. Vapors soon rise and ebullition follows; if we supply an occasional new drop, to replace what is lost by evaporation, the process can be continued indefinitely.

Sections, on the other hand, are apt to be destroyed by such treatment; they fall apart and become useless. Here, then, it is better not to attempt the warming, but to stain for a longer time in cold solution.

If a preparation be now stained with one of the already described coloring matters, we must first remove the superfluous dye-stuff that has not been thoroughly imbibed before proceeding to examine it. In this way alone can a clear picture be obtained, showing distinctly the differences which the staining is intended to develop.

To do this we chiefly employ water and alcohol; both dissolve the pigments, and thus they gradually withdraw the stain from the preparations. In very many cases water suffices, and if the washing is continued sufficiently long, the extra color will be so far removed as to leave what are called "well-differentiated" preparations.

We often endeavor to supplement the decoloring power of the water or the alcohol, and give it a special action by the addition of acids.

Acetic acid, as Weigert has taught, is almost indispensably necessary for producing a decided nucleus staining. This acid has the property of making the protoplasm swell, while it causes a contraction of the nuclei. At the same time, it removes the coloring matter from the former, but fails to penetrate into the latter, which therefore appear more prominently than their pale surroundings, and particularly attract the eye by their staining.

We usually add three or four drops of acetic acid to about 20 cm. of water, and wash the sections in this solution for a considerable time.

The other acids act much more energetically, and some of them in a different manner also. These are muriatic acid, sulphuric acid, and nitric acid, of which the last mentioned is a particularly strong bleaching agent. None of them should be used without caution, as they exercise a destroying influence on most dyes and are capable of neutralizing the fastest stains.

There is more power in alcohol than in water, and therefore it is preferable in the case of strong stains. As may be imagined, too, the acidulated alcohol surpasses the acidulated water in its decoloring power. All the decoloring agents may therefore be set down in a list of increasing strength with water at the bottom and nitric-acid alcohol at the top.

Iodine has a somewhat similar action to those just mentioned, though its application is not the same. Iodine is employed with arsenite of iodide of potassium (iodine 1 part, arsenite of iodide of potassium 2 parts, water 300 parts). This mixture exercises a very peculiar influence on preparations which have been treated with anilin water and gentian-violet. It forms a deposit with the coloring matter, which, however, only adheres to the bacteria, and can be washed out of all other parts of the tissue, the nuclei included.

We thus obtain an isolated bacterial staining which is particularly suited for special purposes. Such a process is that recommended by Gram.

Remember that we possess stains, such as carmine and picro-carmine, which particularly affect the nuclei, and it may readily be imagined that by the union of the isolated bacteria, staining on the one hand and the isolated nucleus coloring on the other hand, we can obtain the most perfect distinction of all the separate component parts of such a preparation. If we have stained the bacteria alone according to Gram's method, we let the second stain act upon the nuclei, and thus we get a contrast in one picture which shows the mutual relations of bacteria and tissue in the most satisfactory manner.

Unfortunately, Gram's method is not equally applicable to all micro-organisms. Typhus bacilli, the bacilli of Asiatic cholera, the bacilli of the cholera of fowls, septicæmia of rabbits, and malignant œdema, the bacteria found in pneumonia by Friedländer, the anthrax bacilli, the gonococci, and the spirilla of recurrent fever, cannot absorb the color sufficiently to retain it; and under the influence of iodine they lose color with the nuclei. For all the other bacteria, however, and especially for the great majority of the micrococci yet discovered, this method has been found excellent.

IV. PREPARATION OF COVER-GLASS SPECIMENS FOR STAINING.

We now know what stains we can employ in the investigation of bacteria, how they are prepared, how they may be varied, and what the chief points are to be considered in the staining process.

But to obtain fairly good results, it is indispensably necessary to prepare the objects in a particular manner for the reception of the stains.

To do this take, with a bent wire previously heated in the manner already described, a small portion of the liquid and spread it in a very thin layer, as evenly as possible, over the cover-glass, or remove a sample of the bacterial film from the surface of the boiled potato, and with the aid of a drop of distilled water spread it over the cover-glass.

The preparation must be completely air-dried and the water must be entirely removed. This process may be accelerated by moving the cover-glass backward and forward at some distance above a gas flame. Yet this must be done with the utmost caution, lest the preparation should be overheated or even burned. It is therefore best to hold the cover-glass between two fingers above the Bunsen burner or spirit-lamp; the fingers are very sensitive thermometers, and will be sure to give warning of any dangerous proximity to the flame.

Now allow the staining to proceed. As a rule, however, we have not such exceedingly simple measures to deal with. The chief value of staining is for cases in which the presence of bacteria is suspected in the interior organism, and we must, therefore, examine blood, tissue juices, pus, or sputum. Such preparations cannot be stained immediately after they have been air-dried.

They contain albumin, which is not rendered insoluble by simple drying. When this albumin comes in contact with the dyes it swells and dissolves, may be washed from the cover-glass, and often causes deposits under the influence of the staining solution, which spoil the preparation.

Albuminous fluids may be fixed on cover-glasses, as recommended by Ehrlich, by exposing the glasses to a heat of 120° C. for about twenty minutes, or, what is much more convenient and equally efficacious, draw them, as recommended by Koch, three times, at a moderate speed, through the flame of a Bunsen burner, with the preparation on the upper surface, to avoid its coming into direct contact with the flame.

When cautiously heated in this manner, the forms of the cells, bacteria, etc., do not alter in the least, nor do they lose any of their capacity for absorbing color; the albumin, however, passes into an insoluble state, in which condition it remains unchanged under the further manipulation of the cover-glass in the process of staining.

A few precautions must not be neglected. The preparation must be completely air-dried before it enters the flame; otherwise the albuminous matter coagulates under the influence of a high temperature, instead of becoming homogeneous. Again, the heating process must not be carried too far. Many bacteria—for instance, the anthrax bacilli—are extremely sensitive to an excess of heating; they change their form, break up into separate granules, or swell up like bubbles, become surrounded with a sort of halo, and lose their ability to stain.

It is generally sufficient if we draw the glass three times through the full flame of a Bunsen burner, at about the same rate as the movement with which we wave a handkerchief in salutation to a person at a distance. It may be that some wave it more rapidly than others, but we can take the average speed. If instead of a Bunsen burner a spirit-lamp is employed, the time must, naturally, be correspondingly lengthened.

This procedure is necessary only for preparations containing albumin. Yet as the heating above described is not deleterious for other objects, and as it cannot always be known whether coagulable matter is present or not, we habitually prepare all cover-glasses for staining in this manner.

The manner of removing bacteria from a liquid or solid nourishing medium to a cover-glass has been described. To examine blood and tissue fluids from an animal just dissected the following procedure is necessary.

Either take up a drop of blood with the bent wire previously heated, and put it on the cover-glass, or press a small piece of any organ gently against the glass, rubbing it over the surface and so

4

spreading out the liquid. Then lay a second cover-glass over the first, and produce between the two a perfectly even, extremely thin layer. Now carefully draw the upper glass away from the lower one, and there are at once two preparations, both of which can be used. It is necessary to wait till they are perfectly air-dried and all traces of moisture have disappeared. Then seize the cover-glass with the forceps and draw it three times through the flame. If we wish to remove any hæmoglobin that may be in the layer before staining it, and thus isolate the bacteria as much as possible, we must lay the preparation for a few seconds in a 1% to 5% solution of acetic acid (as recommended by Gunther), remove this with distilled water, and dry the cover-glass anew before proceeding.

Generally, however, this precaution is unnecessary, and we may therefore drop some of the dilute alcoholic anilin solution from a pipette upon the preparation immediately after the heating process in the flame.

When the stain has operated a few minutes, wash away what is superfluous with distilled water and the process is complete.

The time to be allowed for the action of the stain cannot be definitely fixed. It depends upon the nature of the preparation itself and on the staining power of the dye employed.

Methyl-blue, for instance, always requires much more time than fuchsin or gentian-violet.

The preparation is now ready to be examined in water. Dry the upper surface of the cover-glass with some blotting-paper or with the finger (since it has to receive the drop of oil for the immersion lens), and then the wet preparation is laid on a slide. The evaporating fluid must be added to from time to time, to avoid difficulties in the refractive conditions—the so-called dry blots—which would otherwise occur and make investigation impossible.

In examining such objects in water, too, one often sees Brown's molecular movement, even in the stained bacteria. A few bacilli or cocci which have freed themselves from their surroundings may be seen to merrily dance about in the fluid.

Of course the cover-glasses will bear all kinds of staining matter. We can with equal satisfaction use the simple anilin staining solutions, or the alkaline bacteria stains, or the anilin-water mixtures. If it is desired to obtain particularly intense and rapid stainings, the coloring matter is warmed on the cover-glass, or the latter is allowed to float, preparation side downward, on the surface of a hot solution of the dye.

After the more active staining processes, it is often necessary to employ the stronger decoloring agents, alcohol, acids, etc.

If Gram's method is to be employed, the cover-glasses are stained with a hot, strong, saturated anilin-water gentian-violet solution for one or two minutes; they are next put immediately into iodide of potassium solution for about half a minute, and lastly washed with alcohol till no more stain can be removed.

They may then be examined at once, or they may be completed with a contrast stain such as saffranine, carmine, or a very weak solution of Bismarck-brown. An alcoholic solution of eosin is particularly recommended, since it displays the cellular portions of the blood with rare clearness. The superfluous eosin is removed with distilled water or dried with blotting-paper and mounted in balsam.

The immediate examination of the stained cover-glass preparations in water is particularly recommended in most cases, because it injures the micro-organisms least and shows them to us in a condition more nearly approaching their natural form. It is true, the hollow slide is much superior for this purpose, and that the alcohol, the staining fluid, the iodide of potassium, the acids—in short, the whole process of preparation, affects the appearance of the bacteria, and leaves us only mummies and corpses to examine.

The protoplasmic body of the cell contracts, the membrane alters its appearance, granulations are produced which are quite foreign to the living organism.

But in spite of all this, we can on no account renounce the staining system, which has its advantage and is quite indispensable for our investigations.

In the preparations mounted in water the cells are still expanded and full of juices, the membrane comes out clearly, the separate micro-organisms show body and mass.

All this is changed in a most striking manner as soon as we preserve such an object, whenever it is prepared for comparative investigations or for lecture-room demonstrations. In this case the object, after having been thoroughly air-dried, has to be mounted in Canada balsam.

This medium causes the bacteria to shrink considerably. They take a poor, shrivelled appearance, and when re-examined in Canada balsam, what had before been seen in water will frequently be recognized as the same with difficulty.

Therefore permanent preparations should only be made when necessary for the purposes above mentioned, or when there is some other particular reason for wishing to preserve an object.

Two special applications of the cover-glass staining for particular purposes will be most appropriately mentioned here. The endogenous spores of the bacteria are distinguished by a firm covering, a membrane with great powers of resistance against exterior influences.

This membrane obstinately refuses to admit our ordinary staining matters into the interior of the spore.

If we treat a preparation containing spores with a solution of gentian-violet ever so long, they (the spores) remain as uncolored, brightly-shining gaps, of the well-known egg-like shape, which contrast with the fully-colored residuum of the bacterium not employed in forming the spore.

It is therefore necessary to resort to the most active means in order to overcome the resistance offered by the spore-membrane to the penetration of coloring matter. When this is done, the stain is, as it were, surrounded by a protecting mantle, its retreat is cut off, and it can now only with great difficulty be removed again.

After this explanation, the process of spore-staining will at once be understood. We stain the cover-glass with spore-bearing bacteria for an hour in Ziehl's solution, hot, or still better, boiling.

This is done by keeping a small dish of carbol-fuchsin at the boiling point by means of a gas-flame or spirit-lamp, and from time to time adding of the solution sufficient to make up for the loss by evaporation.

Then the spore is penetrated by the red coloring matter, which is now just as hard to remove as it was to retain in the first place. From the remaining part of the cell, however, it can easily be removed. If the cover-glass be now treated with diluted or pure alcohol, the spore retains its color and the rest of the cell is decolorized.

The spores are clearly visible as bright red, egg-shaped formations, while scarcely anything remains visible of the other part of the cell. But that it really is there, and only requires a little aid to again become distinctly visible, will be seen if we employ a contrasting color (for red, the best is blue), and of course it must be a bacterial stain. A diluted alcoholic solution of methyl-blue is greedily absorbed by the cell, and in such a double-stained preparation one sees the deep red spores lying in rows in their deep blue cells like a string of beads. The picture is really beautiful and well calculated to show the advantages of staining.

All the spores of the different kinds of bacteria do not by any means behave in the same manner under the treatment above described. As regards their powers of resistance to other influences, such as heat, chemicals, etc., a difference of behavior has been observed, and it is therefore not surprising that some spores yield but slowly, others very quickly, to the specific stains. The anthrax bacilli, for instance, take the color very unwillingly, while several saprophytic bacteria, the hay-bacillus, the Bacillus megaterium, etc., are far more easily managed. Lastly, there are a few micro-organisms into whose spores even the ordinary staining solutions readily penetrate.

As regards this last point, it is necessary to be guarded against deceptions. The spore-formation is a gradual process. It has its preliminary stages, which display smaller or larger granules in the body of the protoplasm. These things take the stains sometimes more readily than the surrounding parts, and when they are of a certain size they may easily be mistaken for fully-developed spores. Their nature is not yet properly understood, but they certainly are not fully-formed spores.

They are incapable of the genuine spore-staining; yet that they differ considerably from the other contents of the cell is clearly seen by the fact that we can occasionally succeed in demonstrating them by means of a staining process described by Ernst. He first treats the preparation with warm (not hot) alkaline methyl-blue solution, washes with water, and then stains again with aqueous solution of Bismarck-brown. The sporogenic granules then appear in blue on a brown background.

The second peculiar staining process which it is desirable to consider, because, like the spore-staining, it is only applicable in the case of cover-glass preparations, is the method, already mentioned, introduced by Löffler for displaying the flagella of the motile bacteria.

The principal measures to be employed with the mordant solutions have already been noted, as well as their composition. Let a few drops of the mordant fall upon the preparation on the cover-glass, which must then be warmed immediately over the flame till vapors arise or the solution begins to boil. The mordant must now be washed away from all parts of the preparation with water, and even from the edges wiped away with blotting-paper. Wherever traces of the mordant which have not been absorbed by the bacteria are allowed to remain, troublesome deposits are sure to afterward arise. On this account, in making a preparation it is necessary to be careful that the layers be spread over the glass as thinly and transparently as possible. When this step is accomplished the staining proper begins, for which anilin gentian-violet, or anilin-fuchsin, or better still, carbol-fuchsin, can be employed. Löffler himself prefers to use here his latest coloring mixture, which has already been mentioned. He takes 100 c.cm. of 1% soda-lye solution, and to this 4 or 5 grams of gentian-violet, fuchsin, or methyl-blue, in powder, is added. Two or three drops of this solution must be filtered upon the cover-glass.

To act satisfactorily the color must be applied hot; it is then rinsed with water and the preparation at once examined.

Not only the cover-glasses, but also the tissues in which bacteria are to be sought require special preparation for the staining process. They cannot be used just as they come from an animal immediately after death. For the determination of the simple question as to whether bacteria are present or not, the easily-made "smears" are far better. The bacteriologist, therefore, has no occasion to use fresh tissues—for instance, sections made with a freezing microtome—which indeed, as a rule, do not even show the micro-organisms themselves, much less the details of their structure.

Portions of the organs about as large as nuts, while still as fresh as possible and before putrefaction and decomposition have begun, are placed in absolute alcohol to harden them.

The alcohol must be perfectly anhydrous. It is therefore desirable to lay some blotting-paper in the vessels for the reception of portions removed from the various organs, and to lay such portions on the blotting-paper; for if the alcohol imbibes water out of the air and out of the portions thrown into it, the diluted part of the alcohol, being heavier, sinks to the bottom under the blotting-paper, while the portions to be hardened remain above in the anhydrous strata.

When they have remained about two days in the alcohol, which has been changed once or several times, according to circumstances, they are properly hardened; the alcohol has caused their fluid albumin, glutin, mucin, etc., to coagulate, and has withdrawn the water from their tissues.

The hardened portions are now fastened to small corks with some adhesive substance. A mixture of glycerin and gelatin has been found very suitable for this purpose: one part gelatin, two parts water, and four of glycerin are dissolved by heat and boiled to a thick consistency, after which the mixture is ready for use. A drop of it is placed upon the cork, the hardened portion pressed down upon the drop and allowed to cool for a few moments, and then the whole is once more placed in the alcohol.

After two or three hours the preparation is ready for further manipulation.

To examine such an object, first cut it up into a number of the thinnest possible sections. For this purpose employ one or other of the many kinds of microtomes now manufactured.* He who

* Bausch and Lomb, of Rochester, N. Y., manufacture several styles of these instruments, which are quite reliable.—J. H. L.

still makes his sections with the razor instead of the microtome is like the man who travels by coach when he could travel by train.

In cutting the sections both knife and preparation must be continually wet with alcohol, and the sections must be at once returned into alcohol from the blade. They are then ready for staining.

Pour into a glass dish one of the diluted aqueous anilin coloring solutions and lay a section in it. It makes an important difference whether the section is carried from water or from alcohol into the staining fluid. In most cases the former proceeding is preferable. The diffusion processes, which play an important part in staining, then operate more mildly, the tissue is less strongly penetrated by the coloring matter, and consequently the after-process of bleaching can be effected with more simple agents.

When the section has remained from 5 to 15 minutes in the fluid, according to the nature of the preparation and the coloring matter employed, it must be placed in diluted acetic acid, in which the superfluous stain is washed out. The acid is next removed by a short exposure in distilled water, in which it must also be examined to form a rough estimate of the success of the staining process. With some practice and a careful glance it can immediately be seen whether the staining is successful or faulty, and if it be the latter case, whether the staining or bleaching was too strong, whether the staining solutions were at fault, etc., etc. With a second section endeavor to remedy the defects noticed in the first, and if this proves satisfactory, proceed to submit it to further preparation.

The next step is to render the tissue transparent, in order to enable us to distinguish its component parts and to recognize its finest details. This is done by the aid of ethereal oils and similar means. Those most commonly employed are oil of cloves and oil of cedar; the former of which, however, has the disadvantage that it attacks and removes the stain, while the latter, on account of its extreme sensitiveness to water, must be employed with particular care. Oil of cloves can only bear small traces of water without causing opaque cloudy specks on the preparation, and our aim must therefore be to free the sections as completely as possible from the water before placing them in the oil.

This may be accomplished by placing them for a short time in absolute alcohol after they leave the bleaching solution.

The entire process, if it be desired to produce a typical nucleus and bacteria staining on Weigert's plan, is as follows: The sections pass from distilled water into the staining solution, thence into a weak aqueous solution of acetic acid to bleach them, then into distilled water to remove the acid then into absolute alcohol to make them anhydrous, then into oil of cloves or cedar to make them transparent, and lastly into Canada balsam, one drop of which, in the centre of a glass slide, is sufficient for a moderately large section. The removal of a section from one solution to another is best managed with a needle and a metal spatula, to bear the out-spread slice of tissue.

Others, following the plan of Weigert, perform the whole staining process from the beginning on the slide, thus avoiding the necessity of transporting the section. This plan has undeniable advantages for many cases; the sections do not fold or roll up, do not tear so often, and require very much less time for their treatment. Lay the section on a slide, pour the staining fluid upon it, pour it off again after staining, rinse it thoroughly, first with diluted acetic acid and then with distilled water, remove traces of water by alcohol, let a few drops of oil of cedar or oil of cloves fall upon it, remove excess of oil with blotting-paper, and inclose it in Canada balsam.

As to the choice of a stain in particular cases, either the simple or the compound fluids can be used; for example, either Löffler's methyl-blue or Ziehl's carbol-fuchsin, etc.

The time during which the preparation should be exposed to the action of the various stains depends upon circumstances; in fact, the range of time over which practical experiments have been made extends from a few seconds up to forty-eight hours.

If the bleaching power of water is not sufficient it is replaced by alcohol, and if this is still insufficient, strong acids—muriatic, sulphuric, or nitric—are added to the water or alcohol. The coloring matter then disappears almost entirely from the connective tissues, but it is also apt to be washed out of the nuclei and bacteria, and these strong decoloring agents should only be resorted to in exceptional cases.

The methods above described will be generally successful, but they fail occasionally, especially when a tissue contains bacteria which receive the stain without difficulty, but lose it again with equal readiness when bleaching.

A particularly dangerous rock for these sensitive micro-organisms is the alcohol treatment to which the sections must be submitted before they can be transferred to the oil which is to clarify them.

It is known that alcohol is not only a means for removing water, but also a very efficacious decoloring or bleaching agent, and this latter quality often shows itself to a very undesirable degree. Various attempts have been made to overcome this difficulty, and either to discard alcohol altogether and replace it by some other substance, or to find some method of diminishing its excessive bleaching power. The first of these two plans was adopted by Unna, who treats the sections, after they have been properly decolored, with acetic acid and distilled water over a flame, thus removing all moisture by evaporation, a process which must of course be performed on the slide. Opinions are much divided as to the value of this dry method. If the wet section be held on the glass above a flame till the water is dispelled, then given transparency in xylol as Unna recommends, mounted in Canada balsam, and examined under the microscope, it will be found that, under some circumstances, bacteria can be seen which could not have been made visible by the ordinary process.

But the tissue is seriously damaged. It is full of gaps and cracks; it looks lumpy and curdled. Therefore when we wish to display the typical structure of the tissue, the dry method is scarcely applicable. If, on the other hand, it is only desired to display the micro-organisms that may be in the tissues, it may in some cases prove of advantage.

There is another similar process by which the preparation suffers less and in which the water is not removed by heat, but by a current of air. The section, lying on the slide, is dried by means of a small balloon-bellows, rendered transparent in xylol, and imbedded in balsam.

H. Kühne uses alcohol, but previously adds to it a considerable quantity of the same stain with which the preparation was treated. In the process for removing water almost every atom of coloring matter which is removed by diffusion is replaced by another, and thus the decoloration is reduced to a minimum.

Far better results, however, are gained by another process—that of Weigert, which is the best of all wherever we have to deal with bacteria in tissue that are difficult to stain.

Weigert employs, instead of alcohol, anilin oil, or a mixture of two parts anilin oil and one part xylol. The anilin oil acts similar to the alcohol, but far more mildly; it removes water quickly and decolors comparatively little. If necessary to avoid even this little, like H. Kühne, first color the anilin oil with a little of the stain. To do this, rub as much of the powdered stain as can be raised on the point of a knife into about 10 c.cm. of oil in a little dish. If the mixture be allowed to stand for a time it becomes clear. Then add as much of it to the pure anilin oil as is required to produce the desired tint. The anilin process has this advantage, that the sections on leaving the water are apt to roll up and contract, so much so that they become useless. To meet this difficulty it is best to perform the whole staining process upon the slide.

If desirable to employ Gram's isolated bacterial stain, place the sections, for about twenty-five minutes, in a thin anilin-water gentian-violet solution, and then, for about two and a half or three minutes, in iodide of potassium solution, after which they must be washed in alcohol. If they become quite black in the iodine the stain is now detached by the alcohol and passes off in red clouds from the preparation, which, after about twenty minutes, is bleached. It is well to change the alcohol frequently, and in general not to be too sparing in materials for Gram's process. If the sections which have been colored by Gram's process are afterward to receive a second contrasting stain, they should be placed, for quite a short time, in a very thin solution of Bismarck-brown.

Yet vesuvin is a bacterial stain, and on that account less suitable for these purposes than the pure nucleus stains—saffranin and carmine. Picro-carmine is very useful for double staining, according to Gram.

The best way is to obtain the nucleus coloring first and then stain the bacteria afterward. The sections pass from the alcohol into a strong solution of picric-acid carmine for about half an hour. The superfluous color is then wholly removed in 50% alcohol.

The sections are now rose-colored, and on examination the nuclei will be found of a dark red color, the bodies of the cells of a light red, and the connective tissue pale yellow. Next place the preparations in anilin-water gentian-violet.

Into a small dish of anilin-water pour four or five drops of a concentrated alcoholic solution of gentian-violet till the fluid begins to look opaque, but not till a saturation with coloring matter has taken place—i.e., not until the iridescent film appears. Let the sections remain in gentian-violet precisely half an hour, and then bring them directly, without any previous washing in alcohol, into the sodium solution. After three minutes they must be taken out of the iodine and placed in the alcohol; here the stain is washed out, the red ground color of the tissue becomes more and more distinct, and at length the sections have the same color as before their treatment with gentian-violet.

The results of this process are excellent: the staining is in fact triple; the cells are red; connective tissue, yellow; bacteria, blue; and the latter stand out from their surroundings with remarkable distinctness. In all staining by Gram's method the precipitates are troublesome, and often settle upon the preparation in large masses. In order to avoid them, Gunther has proposed a slight alteration of the original method of Gram, which often proves serviceable.

Günther stains the sections in a fully-saturated opaque solution of anilin-water gentian-violet for a minute; then two minutes in iodide of potassium solution, half a minute in absolute alcohol, ten seconds in a 3 to 5% solution of muriatic acid in alcohol, then once more absolute alcohol, then contrasting stain, etc. The addition of muriatic acid gives the alcohol the power of dissolving and removing the particles of coloring matter which adhere to the tissue.

Also in the double staining by Gram's method the strong decoloring influence of alcohol is often an inconvenience. The bacteria lose their color and disappear completely from view.

It is, therefore, frequently advisable to take anilin oil instead of alcohol. Stain the section first with picro-carmine, etc., then place it in anilin-water gentian-violet, next in iodide of potassium solution (Gram's solution), and then it must go immediately, without any previous immersion in alcohol, into the anilin oil, then comes oil of cloves, or xylol, and Canada balsam. All this is best done on the slide.

Never stain a large number of sections at once without having proved the correctness of the whole process by one or more successful trials. For almost every separate case and every staining solution one should endeavor personally to find out by experiment the right conditions of time, strength, heat, etc., etc.

Precise directions as to the time, etc., for the different processes have, therefore, been intentionally avoided as far as possible.

It may seem that this is a very strange way of beginning, and that if we worked with No. 1 solution, No. 2 solution, etc., and always knew their exact percentage of coloring matter, we might have a little chance of success. Such directions, however, would only be found infallible in a restricted number of cases, and the mechanical observance of such rules would sink the art of staining into a mere trade. If one only knows rules for staining and not why he must act so and not otherwise, he knows in truth very little.

And yet many a person thinks he has all the secrets of staining, or even the whole science of bacteriology, in his pocket, when he carries home with him a few pretty red and blue double stainings. Yet the immense advantages which experimental science gains from the use of stained preparations will not be renounced on that account.

The finer varieties of form in the bacteria, the slight differences of length and thickness, the extremely characteristic shapes of particular species—all this can only be demonstrated by staining.

It supplies us with durable preparations which permit of comparative examination.

It is staining alone that gives us an immediate insight into the micro-organic life that exists within the tissues, and it is double staining that enables us to differentiate between tissue and bacteria with such extraordinary clearness and precision.

By means of staining—that is to say, by means of the peculiar relations existing between particular stains and particular species of bacteria—attention has been directed to the special importance of many kinds of micro-organisms, and a successful staining has thus been the first step on the way to important discoveries.

Coloring is an invaluable aid in the hands of those who know how to employ it, but it is an art that must be studied, and the great number of the so-called "faults of investigation"—defective observations of different kinds—show that here also apprenticeship must precede mastership.

The most common of these faults of investigation are as follows. A part of them arise from an insufficient preparation of the object.

Glass covers are heated before they are completely air-dry, or they are exposed too long to the action of high temperature. This causes the formation of those peculiar alterations of shape which have already been mentioned. The bacteria swell up, or they collapse, or they become surrounded by a halo, and quite lose their characteristic appearance.

There is another fault, too, referable to an insufficient previous treatment of the preparations. If portions of organs are left too long before being put into alcohol, they begin to putrefy—i.e., the septic bacteria gain access to them. If these objects be hardened and the sections made from them be colored, these foreign micro-organisms are of course stained with the rest, and thus give rise to all sorts of deceptions.

To avoid these, put the portions of tissue, as fresh as possible, into alcohol, but besides this, be particular to notice the distribution of the bacteria within the section.

The septic bacteria, of course, penetrate the organs from without; their numbers are therefore found to diminish in proportion to the distance from the outside surface, while the inner portions are usually quite free from them.

Another error proceeds from the fact that many of our staining solutions, especially hæmatoxylon and carmine, but also the anilin colors and even distilled water, are often enough the home of numerous bacteria, which find there a field for their development. When staining our objects, these micro-organisms are deposited on the sections or are even floated into the tissues, and we may easily fall victims to deception if their somewhat superficial position be not carefully noted.

Some faults of investigation arise from faulty staining.

Very frequently in Gram's process, and also in other processes, the coloring matter of badly-filtered solutions is deposited upon the preparation, usually in the form of very small, roundish bodies, which lie together in masses and are apt to be mistaken for micrococci. The irregular forms of the particles, their peculiar shining appearance, and the want of order in their distribution over the different parts of the tissue will suffice to prove that they are not what they seem.

The influence of the iodine solution on bacterial preparations sometimes shows itself in a very remarkable manner. The rod-bacteria break up into a bead-like row of granules, reminding one of a chain of globular bacteria. The more strongly-acting acids also produce such appearances now and then, and in fact this has led to the mistaking of indubitable bacilli for micrococci. Yet the true state of the case is nearly always recognizable by the occurrence of unaltered rods in the preparation, as also by the presence of intermediate, only partly-altered, forms.

Some faults of investigation lie, not in any defect of the object, but in the observer, who wrongly interprets what he rightly sees.

In making preparations of blood to search for bacteria, we place a small quantity of the fluid on a cover-glass, lay a second glass upon it, and then pull the two apart again. This proceeding does not fail to produce its influence on certain ingredients of the blood.

By the attraction of the two glasses some white blood-cells are burst, their nuclei are crushed, and then drawn apart when the glasses are separated. As these consist of nucleus substance, they are naturally stained by the anilin. In this manner one gets the strange-looking, comet-like figures in the picture—thick heads with long tails; in others, all remains of nucleus form are destroyed and one sees only long stretched threads, which have more than once been taken for the mycelia of blue-mould fungi. Lastly, such a thread will sometimes break up into pieces or resolve itself into a row of little granules, and then we see "bacilli" and "cocci," but with a little attention and experience we learn to be on our guard against such mistakes. The list of the commonest faults of investigation will be closed by warning against one which is made by almost every beginner.

There is in tissue a particular kind of cells, the so-called plasma-cells or granule-cells, whose behavior under the influence of anilin stains is precisely the opposite of that of all other cells. They are generally found as large, flat formations of the outer walls of the vessels, and consist of a nucleus and a very fine-grained protoplasm. Now, with these cells it is only the protoplasm that stains; the nucleus remains uncolored and therefore escapes all but a very attentive observation; the cell body, however, presents an even, deeply-colored heap of granules, which in fact has a very strong resemblance to a colony of micrococci.

Often, indeed, these cells have been taken for such a colony, and more than once they have been taken for the long-sought cause of some particularly interesting disease. From the insignificant cold in the head to the terrible hydrophobia they have, perhaps, for a longer or shorter time, been taken for the source of all the diseases which could be thought to proceed from bacteria, and it is hardly to be expected that they should soon cease to play this deceptive role.

Their true nature may be recognized without much difficulty by their granules being of unequal size, by the nucleus being generally recognizable when carefully sought, and by there being usually several such cells, of exactly the same size and appearance, together. It is, further, a remarkable fact that these plasma-cells often stain with a different tint from the surrounding tissue. This is most clearly seen when the sections are treated with methyl-blue; as a rule the plasma-cells become deep violet, so that it seems as if a special chemical combination took place between them and the coloring matter. The plasma-cells are also sometimes accessible to Gram's method, which leads to the increased possibility of their being mistaken for swarms of micrococci. The peculiar nature of these strange forms is as yet but very imperfectly understood.

CHAPTER III.

Methods of Breeding; Sterilization; Liquid Culture Media; Preparation of Beef-bouillon; Potato Cultures; Beef-bouillon Gelatin and Agar-agar; Uses of Food Media for Obtaining and Maintaining Pure Cultures; Plate Cultures; Petri Dish Cultures; Esmarch's Roll-tube Cultures; Pure Cultures; Culture of Anaërobic Bacteria; Incubators; Thermo-regulators and Safety Burners.

METHODS OF BREEDING.

WERE the microscopical examination of the bacteria as they occur in their natural state the only means at our disposal for studying them, our knowledge of bacteriology would never get beyond certain very narrow limits. We might know something of the wide-spread existence of these micro-organisms, we might observe their frequent presence in connection with certain forms of disease, and we might perhaps even be able in some cases to prove the regular presence of the same species (as judged by form and appearance), and then, by jumping at a conclusion, maintain it to be the source of the pathological conditions in question. But this would be all, and even this little would stand on weak feet. It is always unsatisfactory to form a judgment from the mere appearance of the smallest living organisms, whose forms are the simplest imaginable, and experience has shown to what great errors one may be led by so doing. Bacteria, which seemed to agree fully as to appearance, turned out on more careful examination to be altogether different species, which had nothing in common but their similarity of form.

The disadvantage of this way of proceeding was soon recognized, and attempts were therefore made to render the bacteria as independent of their natural conditions as possible, in particular to sever the parasitic kinds from the organisms whose parasites they are, to bring them into conditions which allow of more easy examination—in a word, to breed them artificially.

In this we have succeeded with a great number of species, with some easily, with others after much difficulty.

It will at once be apparent what an immense stride was made in our knowledge of the bacteria by this success. We were now enabled to study the micro-organisms freed from the many adventitious circumstances of their existence. All the hindrances to investigation which had before existed in the intimate mutual relations between the bacteria and their natural nourishing soil were swept away. We were even enabled to vary at pleasure and in detail the conditions under which the bacteria were allowed to develop, and by observing their behavior under the conditions thus altered very valuable criteria were obtained.

New, hitherto unknown peculiarities were discovered, and the rich abundance of added facts made it possible to distinguish between clearly separate species formerly thought to be identical.

The artificial breeding of bacteria has been of the highest importance in enlightening us as to their agency in producing diseases. Although from the regular occurrence of the same bacteria in the same disease the former might be regarded as the probable cause of the latter, and although this probability was raised to something like certainty by transferring portions of diseased organisms to a healthy one, which then developed the same form of disease and showed the same bacteria—while this experiment could be successfully repeated through a whole series of cases, yet all these demonstrations were open to serious objections.

The existence of a special kind of bacteria was granted when it could no longer be denied, but the micro-organisms were not acknowledged to be the direct cause of the morbid changes; they were regarded merely as an accompaniment, a consequence of the disease, as uninvited yet harmless guests, for whose development the pathological conditions had proved particularly suitable. The successful inoculations were explained by the supposition that the disease produced a specific organic pathogenic virus, with the power of reproducing the affection and calling forth the same changes, in the course of which the bacteria found their way into the soil thus prepared for them.

This view of the case could not be disproved till the parasites had been entirely removed from the diseased organism and freed from all surroundings to which disease-producing influence could be attributed—i.e., not till they had been bred in isolation under artificial conditions, and then tested as to their pathogenic power. Should it now be possible by their aid to generate the same symptoms, there could no longer remain a doubt that they and they alone were the cause of those symptoms.

This experiment has been performed with success repeatedly, and it is to the breeding of bacteria that we owe the most important discoveries of recent times with regard to the origin and nature of diseases in general. Here far more than in other spheres is to be found in the recognition of its true cause the key to a right understanding of all the different symptoms under which a disease appears.

I. STERILIZATION.

To obtain the most advantages possible from the artificial breeding of bacteria, certain precautionary measures must be employed.

To get a clear, definite idea of the peculiar properties and general behavior of a particular micro-organism, it is above all necessary to bring that particular species under observation alone and free from all admixture with other species. A medley of bacteria is useless for exact investigations. It is only when one species is obtained in pure culture, as it is called, that it is possible to rely on obtaining safe, unobjectionable results.

The characteristic marks of a species which are perhaps too slight in the individual to be easily noticeable are in cultivation made clearly evident, and all that was characteristic of the species in question now shows itself in a greater degree with numberless repetitions. Attention was early directed to these advantages, and investigators endeavored to find some means of obtaining pure cultures of the various species of bacteria. But it was soon found (the more so that all attempts were made with liquid media) that the problem presented great difficulties—difficulties arising from the immense diffusion, it might be said the omnipresence, of the bacteria and the high power of resistance of their spores.

Of course, in order to breed a particular species of bacteria artificially in pure culture on any nutrient medium, the latter must first be freed from all other micro-organisms before employing it, and further, every pure culture must be protected during its growth from the invasion by foreign bacterial germs—i.e., from contamination.

And this is no easy task, especially the first destruction of germs in the culture media—their "sterilization," as the French call it and as it is now universally called—requires particular attention.

The bacteria in their common form are not particularly able to withstand exterior influences. A number of them are, however, provided with a peculiar protective arrangement, destined to preserve the species in the face of hostile exterior influences, and, as already mentioned, the fruits or spores which serve this purpose are perhaps the most tenacious of life among all the organized beings of our world. To free a nourishing solution or anything else from germs, it must be borne in mind that it may contain some of the spores which it is so very difficult to annihilate. In other words, to effect sterilization, only such means must be employed as experiment and experience have shown to be capable of regularly and certainly killing even the most tenacious spores.

This is no requirement based on theory and useless for everyday practice. The incorrectness of such a view would soon be proved by very unpleasant experiences. Bacilli and their spores exist, in fact, everywhere, and the majority of failures made in working with bacteria is traceable to the insufficient sterilization of the substances and utensils employed.

What means are at our disposal to attain the ends in view? It is not easy to arrive at a just appreciation of them, because for a long time sterilization, in its proper sense, was not sufficiently distinguished from other similar procedures. A process was regarded as fully sufficient if it only checked the development of the bacteria, and the fact was overlooked that as soon as the means was removed its operation was at an end, and the checked bacterial growth was resumed. We demand from a truly disinfecting measure that it should once for all destroy every trace of life in the micro-organisms; that it should also infallibly annihilate their most enduring forms.

This requirement must, it is true, be taken "cum grano salis." The different species of bacteria are by no means all alike as regards the tenacious vitality of their spores. Some are known which yield to slight attacks, and others that can only be destroyed with the greatest difficulty. If the sterilizing process be applied to these latter, requirements are met with which cannot be carried out in practice.

Fortunately, however, the varieties which it is so very difficult to destroy are by no means frequent. It is sufficient to know where to expect to meet with them; for instance, in garden mould, in manure, and in decaying mixtures. In such cases it is, of course, necessary to proceed with extra care.

In ordinary cases, however, sterilization may be accomplished with less trouble, and the principles which sterilization should follow for purposes of culture have, as a rule, been determined by careful investigations, and are as follows:

It has been found that a small number of chemical agents are capable of accomplishing what is required. Above all, as specially important must be mentioned concentrated carbolic acid, corrosive sublimate in solution of 1 to 1,000, and also quick-lime.

But the use of these substances in connection with the breeding of bacteria is of little importance. For if one of these substances be added to a culture medium, it (the solution) will indeed become germ-free, but as it is impossible to remove the germicide again, the culture soil becomes permanently unsuited for bacterium culture; besides all of which, the further changes which the action of the chemical agent may produce in the solution must be considered. The effects of these disinfecting substances are such that neither the vessels in which the culture media are held nor the implements with which we inoculate bacteria can be brought into contact with them without incurring the danger of failure in the attempted breeding experiment.

It is necessary, then, to adopt some plan of destroying bacterial germs without so affecting the substances containing them as to make them valueless for further use: the culture media must be germ-free, but not barren—i.e., they must retain their nourishing qualities.

The only means which can fulfil the requirements is heat in its various forms. Even the spores cannot resist the continued action of high temperatures, and the modern process of sterilization is based on the utilization of this fact.

Dry heat may be employed. Place the object which it is desired to sterilize in a flame, and in a short time every trace of organic life will be destroyed.

Of course this process is only applicable now and then, and solely in the case of incombustible matters. The platinum wire, for instance, with which the bacteria are transferred must in each and every case be heated in the flame of the Bunsen burner or spirit-lamp before being used.

For the other metal implements—inoculating needles, knives, scissors, etc.—a direct heating in the flame will also be found the quickest and simplest way of insuring thorough sterilization. It is by no means necessary to make the instruments red-hot; if they are moved backward and forward perhaps for a minute over the Bunsen burner, it is all that is necessary. Even such a heating destroys the temper and sharpness of the blades quickly enough, but this disadvantage is submitted to willingly for the certainty with which sterilization is accomplished.

Wherever, for any particular reason, the direct application of the flame is impossible, whether the objects are too large, too numerous, or too combustible, or whether they are in some other way difficult to manage, we have special apparatus for the application of dry heat which insure the equable and regular action of high temperatures.

They are double-walled cases of sheet iron, and are heated from below by a powerful gas-burner. It may be seen by the thermometer in the lid, after about ten minutes' warming, that the air inside reaches a heat of about 150° C. and remains at that temperature.

Experience and experiment have shown that spores even of very high powers of resistance cannot outlive an exposure of more than half an hour in such a high temperature. For half an hour or more, then, put the larger glass and metal instruments, the vessels for the culture media, pipettes, and syringes—in short, whatever withstands the action of a high temperature without damage—into the dry box or hot-air oven.

The nourishing liquids—the most important item to be considered in the entire chapter of sterilization—will not bear this treatment. Fluids and substances which melt under the influence of heat must not be exposed for any length of time to such a temperature, since they would be affected very prejudicially, or even destroyed by it.

Fortunately, however, heat is able to operate much more quickly and powerfully in fluids than in a dry state, as in the case of hot air. Even spores which resist a temperature of 150° C. in the air for an hour or more lose their vitality in a few minutes in boiling water.

Advantage is taken of this energetic action of moist heat for the sterilization of our nourishing media. Of course they can be freed from any germs that they may contain by simply boiling; yet certain precautionary measures are necessary in order that all portions may be equally submitted to the boiling process. And even then it has, in the case of numerous liquids, its decided disadvantages. Other methods were therefore resorted to. Either the vessels were plunged into boiling water and kept there for a long time, or steam at above 100° C.—under pressure therefore—was employed.

The first-mentioned plan frequently proved inefficient, because it is difficult to keep somewhat large vessels so protected from exterior cooling—from loss of warmth in contact with the air—that the desired and required temperature is kept up all the time and throughout all parts of the liquid. The use of steam under pressure has also its disadvantages. In the first place the apparatus called an autoclave, intended for its application, is a very complicated and expensive instrument, which requires particularly careful treatment and attentive management, and, moreover, the distribution of heat within it, as Koch has shown, is by no means always equable. It is apt to have "dead corners" containing air that is not reached by the motionless steam, which endanger the success of the whole operation. Thus the steam may show a temperature of 130° C. while the fluid inclosed in it is, in some portions, not more than 70° or 80° C. It is true that such differences will not occur as a rule. In dealing with small articles—for example, with thin silk threads—on which are dried spores and which are usually exposed to the action of this apparatus, such variations can scarcely occur. In such cases the operation of saturated vapors under pressure are unexceptionable, and there is no doubt that with increased pressure—that is, with increased temperature—the results will be better. Thus Globig found that the spores of a particular species of bacteria belonging to the class of potato bacilli, and which are of specially tenacious vitality, were destroyed by steam at 100° C. in five and a half to six hours; by steam at about 110° C. in three-quarters of an hour; at 113° C. in twenty-five minutes; at 120° C. in ten minutes; at 126° C. in three minutes; at 127° C. in two minutes; and at 130° C. in a moment.

Nevertheless, if compressed steam be but seldom employed in our laboratory and only in exceptional cases, the reason is to be sought partly in the disadvantages already mentioned, but chiefly in the fact that we possess another, more convenient, less expensive means, which is easier to manage, fulfils all requirements in the great majority of cases, has been in use for years, has stood the test of many thousand experiments, and rarely proves inefficient.

This process is sterilization by the freely-escaping steam of boiling water. Koch and his collaborators, Gaffky and Löffler, found that such steam, when properly confined and protected from the cold outside air, keeps the boiling temperature permanently, which it also speedily and perfectly communicates to liquids exposed to its action, so that an exposure of from half an hour to an hour suffices, as a rule, to render them germ-free.

On this extremely important fact depends the construction of the "Koch steam generator," which is almost exclusively used in sterilizing our food solutions.

A cylinder about 3 to 4 m. high and 30 cm. in diameter, of plain tinned iron, or still better of sheet copper, is enveloped in a close layer of felt, to protect it from loss of heat. At the top it has also a felt-covered lid, called the "helmet," which fits loosely and must not be air-tight. In this helmet a thermometer is usually fixed. The cylinder itself has a grating in its interior, at the boundary of the lower third, above the water which is made to boil in it. The height of the water can always be seen by a gauge-pipe at the side.

The grating, therefore, divides the cylinder into a lower water space and an upper steam space.

If such an apparatus be heated, it is certain that as soon as the water fairly boils and the full steam development has commenced the thermometer will rise to 100° C. and will continue at that point. An exposure of fluids for from half an hour to an hour, reckoning from the moment of full steam development, will, as already said, suffice in the great majority of cases to effectually sterilize them.

Of course we can also sterilize by this means all substances which bear the high temperature without damage—for instance, India-rubber stoppers, paper filters, etc.

The peculiar efficacy of steam, the undoubted superiority of wet over dry heat, is a very striking circumstance. It is explained by the fact that the tough covering of the spore, the very touchstone of all disinfecting processes, swells and softens on coming into contact with the moisture, and thus becomes permeable, as has been proved by the experiments of Esmarch.

He found that if the steam which has been developed by boiling water under atmospheric pressure is afterward superheated by passing over hot metal plates, its germ-killing power diminishes. This is because the steam is dryer, further removed from its condensation point, less disposed to set water free; it has a longer way to go, must lose more heat than steam of 100° C., before it can deposit moisture and effect the softening of the spore envelopes. It is a different case, naturally, when the increase of temperature goes hand in hand with the increase of pressure which results from increased tension. In that case the steam remains close to its condensation point, and is capable of losing water, of becoming water, at the moment when it meets with the bacterial spores intended to be sterilized.

But the heat of freely-escaping steam cannot be employed when dealing with substances which cannot support the boiling temperature. Strongly albuminous fluids, for instance, cannot be heated to 100° C., since their albumin curdles and the solution is essentially altered in its composition. For such cases a method introduced by Tyndall and afterward considerably improved by Koch, called "discontinuous sterilization," is employed.

Most bacteria, in their usual forms, cannot survive a temperature of about 60° C., while their spores are in no way affected by this heat. If a culture fluid be warmed, therefore, for some time to 60° C., as a general rule only the spores remain alive. But when the temperature diminishes these very soon begin to germinate, the bacilli lose their protective sheath, and if on the following day heat is again employed to 60° C., the newly-formed rods are destroyed; if this process is often repeated, it is certain that all the spores will have become bacilli and that all the bacilli will have been killed. Experience has shown that it is best to warm the solutions to 56° or 58° C. every day for about four or five hours for a week. The result is usually satisfactory. Of course this method can only be used for culture fluids in which the spores would in any case proceed to germination.

These are the general principles of sterilization, and are to be applied in special cases.

We heat and thereby securely sterilize—

First. All glass and metal instruments which do not suffer from the prolonged effects of high temperature for about half or three-quarters of an hour in the hot-air oven at 150° C.

Second. All culture media and similar substances for about half or three-quarters of an hour at 100° C., by means of a current of steam in the generator.

Third. All substances containing albumin which do not support 100° C., three or four hours daily for a week at 56° or 58° C.

When the culture solutions have been sterilized and thus the first foundation laid for a pure culture, endeavor as far as possible to protect them from all foreign impurities. This may be done by protecting all the vessels containing the culture media from the germs of foreign micro-organisms.

A good plug of cotton-wool has proved the best and simplest safeguard. The cotton-wool is a good germ filter without any special preparation, which prevents almost to a certainty invasion on the part of micro-organisms.

Only the mould fungi can get through a plug of cotton-wool. These disagreeable visitors are found particularly often when one has put an India-rubber cap over the cotton plug to prevent evaporation of the fluid in the culture vessel. Then a sort of damp chamber is formed under this cover, and any mould spores that may have lain on the surface of the plug begin to germinate, send their mycelia through the close complex of wool fibres, and after a short time mould fibres may be seen on the under side of the plug. The best protection against such uninvited guests is either to carefully drop a little solution of corrosive sublimate on the plug or burn off the top end of the plug before applying the cap.

Generally speaking, however, the simple plug of cotton-wool is quite sufficient. The test-tubes, Erlenmeyer's flasks, and larger glasses are provided with such plugs in advance, and they are sterilized together with the glasses in the dry heat. The slight scorching of the wool also serves as evidence of the sufficient sterilization of the objects in question. In addition to these precautions, care must be taken to exercise the greatest cleanliness and nicety in all the operations with the culture media.

The cotton plug should be taken out as seldom and for as short a time as possible; the mouth of the vessel, when opened, should not be held directly upward, for fear of receiving germs from the air, and every instrument which is employed must be thoroughly sterilized.

All our surroundings swarm with micro-organisms, and a single one falling in the wrong place may suffice to spoil everything. The reproductive powers of the bacteria are so unlimited that in a very short time they can multiply almost immeasurably, and expel all the former inhabitants from their field of development. In this manner it is easily possible to obtain "transformations" of one species into another—for instance, the harmless hay bacilli into the virulent anthrax bacilli, and vice versa.

Among the bacteria, as among other creatures of the organic world, there is a "struggle for existence."

If two germs of different species and different origin alight upon the same nourishing field, both will develop side by side for a time in peace. But, from some cause or other, the conditions are more favorable to one species than the other, it soon gains the mastery, and often enough completely exterminates the other.

Thus it is seen what dangers for the pure cultures this exercise of the right of the strongest may bring with it. One foreign germ is able in a short time to alter and destroy such a culture, in either case to rob it of all the advantages which constitute its real value; and those who talk of "tolerably" or "almost" pure culture show clearly that they have as yet comprehended but little of the true rules and principles of bacteriology.

II. LIQUID CULTURE MEDIA—PREPARATION OF BEEF-BOUILLON.

The bacteria are not particularly dainty. An organic mass containing nitrogen and carbon, especially if it has a slight alkaline reaction, together with favorable atmospheric and thermal conditions, is fully sufficient for most of them. Others are harder to satisfy; particularly among the pathogenic species a great many are found whose taste is much more circumscribed.

Efforts have been made to discover a food solution whose composition might fulfil most, if not all, requirements. Pasteur and Cohn made experiments in this line, and have given directions for making two artificial food solutions which are now but little used, but which, being of interest in the history of bacteriology, should be mentioned. Pasteur's consists of one part of tartrate of ammonium, ten parts of sugar candy, and the ashes of one part of yeast, with 100 parts of water.

Cohn's consists of ½ gram of phosphate of potassium, ½ gram of basic triple phosphate of lime, and 100 grams of water. To the whole is added 1 gram of tartrate of ammonium.

But it was soon observed that it was better to breed the bacteria under conditions resembling those of their natural state as closely as possible.

Therefore, for the purely saprophytic species, which occur principally on vegetable substances, infusions of wheat, pea-straw, potatoes, and decoctions of fruit were prepared; the species observed in animal excrements were to thrive on extracts of manure, etc. For the benefit of those whose habitat is the living organism, efforts were made to produce a liquid that should approximately answer to the juices of the body without being too complicated in its composition. It was to contain liquid albumin and extractives in about the same quantities as they occur in the blood, and were to show a decided alkaline reaction.

The simplest and most satisfactory medium in most cases was found to be a decoction of chopped meat, made slightly alkaline by an addition of solution of soda. Thus Pasteur early employed his bouillon of fowl's flesh with success, and we still often employ it, or at least a similar liquid food, at the present day.

We prepare our bouillon according to Löffler's directions. We take a definite quantity (500 grams) of finely-minced beef, as lean as possible, and mix it up with about a litre of ordinary water and allow the mixture to stand about twelve hours. In summer this should take place in the refrigerator to prevent putrefaction. A number of soluble albuminous and extractive substances pass into the liquid part, which is then separated from the flesh. This is best done by turning out the whole mass into a loosely-woven cloth ("cheese-cloth") and squeezing with the hands until the 1,500 grams of mixture have yielded 1,000 grams of meat extract.

If this liquid be heated, most of the albuminous substances contained in it are precipitated and lost. To make good this loss and to insure the food solution as much albuminous matter as possible, we previously add a certain quantity of peptone, which cannot coagulate in the warming process. We take about 1% for the quantity above mentioned—therefore about 10 grams—and to facilitate the solution of the peptone we add ½% or 5 grams of common salt.

The meat extract, with the peptone and the salt, is next boiled for about three-quarters of an hour in the water bath over the open flame or in Koch's steam generator. Then follows the neutralization of the liquid (which is as a rule strongly acid) by the cautious addition of a saturated solution of carbonate of soda, changing its reaction till a drop taken out with a glass rod no longer reddens blue litmus paper, but gives a slightly blue tinge to red test paper.

After this is done the liquid is boiled for about an hour longer. The coagulable albuminous substances are now curdled, one part floating on the surface of the clear broth as an opaque conglobated scum, the rest lying as a solid mass at the bottom. When the liquid is cool pour it through a filter which has been moistened with distilled water, and let the clear, almost colorless bouillon run out below. After this filtration it must still be alkaline, or at least neutral, and must not become in the least turbid when repeatedly boiled.

If it fails to answer these requirements, the faults must be rectified and the filtration repeated.

If, in spite of all efforts, the clearing remains unsatisfactory, the white of a hen's egg should be added to the filtered solution. If we boil it again for half an hour, the white of egg coagulates and carries off with it the slight turbidity that was present.

When the bouillon is thus satisfactorily perfected, pour into well-sterilized test-tubes, Erlenmeyer's small flasks, etc., about 10 cm. each, and provide each with a good plug of ordinary cotton-wool. But before proceeding further the food solution must be made sterile, and with the bouillon this requires special care, since it possesses, in common with all liquid foods, the disadvantage of presenting but little opposition to the invasion and unlimited growth of bacterial strangers. It is, therefore, best to expose the vessels, with their contents, for about an hour to the sterilizing action of the current of steam in the generator, and to make quite sure it is best to repeat the operation once more on the following day.

The bouillon is then ready for use and forms a very valuable, often useful, form of nourishment, originally intended for the breeding of parasitical bacteria. It is also very suitable for the saprophytic species, and as yet we know but few micro-organisms which are able to thrive on other culture media and not in bouillon.

It is necessary in some cases to increase its nourishing power by certain additions. Thus the supplement of 1 or 2% of grape sugar or from 3 to 5% of glycerin has sometimes proved advantageous.

The object of all our bouillons and artificial foods is to imitate the natural nourishing media. If these in particular cases have their peculiar nature, we must note the fact and proceed accordingly in our experiments.

The bouillon is specially useful in cases where we can utilize one of the chief advantages of liquid nourishing media—namely, the very exact, thorough, and equable distribution of the germs to be introduced.

If, for instance, we place a trace of any given species of bacteria into a test-tube of bouillon, if the tube be "inoculated," as we say, with the given species, an equable development of it throughout all parts of the nourishing solution will soon take place. If a little of the fluid be then removed with a pipette, we find, as experiment has always proved, that each drop contains almost exactly the same number of micro-organisms.

We are thereby enabled to work, if necessary, with well-measured, exactly-defined quantities of bacteria, and to compare the results of different experiments in a very simple manner.

Another case in which bouillon is quite indispensable is the breeding of bacteria in the hollowed slide.

We can watch their development, their vital functions, step by step under the microscope, just as in the hanging drop, if, instead of water, we use a liquid in which the germs will grow and develop. A drop of bouillon is inoculated with the desired species, brought upon the hollowed slide in the manner formerly described, and examined with the highest magnifying power. If too much material has not been taken, if the temperature is not unfavorable, and if sufficient patience be possessed for the task, it will not be difficult to watch the details of cell-development. The growth can be seen, the division of the cells, the origin of the simple groups, sometimes also the formation and budding of the spores as they take place under the eye. In this manner many of the most important secrets of bacterial biology are unfolded.

Lastly, the bouillons are extremely useful in cases where it is necessary to obtain large quantities of one definite species of bacteria—the so-called cultures *en masse.* The germs are placed in one or more litres of ready-prepared sterilized bouillon, and in a few days the desired material is obtained.

Here, however, for all practical purposes, the usefulness of the bouillons for experiments ends. The undoubted superiority of solid nourishing media has prevented all need for further extending the employment of liquid media.

It is scarcely necessary to go into particulars to show the disadvantages of breeding in liquids. The basis of bacteriology may be said to be the obtaining of undoubted pure cultures, the determination of the distinguishing qualities of one species. This is extremely difficult as long as we confine ourselves to fluids.

For instance, given a mixture of bacteria—a putrid vegetable infusion: separate the various species which it contains, isolate them, and breed each separately. It will scarcely be possible to do so. However carefully we go to work, however frequently we transpose from glass to glass, however small may be the portions with which we inoculate in order to obtain the greatest separation of germs, we will scarcely ever succeed, even by a lucky chance. Indeed, as a rule we will fail, and our investigations will be altogether wanting in reliability.

The case is almost more unfavorable still if it is necessary to get one particular species, and this alone, out of the mixture. The only way is to work on at a venture till good luck places the desired bacterium at our disposal. If we now breed this artificially, and if but one single foreign germ has found access to it, it may be that the stranger finds the conditions particularly suitable to him, he multiplies infinitely, pushes aside the lawful occupants of the soil, overgrows them completely, and gives the culture quite a new appearance. It can, therefore, readily be understood how this striking phenomenon caused the serious belief in the transformation of one species into another.

The fact is, we want some reliable means for dictating to the germs the definite, controllable ways and limits which they are not to leave or dispute without consent and aid.

These facts are clear enough, and the great majority of investigators have long since acknowledged their truth. The hindrances and difficulties which are met at every turn in the use of liquid media are, in fact, not less weighty than numerous, and although the French school has not yet renounced this method of investigation, one must, nevertheless, pay the tribute of admiration to the ability and dexterity which have enabled them to attain such great results with such imperfect means.

III. POTATO CULTURES.

It is not surprising if even the most ingenious methods fail to conquer the difficulties just described, since they lie in the nature of the media themselves, yet all obstacles were removed, as it were, by a single blow as soon as the employment of solid food bases took the place of the liquid ones.

The way in which the former came to be employed is sufficiently curious to deserve mention.

It was observed that when slices of boiled potato were left exposed to the air for a time and then kept a day or two where they could not dry up, there appeared on the surface a series of small, white-colored points and drops, which increased quickly in extent till they at length covered the whole slice. Microscopic examination showed that these spots were nothing but collections of micro-organisms, and further, that such a point contained only bacteria of one and the same species. This last fact was, at the period of its discovery, something very striking, but its explanation was not very difficult. The little aggregation had proceeded from germs in the air, which, settling on the potato, found there a suitable place for their development and growth; they were, however, compelled by the existing conditions to develop and grow at that point of the solid culture medium on which they had first fallen, so that here only cells of the same species could arise. If, instead of falling on the potato, they had fallen into a glass of bouillon, they would also have developed, but in a short time the different species would have got mixed, and very soon all would have become a lawless, inextricable chaos.

On the potato it was impossible for one species to gain the ascendency, repressing the others and preventing their further development; for here each germ found a quiet spot where it could proceed to the reproduction of its kind.

This collective grouping of individuals of one and the same species only enabled us to observe these species more accurately and to determine distinguishing qualities with increased certainty. With the herd before us, we succeeded in discovering many peculiarities whch had altogether escaped observation when we had only one specimen to examine; in a word, we had in this first collection of bacteria on a potato (in a colony of bacteria, as it is called) the point of departure for a pure culture with all its advantages, and it proved easy to utilize this fact.

In addition to this, the relations of the bacteria to their solid food medium led to the discovery of a whole series of hitherto unknown properties in these micro-organisms; properties which could not be seen in the liquid food solutions. This gave rise to an abundance of new ideas, and yielded extremely valuable criteria for the comparison and discrimination of the separate species.

Thus the potato all at once opened our eyes to the advantages of the solid culture media—the facility which they offer to the obtaining and preserving of pure cultures, the certainty with which they afford to each species an undisturbed and unfavored development, and the revelation of a never-dreamt-of multitude of new qualities in the bacteria which could not otherwise be recognized.

The acuteness of Koch enabled him to fully appreciate these advantages, and also to utilize them throughout the extent of their applicability.

With the culture of bacteria on the potato as his point of departure, he was soon able to make improvements in solid food media, and thereby to give the start to the surprising progress made in bacteriology within the last few years.

The first place is given to the potato in treating of solid food media, not alone to fulfil the duty of gratitude in a matter of history, but also because it still continues to be, in many cases, a valuable means for the breeding of micro-organisms. It has, indeed, been found that the greater number of all the bacteria as yet known to us (not alone the saprophytic ones) are capable of thriving upon it, and thereby displaying very characteristic appearances, so that the potato culture sometimes offers the most essential service in distinguishing species which, under other circumstances, are easily confounded with one another.

Take good medium-sized potatoes and cleanse them from the dirt that cleaves to them by repeated vigorous brushing, for it is evident that to prepare the potatoes for serving as nourishment to pure cultures they must be freed from foreign germs. In the upper strata of earth from which the tubers were taken there are always great numbers of bacteria in the form of spores with a high degree of tenacious vitality. These cling to the surface of the potatoes and are particularly fond of hiding in those little recesses called "eyes," whence the shoots proceed, or in the so-called "bad spots"—i.e., portions of dead substance. These latter should be removed.

With the point of the knife dig out suspicious-looking portions, taking care to go deep enough to reach the pure, unaltered "flesh" of the potato. Only those parts of the skin should be removed from which an undesired swarm of bacteria is feared, for the healthy, smooth skin must be regarded as a valuable protection against exterior pollution, and therefore should be left on.

To insure the thorough destruction of all and any germs that may remain, put the potatoes for half an hour into a solution of $\frac{1}{10}\%$ corrosive sublimate.

The potatoes are then boiled, for it has been found that in their raw state they are not a suitable food for bacteria. This is best accomplished by putting them in a tin vessel with a grating at the bottom and placing this for about three-quarters of an hour in the steam generator. The time required for thorough boiling will, of course, differ somewhat according to the size and the kind of potatoes used. They are next halved with a heated and recooled knife in the direction of their length, being held by the finger and thumb of the left hand, finger and thumb to be previously dipped in 1% corrosive sublimate solution.

Next, with a sterilized scalpel or a platinum loop take the matter to be inoculated, and by spreading and rubbing extend it as evenly as possible over the surface of the slice. It is well to have a border or margin of about one cm. around the edge; the culture will afterward show a clearly-limited, definite growth, upon which is best calculated to form a sound judgment. Foreign growths, too, proceeding from germs that have by some chance escaped destruction (if such there be) are generally found to start from the edges, so that it is desirable to avoid contact with them. When the slice is properly inoculated it must be kept from drying up and from pollutions from the air. A number of such slices are, therefore, placed in large glass dishes, with a layer of damp filter paper at the bottom, the lid only being removed in case of necessity.

This "moist chamber" must not be fairly wet, for in that case drops of water are apt to form on the inside of the lid and fall down on the slices, disturbing the quiet development of the culture.

It is unnecessary and even dangerous to use sublimate for the cleaning of the dishes or the moistening of the blotting-paper. The inoculated surface of the potatoes does not come into contact with anything dangerous to it, and the sublimate solution, if it gets to the slices, can only do harm and render the culture useless.

There is another process of preparing potatoes for bacterium culture, very much simpler and for many purposes equally sufficient—that of E. von Esmarch.

Sterilize a number of small double glass dishes (such as are often used for holding staining fluids) in the hot-air oven. Some potatoes are peeled with an ordinary kitchen knife and washed clean under the faucet. Then carefully cut out the eyes and the "bad places," and cut the potatoes into slices about 1 cm. thick, placing one such slice in each of the dishes. These dishes, with the potato slices, are then exposed for about one and one-half hours to the action of the steam generator, in which they are well boiled and sterilized. The skin, as the protector against foreign germs, being here removed, the danger of after-pollution is far less, and as the evaporation proceeds very slowly in such small dishes, the danger of the slices drying up is small. General use can, therefore, be made of the surface; it can be inoculated and the developed culture kept for months. Kral in particular has made this fact subservient to his purposes in the preparation of his magnificent cultures for demonstrations.

If Esmarch potatoes are not recommended entirely and exclusively, the reason is that a few bacteria only display their most typcial, characteristic manner of growth when cultivated on the fresh halves prepared as previously described.

There is a third method of preparing the potato for the use of the bacteriologist, which is in many ways very useful, but which has not proved reliable in all cases. This method, with slight variations, was recommended, almost at the same time as the former, by Globig, Bolton, and E. Roux. It consists in boring out a cylindrical plug from the flesh of a good large potato (which requires no previous preparation) with a cork-borer or some similar instrument. The skin at either end is cut off, and then, with an ordinary scalpel, the cylinder is halved diagonally by an oblique longitudinal cut, each half being then pushed into a test-tube that has been previously sterilized. If the plug of potato is too thick it must be thinned down with a knife. Next proceed to sterilize the test-tube with its contents and cotton-wool stopper by the aid of the steam generator, in which it remains one or two hours. This suffices to kill any spores of potato bacilli that may have been left, and the culture medium is now entirely ready.

With a platinum needle or loop transfer the material for inoculation to the oblique surface of the potato plug, and spread it out properly. The further development must now be awaited.

These cultures can also be preserved indefinitely without any fear of pollution. The test-tube potatoes share this advantage with those prepared in little dishes on Esmarch's plan, with which they also have a second advantage in common, namely, that a great quantity of them can be prepared at once and kept ready for use without spoiling.

The blood-red film made by a well-known bacterium—the Micrococcus prodigiosus—is a striking example of growth on the potato; also the blackish-blue layer produced by a bacillus which is sometimes found in river-water, the dirty green culture of the bacteria of green pus, the grayish-blue one of the blue-milk bacillus, etc., etc. A hick white skin is found in a brood of Bacillus subtilis, and a dull white, granulated mass is produced by the growth of anthrax bacilli. Again, the typhus bacilli, when grown on potato, produce a slightly moist, shining appearance. In fact, the growth is almost invisible to the naked eye, and this peculiarity is taken advantage of to distinguish them from all other species of bacteria.

It is true that the growth of the individual micro-organisms on the potato is not always quite uniform. The reaction of the culture medium is of decided importance for the development of the bacteria. Now the flesh of the potato is often slightly acid, though usually decidedly alkaline, and this circumstance occasionally gives rise to very striking variations in the appearance of one and the same species of bacteria on the potato cultures.

Even such bacteria as always show a preference for alkaline reaction in their culture media will thrive on sour potatoes. It seems, therefore, as if the vegetable acids, particularly the malic acid, which exists in the potatoes are much less unpleasant to the bacteria than the mineral acids and the lower acids of the sebacic group.

The potato process can be used for dividing a mixture of bacteria into its separate component parts.

If an exceedingly small portion of a mixture of bacteria be spread over the surface of a slice of potato, the germs would be too numerous and lie too close together. A uniform layer, in which no distinction could be made, would cover the surface. Therefore with continually-changed sterilized knives remove a little from the first potato to a second, and from this again to a third, from the third to a fourth, and so on to a fifth and sixth, always with smaller quantities, so that at last, by this repeated thinning of the inoculated matter, a very wide diffusion of germs will be obtained.

The different kinds of bacteria will develop in small pure cultures each by itself, clearly distinguishable by their color. The germs will be so far separated from each other that each will be able to reproduce its own species on the solid medium without coming in contact with other species.

This process lies at the foundation of our whole present method of resolving mixed masses of bacteria.

Occasionally the potato is employed in a form which differs considerably from those already described. The potatoes are peeled as if for the dinner-table and boiled in the steam generator, without any particular previous sterilizing. When they are sufficiently boiled they are mashed to a stiff mass in an earthenware dish, with a little distilled water. The mash is then put into Erlenmeyer's small flasks—a troublesome task—and thoroughly sterilized on three successive days by an exposure for an hour each time to a current of steam.

The bacteria grow just as well upon this medium as upon the slices, and being well guarded against pollutions from without by the cotton-wool plug, this system is extremely well adapted for all those cases in which it is desired to at once obtain large quantities of some particular micro-organism, to get a so-called potato culture *en masse*.

In a similar way another solid food medium is made which is

very useful for some purposes. We dry ordinary bread—rye bread is the best—at a moderate heat and reduce it to a fine powder. This is put into Erlenmeyer's flasks (about 20 grams in each), and enough distilled water added to make the whole into a uniformly soft, moist pap. This is sterilized an hour a day for three days in the steam generator.

The bread powder has a slightly acid reaction, and is, therefore, greatly favored by the mould fungi, which, indeed, are generally bred upon it in pure cultures. To make it suitable for bacteria, after the softening with distilled water a solution of soda must be added till alkalescence is produced.

IV. LIQUID AND SOLID CULTURE MEDIA—BEEF-BOUILLON GELATIN AND AGAR-AGAR.

The solid food media already considered have a decided superiority over liquid media, and yet every one of them had a decided fault about it. They were all opaque and therefore unsuitable for direct microscopic examination, which had yielded such valuable revelations in the case of the liquid media. It was a brilliant idea of Koch's that found a means of overcoming this difficulty also: "He changed the liquid media into solid ones by the addition of transparent substances capable of consolidation." This was the grand secret whose discovery was destined to disclose to us a world of new phenomena.

As liquid food Koch employed beef-bouillon; as consolidating substance he employed gelatin, which gives the liquid solidity while leaving its transparency undiminished.

Gelatin is a peculiar mass, chiefly obtained from calf's bones or from sinewy and cartilaginous substances. Its chemical composition is not yet exactly known, and probably varies according to circumstances; yet we do know that chondrin and gelatin, with the albuminoids allied to them, are its chief constituents and give it its characteristic properties. The gelatin * which we commonly employ is a French article and is sold in thin transparent leaves. If placed in water or watery fluids—for example, in beef-bouillon—it swells up and melts, at a temperature of about 24° C., into a homogeneous solution. It is then perfectly liquid and boils at 100° C., passing readily through filtering membranes, and remains transparent and unaltered for any length of time. On the other hand, it returns to the solid state at temperatures under 24° C., and

* This gelatin is sold under the name of "gold-label gelatin." J. H. L.

forms an almost colorless mass, of glassy transparency and jelly-like consistency, which when protected from drying up may be preserved a long time without deterioration. In consequence of these properties we can employ gelatin, both in its liquid and in its solid forms, for our culture media. A comparatively small amount of raw gelatin suffices to change our liquid media into solid substances. The more of it used the more solid is the new combination, and the better it is able to withstand the softening influences of heat and other agents.

We generally use a bouillon with 10% of gelatin, a mixture which gives the desired degree of solidity without causing too much difficulty in its preparation or in its use.

The way of making this culture solution is as follows:

The "meat-water" is prepared precisely as for the bouillon already mentioned. Add 1% of peptone, ½% of common salt, and in addition to these 10% of gelatin.

That is to say, add to the 1,000 grams of meat-water 10 grams of peptone, 5 grams of salt, and 100 grams solid gelatin.

This mixture is to be well shaken in a large flask, in order to distribute the peptone, and then heated for about half an hour, i.e., till the gelatin is fully dissolved, over an open flame or in the water bath or steam generator. It should not be heated more than this, since otherwise a premature coagulation of the albuminous matter might easily take place.

Next follows the neutralization, for gelatin has a decidedly acid reaction. Concentrated solution of carbonate of soda is added till the blue litmus paper is no longer reddened, but the red appears slightly blue; an operation which is frequently a good test of patience and requires a good deal of test paper. To get rid of the coagulable albuminous matter boil for about another hour, at the end of which time the liquid is found to have become clear and some of the albuminous matter is found floating on the surface as a dirty gray scum, while some also has sunk to the bottom.

Next comes the filtering. A folded filter, such as is used in chemical experiments, is put into a glass funnel and slightly moistened with distilled water; then the hot mixture is carefully and slowly poured through it. Avoid pouring in too large quantities at once; the gelatin would otherwise cool in the funnel and could not run through. In order to avoid this difficulty there exist so-called "hot-water funnels." A glass funnel is surrounded on all sides with a covering of copper; between the glass and copper is a closed space which is filled with water. Round the outside copper wall runs a perforated metal tube which can be placed in connection with the gas-pipe and feeds about thirty little flames. These make the water boil, and the inner glass funnel is thus surrounded by boiling water. This, of course, very considerably accelerates the filtration of the gelatin, and if one takes care always to have a sufficient quantity of water between the two walls of the funnel, the apparatus gives no occasion for complaint. Yet the filtration can easily be performed without its aid if the unfiltered gelatin in the flask be kept warm in the water bath. In Koch's laboratory the hot-water funnel has hardly ever been used for years past.

The filtered gelatin must be perfectly clear, transparent as water, and but slightly tinged with yellow; it should contain nothing flaky, should not become turbid when boiled or cooled, and should show a clearly alkaline reaction.

One may test the two last-named qualities by letting the first liquid that passes through the filter run into a test-tube, testing its reaction, and heating it until it boils. If the gelatin proves acid the filtration must be interrupted and the requisite quantity of carbonate of soda solution added, after which it should be boiled again for about a quarter of an hour. This acidity is sometimes found when the solution was decidedly alkaline after its first neutralization. The chief cause for this change in reaction is that during the boiling, meat acids and acid salts which were not present before have been set free.

It is more difficult to get rid of any turbidity that may appear in the gelatin, since the causes are various.

Perhaps the filter was not fine enough, may have been torn at the point in twisting it at first, or it may have burst when the gelatin was poured into it. Carefully examine the filter before using it, and only admit the liquid slowly and cautiously. It is desirable to increase the strength of the filter by putting a small protecting cap of cotton-wool at the bottom of it.

It sometimes also happens that the gelatin is turbid only when it first begins to run through, while that which follows is clear; that is, when the filtering paper is poor and its pores have to undergo an alteration before they can hold back the little coagula. Or it may be that the gelatin is too strongly alkaline. This is frequently the result of its being too rich in carbonates.

If the mass has to be treated anew, the carbonic acid is driven off, the combinations are dissolved, the solution is made turbid. In all such cases the evil may easily be removed: readjust the reaction, employ good filters, and recommence the filtration.

Yet it sometimes happens that none of these means prove efficient. Filter as often and as carefully as we may, the gelatin remains opaque. Sometimes the reason is that it was boiled too long or too violently after the neutralization, so that substances have been dissolved which remain in the liquid as obstacles to transparency, and cannot be held back by the filter. The same fault is also seen when the opposite mistake has been made, when it has been heated for too short a time—i.e., not long enough to separate all the coagulable albuminous matter.

It is not very easy to overcome these difficulties. The best way is to let the gelatin run through the filter, cloudy as it is, and then endeavor to make it clear. To do this, add some uncoagulated albumin [the best being the white of a hen's egg] to the solution and then boil again. Soon the white of egg curdles, it conglobates and carries away the suspended particles with it. Now filter once more, and a clear and beautiful gelatin will be obtained. The success of this measure is, in fact, so sure and perfect that, as a rule, we do not wait to see whether the gelatin clears unaided, but in all cases, about one and a half hours after the neutralization, we add the white of an egg and then boil again for another half-hour.

One other case may be cursorily mentioned in which the gelatin may be spoiled. If the solution be poured into new, unused glasses, sometimes it will be found that it afterward becomes cloudy again. This is because the glasses, as they come direct from the glass-works, often contain some traces of alkali; their surface is not chemically neutral, as may easily be proved, and this small quantity of alkali suffices in certain cases to cause loss of transparency in the gelatin. It may happen, for instance, that of 100 test-tubes with gelatin, all prepared in the same manner, 20 or 30 afterward become cloudy, while the rest remain clear. It is therefore desirable to clean the glasses, for the first time, with acidulated water.

When all these difficulties above enumerated have been overcome, pour the clear, transparent gelatin into sterilized test-tubes, not allowing the gelatin to touch the mouth of the tube where the stopper comes, else the cotton-wool sticks to the glass and gives trouble afterward.

Of course the gelatin can be used in bottles, in Erlenmeyer's flasks, or in dishes, as well as in test-tubes. In all cases, however, the gelatin must be thoroughly sterilized. This, of course, is done in the steam generator. Yet it has been found that gelatin will not bear the prolonged heating which would be necessary to make it germ-free by one single operation; it is apt to lose its capacity of solidification, in which case it becomes useless for our purposes. We therefore put it, for three successive days, into the steam generator, each time for twenty-five minutes.

This precaution should not be neglected, for to obtain a good medium, perfectly germ-free, it must be carried out to the letter. This interrupted heating, which reminds us of the discontinuous sterilization, is also the best method of killing the spores which possess the most tenacious vitality, since they germinate in the intervals and are then sure to perish at the next heating.

When the tubes have been sterilized three times they may be used.

The kind of gelatin whose preparation has just been described in detail, and which is chiefly employed on account of its composition, is called "ten-per-cent meat-water peptone gelatin," and when in future nutrient gelatin is spoken of, it is always this mixture that is meant. But of course it is possible to employ a variety of compounds differing from the above

On the one hand, the quantity of gelatin put into the different food solutions may be largely varied, and some, indeed, prefer to work with a 7½% or 5% mixture. Still further diminution of the quantity of gelatin is not advisable, since we then have something approaching too nearly the liquid foods, which softens with a slight degree of heat and does not offer sufficient resistance to the peptonizing influence of many kinds of bacteria.

On the other hand, gelatin may be mixed with some other nutrient solution instead of with the usual meat-bouillon. Some, for instance, do not employ fresh meat for their bouillon, but take beef extract. The following is a recipe: 1,000 grams water, 30 grams peptone, 5 grams extract of meat, 100 grams gelatin. The sterilization here requires particular care, since the extract of meat is extremely rich in very tenacious germs.

Besides the different varieties of meat-bouillon, other kinds of nutrient solution—as, for instance, blood-serum, milk, spices, and urine—have been consolidated by the addition of gelatin. It would lead too far to go into all the particulars concerning them, yet a few words must be written about a number of special admixtures employed for particular purposes, and which are usually added to the bouillon gelatin when it is fully prepared and ready to be poured into the test-tubes.

Thus, for instance, 4 to 6% of glycerin is added to the gelatin, because it has been found that certain bacteria thrive better with this addition. An increase of nourishing power may also be obtained by adding from ½ to 2% of grape sugar (dextrose), which is, indeed, frequently done.

But the grape sugar has another valuable property: it is a reducing substance which partly absorbs the oxygen, and thus renders the culture media peculiarly suitable for the development of oxygen-avoiding, anaërobic species. There are also other substances which act in the same way, and have been recommended by Kitasato and Weyl, namely, formate of soda, which by oxidation changes into the carbonate, resorcin, etc.

As an indicator for certain chemical changes within the food medium caused by the growth of bacteria in it, we use an almost saturated solution of blue litmus. Buchner and Weisser, and of late more especially Petruschky, have shown that the changes of color in gelatin tinged with this solution give a very exact key to the amount of acid, or alkali, formed in the food mass.

Quite different from this is another use of the litmus solution. The litmus coloring matter is easily reduced and loses its color; it discolors completely and seems to have vanished until oxygen finds its way back to it and instantly gives it back its color. This property has been taken advantage of to study any reduction processes that may take place in the substratum; and by the labors of Cahen, and more especially Behring, very important facts have been elicited in this way. In many cases, it is true, we find it desirable to replace the litmus (whose chemical composition is not yet precisely ascertained) by other dyes, which are also discolored by reduction; for instance, sulphate of indigo in 1 to 10% solution, etc.

All these modifications of the original simple bouillon gelatin share with it in its chief properties: they soften at a somewhat elevated temperature and have then the qualities of liquid food media, are easy of application and distribute the germs equally, and at lower temperatures they return to the solid state, where they possess the important advantages of the solid food media.

It is, on the whole, immaterial as to which sort of gelatin should be given the preference, but one thing should be specially noted: that for comparative investigations always one and the same food medium should be employed. For even slight differences in the composition of the food solutions often cause very great differences in the appearance and behavior of the bacteria.

Gelatin is an excellent substance to give to our food liquids the necessary stiffness, and the ease with which it may be prepared and handled makes it permanently valuable to us; but it has one fault. Under the influence of heat above 25° C., and in consequence of the peptonizing action of many species of bacteria, it softens and becomes a liquid instead of a solid medium. The food gelatin, therefore, cannot be used with advantage for breeding experiments at a high temperature, nor for micro-organisms which have a strong decomposing tendency.

Instead of gelatin, another substance may be employed, called agar-agar, which is also transparent and capable of solidification. In contrast to gelatin, which is' an animal product and related to the albuminates, agar-agar is a vegetable jelly, obtained from certain species of sea-tangle on the coasts of India and Japan.

It is sold in the form of dry transparent strips and as a white powder. In both forms it is able to give a completely solid consistency, without injuring the transparency. It melts at about 90° C., stiffens again at about 40° C., and is not affected by the digestive action of the bacteria.

If, in spite of these advantages, agar solutions have not entirely replaced gelatin, the reason is that the agar is much more difficult to prepare as a culture medium.

In principle its preparation quite resembles that of gelatin. It is mixed with bouillon, and the mixing is performed as follows:

First of all, make an ordinary food bouillon, in the manner heretofore described, with 1,000 grams meat-water, 10 grams peptone, and 5 grams common salt, and then add to the filtered, clear, alkaline liquid, in proportion to its quantity, 1½ or at most 2% of the solid agar-agar. It is not advantageous, as in the case of gelatin, to put the hardening medium in before the filtration, since the conglobated masses of albuminous matter unnecessarily add to the length of the always tedious process of filtering agar-agar.

For the same reason do not employ more than the above-given proportion of agar-agar; if too much is used, its solidity and consistency do not allow it to pass through the filter at all.

When the agar, cut up into as small pieces as possible, has been put into the bouillon, it is left there for a few hours to swell.

Then begin to heat the mixture, in order to dissolve the agar. The longer this is continued the more patience must be exercised, the better will the food medium turn out, the more easily will it filter, and the clearer will it be. The boiling must be extended to five or six hours at least; whether it is done over an open flame, or in the steam generator, or in a well-filled water bath is immaterial. When the agar-agar has become so far dissolved that no more coarse portions are perceptible in the bouillon, the reaction must be adjusted.

If the bouillon was distinctly alkaline a few drops of carbonate of soda solution will suffice, the agar being, in contrast to the gelatin, almost completely neutral in its reaction. If the liquid after another hour's boiling gives no sign of becoming clear, one can here, as with the gelatin, add the white of a hen's egg.

Then filter. The great number of proposals which have been

made, in the most different quarters, to facilitate this part of the process, is a clear proof that a thoroughly satisfactory manner of proceeding has not yet been found. By experience, we have found it best to pass it through filter paper—through an ordinary folded filter. All other means—cotton-wool, glass-wool, etc.—are of little use. The filtration takes place in the steam generator or with the hot-water funnel. Both these aids may be dispensed with if only the precaution is taken of "previously wetting the filter with boiling water." If this is not done, the inside of the paper becomes covered with a thin coating of solidified agar and the filtration is stopped in its very beginning.

The process of filtration is one that always requires a great amount of patience and perseverance. It is better, in many cases, to avoid filtration altogether.

The food agar is used in a manner somewhat different from gelatin; it is chiefly its surface that we employ as a basis of development for bacteria, and it is not necessary to be so particular about its complete and faultless transparency. Pour the agar out of the flask in which it has been boiled and cleared from the sediment into smaller vessels, whence it can be poured out for use as occasion may require. Or, again, pour the whole mixture, after several hours' boiling, into tall, narrow cylinders—measuring glasses—which are then placed for some time in the steam generator. Here the opaque particles settle down more or less completely, and portions can now be removed from the upper transparent strata with pipettes for immediate use.

The agar obtained in this manner, or the one previously described, is next poured, in quantities of about 10 c.cm., into sterilized test-tubes, and made thoroughly germ-free by three heatings of half an hour each in the steam generator.

The agar is allowed to solidify in an oblique position, in order to obtain a large surface for use. A small quantity of water always separates and remains at the bottom of the glass, and does not disappear by evaporation for some time.

Even the best, most carefully-filtered agar, which in its liquid state was perfectly transparent, generally becomes somewhat cloudy and opaque as soon as it hardens, but this appearance, if it occurs, is of no moment as indicating a fault in the process of preparation.

As in the case of the bouillon gelatin, so also in the case of bouillon agar, numerous additions are made for special purposes. Such are grape sugar, formate of soda, resorcin, litmus, etc. For observing the process of reduction, for instance, add to a litre of prepared agar about 40 c.cm. of a saturated solution of litmus.

By far the most important, however, is a modification of the ordinary bouillon agar, which contains from 4 to 6% of pure neutral glycerin. By the investigations of Nocard and Roux, the very important fact has been elicited that on a food agar thus prepared, those micro-organisms are able to thrive which formerly required far more complicated mixtures to live upon. Such, for instance, are the tubercle bacilli. While on agar without glycerin their development is very insignificant, they form a particularly luxuriant growth on glycerin agar, and the growth of many other parasitic bacteria may be greatly aided by a little glycerin.

In consequence of this discovery, glycerin agar is very largely employed, and has superseded the use of solid blood-serum, which was formerly much used in these cases.

As is well known, blood when it is removed from the influences of the living vascular wall coagulates, and that in the further development of this process a separation takes place between the red clots of blood and the liquid, almost colorless or light, amber-colored serum.

The latter is rich in albuminous substances, which coagulate when heated. The principal mass of the serum albumin consolidates at about 70° C. If we do not heat much above that temperature and do not allow it to continue too long, the serum is transformed into a solid, uniform substance, which is scarcely less transparent than the common food gelatin, and which, in like manner, can be employed as a culture medium for bacteria.

Koch had introduced its employment because he saw that certain strictly parasitical micro-organisms were unable to thrive on those rough imitations of animal juices which our liquid and solidified bouillons were supposed to give. He therefore endeavored to procure for these bacteria a nourishment which might more nearly resemble their natural conditions, and in this way he came to employ serum.

Without doubt, this ingredient of blood is an excellent medium for the culture of bacteria, and if at the present day it is used comparatively seldom, the only reason is that its preparation and application are attended with considerable difficulty.

The serum is best prepared in the following manner:

The blood which flows from the jugular vein of animals when slaughtered is collected in large previously-sterilized glass cylinders. They must stand about forty-eight hours undisturbed, if possible in the refrigerator; here the separation of serum and coagulum takes place. The yellowish, sometimes also slightly-reddish, serum is removed with sterilized pipettes and placed in sterilized test-tubes—about 10 c.cm. in each. In these it is made to consolidate by heating, and it is generally desirable, as in the case of agar, to heat the fluid in an oblique position, so as to obtain a large surface.

For this heating, double-cased tin boxes are used whose bottom is moderately inclined. Between its walls there is water, which is heated from below by a gas-burner. Into this apparatus the filled test-tubes are placed, and with them a thermometer, to give at any time the temperature of the interior space. Warm slowly up to about 68° C. and take care that the limit of 70° C. be not exceeded.

The serum consolidates more or less quickly. Those tubes in which it is fully coagulated are at once removed, for as we must not employ more than 70° of heat, so also we must not expose albumin too long to the action of heat after it has once coagulated —in both cases the liquid would harden into a dirty-gray opaque mass.

To be faultless the prepared serum should, as already stated, be transparent, yellowish, and of a jelly-like consistency. The serum in the tube has well-defined, smooth edges; at the bottom of the tube a little clear water always condenses, and for months prevents the substance from drying up.

Were the serum to be used at once, a grave fault would be committed: the substance has not as yet been sterilized.

This cannot be done afterward in the usual way, by heat in the steam generator, for that would cause a complete loss of transparency and render the serum useless. It must, therefore, be sterilized before its consolidation, by Tyndall's method of discontinuous sterilization, the principles of which have been mentioned. The liquid serum must be warmed for about eight days, each day for about two hours, to 54° or 56° C., and then brought to consolidation at 68° C.

It is better still to avoid this troublesome and tedious process altogether by taking all possible care in obtaining the blood, in pouring it from vessel to vessel, and to so manage that no germs get into the serum, in which case sterilization is superfluous. If done in this manner, it will be found that the greater part of the tubes remain unaltered. To make sure, keep the consolidated serum for three or four days at breeding temperature. Any germs of bacteria that may have been present will, by the end of that time, have arrived at a distinctly-visible point of development, and the tubes containing them can be removed. The others may be regarded as germ-free and can be used.

With the other kinds of food media this very convenient method cannot be employed, since the chances of their being polluted from the beginning are far greater.

Serum once consolidated does not lose its solidity again. It is therefore particularly well adapted for the culture of bacteria at high temperatures.

But in consequence of this peculiarity it has not the advantage possessed by other solid and transparent media—that of being able to return to the liquid state under certain conditions, and thereby becoming easier to manipulate. Under the influence of certain bacteria, however, it does soften in the same way as gelatin, and then it is usually peptonized to a very considerable extent.

On this account the employment of blood-serum is restricted to somewhat narrow limits, and we can in many cases greatly facilitate our labors by employing glycerin agar instead. For certain purposes serum of human blood and of substances similar to it—as, for instance, the fluid from dropsy (ascites), from hydroceles, and from the ovary—has been prepared as a culture medium. Human serum is chiefly obtained from the placenta, which generally yields a considerable quantity of blood.

The whole series of solid transparent food-media which are commonly employed have now been considered, and also the special advantages offered by particular kinds: the ease with which gelatin can be prepared and employed; the power of agar to resist not only high temperatures, but also the decomposing action of bacteria; and the importance of glycerin agar and of blood-serum for the breeding of strictly parasitical micro-organisms. But it would be a great mistake if it should be considered that these constituted a wealth of artificial culture media with which could be obtained the desired object at all times and under all circumstances.

By the addition of gelatinizing substances to the bouillon, for instance, the latter is not in itself rendered more suitable to serve as food for the micro-organisms. It is true that bouillon was chosen with a view of offering the requisite conditions for development to as many kinds of bacteria as possible, and up to a certain point its success was undoubted, for great numbers of bacteria are known which thrive in the ordinary culture media.

Yet its nourishing powers must not be overrated. Numbers of bacteria, perhaps the majority of all existing species, do not find in it the conditions necessary for their existence, and therefore resist all efforts to breed them artificially. A simple example may suffice to prove this.

If saliva be examined under the microscope in a cover-glass preparation or as a hanging drop, generally an abundance of micro-organisms will be found in it: cocci, bacilli, separate or hanging together in long threads, but whose form alone shows them to belong to different species; here and there, too, beautifully-twisted, lively, motile spirilla are seen. If some of this saliva be placed on our food media—for example, on bouillon gelatin—it will soon be seen that very few germs develop, and that the number of colonies is very much smaller than might be expected, after the number of bacteria which were seen in the inoculating material under the microscope.

Though these conditions may not be found everywhere to such a striking extent as in the experiment just mentioned, there can be no doubt that many micro-organisms cannot thrive on our culture media.

One more point is to be noticed here. On reading a number of French treatises on bacteriological subjects, passages will surely be met with in which the exclusive use of liquid food media is warmly recommended.

The use of the latter is evidently considered by some of our colleagues beyond the Rhine as a national affair, which, were it only as a matter of patriotism, must be strongly defended against German innovations.

They maintain, for instance, that in the bouillon a great number of bacteria can thrive which fail to do so on a solid medium. This, however, it may be positively asserted, is not the case; such differences do not exist, and would, indeed, be altogether incomprehensible in view of the manner in which the solid food media are prepared.

Our culture media, whether solid or liquid, are all essentially similar, and the monotonous method of proceeding is doubtless sometimes the cause of failures in our breeding experiments.

It would be an interesting and doubtless profitable research to study the necessary conditions of life for those bacteria which have hitherto resisted all efforts to breed them. This would necessitate a methodical, systematic variation of the food solutions, for with many species the question of their cultivation is certainly a mere question of their proper nutrition.

V. USES OF FOOD MEDIA FOR OBTAINING AND MAINTAINING PURE CULTURES.

How are the food media to be used for obtaining and maintaining pure cultures?

It will be remembered how we proceeded with the first solid food medium which was employed: that out of a mixture of bacteria which were to be separated into the different species contained in it a small portion was taken and spread as evenly as possible over the surface of a sliced potato. Then from the first potato to a second, from the second to a third, and so on, the inoculating material was transferred in continually smaller, "more diluted" quantities, so that the germs were at length so widely separated, so far isolated from each other, that they developed separately and singly, yielding pure cultures which could not mix together on their solid medium.

In precisely the same way the gelatin was at first employed, and Koch originally endeavored to imitate the form and other particulars of the successfully-tried potato culture by liquefying the culture gelatin, pouring it into watch-glasses, on slides, etc., allowing it to harden upon them, and inoculating the now solid surfaces by dipping a platinum needle into the mixture of bacteria and then drawing it repeatedly across the gelatin.

Stroke by stroke the number of the germs planted becomes smaller, and when later they develop along the strokes made by the needle and form colonies, the separate species will appear isolated, and it will not be difficult to divide them with certainty and precision.

At the present day the "slide cultures" are employed in exceptional cases only, for it was soon observed that by this procedure one of the advantages of gelatin was lost—the advantage of its permitting, before it assumes the solid form, the same complete and equable distribution of germs as do the liquid media.

We therefore perform the dilution of the inoculating material in the dissolved gelatin, and after this the now universally-employed method for obtaining pure cultures will at once be apparent.

An experiment with human fæces, a substance extremely rich in the most varied species of bacteria, can be made as follows:

First dissolve the gelatin in a number of test-tubes. The best way is to employ the water bath at about 35° C. Take one of the test-tubes, taking care that the gelatin is completely dissolved, and with the platinum needle introduce a trace of the material to be inoculated. With such a tough consistency as that of the material in question it is well to rub it and squeeze it against the glass, in order to mix it well into the liquid; in all cases, however, and under all circumstances, endeavor, by repeated slight variations of the slant of the test-tube, to produce a very thorough and equable distribution of bacteria throughout the gelatin.

This being done, the "dilution" begins. Were the first tube only to be employed, the number of germs that would develop would certainly be far too great to think of separating them.

In making these dilutions, there is one special method which practice has shown to be the best and which usually proves successful: from the first tube—the "original"—remove with the platinum loop three successive drops of the inoculated gelatin and place them into a second tube—the "first dilution;" out of this again, three drops into a third tube—the "second dilution." In the one or the other of these three tubes the germs will generally be found so distributed that the contents can be taken as the starting-point for further researches.

It is not necessary to follow these directions literally under all conditions. On the contrary, the number of dilutions and also the number of drops apportioned to each will have to be varied according to circumstances.

For the transmission of material by means of the platinum loop, there are certain special manipulations which have proved practically useful and are universally employed.

With the left hand opened and the palm directed upward, place the "original" tube between the thumb and fore-finger. The mouth of the tube is turned toward the operator. It should always be held, as much as possible, in a horizontal position, in order that when the cotton-wool plug is removed no germs may fall in from the air. Now place beside this first glass a second, which is to be used for the first dilution; here, too, the opening is directed toward the operator, and is about in the centre of the palm of the hand. Remove the cotton-wool plugs by carefully twisting them, first from tube II., then from tube I.; the former is placed between index and middle finger; the latter between the ring and little fingers of the left hand. By always observing these directions we may avoid confounding the plugs. Next put the sterilized but cooled platinum loop into the first tube, take out a drop of gelatin and transfer it to tube II., and by moving the wire up and down endeavor to distribute the contents of the drop throughout the liquid gelatin. Repeat this process a second and a third time (it is unnecessary to heat the platinum loop anew each time). When this is done the cotton plug is placed in tube I., which is set aside, preferably into an empty tumbler. Then, just in the same way—the palm of the hand remaining always directed upward—transfer from the first dilution to the second.

It is desirable to always know which tube is the original, which the first and which the second dilution. Therefore we mark them from the beginning, either with labels or with a Faber wax-pencil, or by twisting the cotton plug of the first dilution into one, that of the second dilution into two long ends.

Now, the liquid gelatin in the test-tubes must be suitably spread out and solidified. The larger the surface on which it is spread the further the germs are separated from each other and the easier it is to keep the different species apart.

We therefore pour the contents of the test-tubes upon plates of glass to consolidate. These plates, from which the widely-celebrated Koch's "plate process" takes its name, are cut out of moderately thin glass, and must, of course, be thoroughly sterilized before being used. This is best done in a box of sheet-iron, arranged to receive about twenty plates at once. Packed in this manner, the plates are heated for half an hour in the hot-air oven, after which the lid of the plate-box is removed, and with the thumb and finger of the right hand a plate carefully drawn out without touching its surface. If the attempt be made to pour the gelatin over it at once, it would be found impracticable. Gelatin hardens but slowly at the ordinary temperature of rooms, and unless it can be managed to have the plate perfectly horizontal, the gelatin will soon flow over the lowest end.

Koch has, therefore, invented a special "plate apparatus." A vessel is filled with pounded ice and water to the edge, and covered over with a sheet of opaque glass. The whole is placed on a levelling stand with adjusting screws and provided with a spirit-level, so that a horizontal position is easily obtained.

When the plate is placed on this instrument the gelatin may be poured out without difficulty, and on the horizontal, strongly-cooled surface it spreads evenly and hardens quickly. In spite of this aid, it requires a certain degree of practice to spread the gelatin properly over the plate. The process may be made much easier by quickly sterilizing the mouth of the test-tube in the flame of the Bunsen burner immediately after the inoculation, and afterward directing the flow of the gelatin by using the mouth of the test-tube as a "lip." Remember, however, to push the cotton plug a short distance into the tube, that it may not be scorched or burned, and take care that the mouth of the tube is fully cooled again before proceeding to pour out the gelatin.

The whole operation of the "plate process" may be summarized as follows:

Three test-tubes with gelatin are melted in the water bath at 30° to 40° C.; inoculation and dilution are next performed, the original being first prepared, and the first and second dilution by the transfer of three drops with the loop. Next sterilize the mouths of the tubes; while they are cooling, fix the levelling apparatus to an exact horizontal by means of the spirit-level; open the plate-box, take out a plate, and lay it on the cool surface of the glass-plate covering the iced water. A bell-glass prevents germs from falling upon it from the air. Next take the first test-tube, slant it a few times up and down, to distribute the germs as thoroughly as possible in the liquid, remove the cotton plug with a pair of forceps, raise the bell-glass, and pour the gelatin over the plate, aiding its distribution by means of the edge of the test-tube.

Under the protection of the bell-glass, which is immediately replaced, the gelatin consolidates in a few minutes. It should be evenly spread in a thin, uniform layer over the surface, stopping within about 2 cm. of the edge on all sides.

The prepared plate is now removed and placed in a moist chamber, to prevent it from drying up. Spread some blotting-paper over the bottom of a glass dish (as in the potato cultures), moisten it slightly with water, and place the plate preparation upon it. It is desirable, for the sake of convenience, to put several—say half a dozen—such plates together in one dish, one above another, separated by little glass bridges or benches. Each of these bridges should, of course, be labelled with name, source, and date. In the present case, for instance, "II., VII., Fæces, O." (February 7th, Fæces, Original); then follows the glass plate; next immediately over it a "bridge" with its label, "II., VII., Fæces, I." (February 7th, Fæces, 1st dilution); then the second glass bridge, etc., etc. It is not necessary to sterilize the glass bridges, as they do not come into contact with the gelatin and have therefore no opportunity to pollute it.

Keep the bell-glasses in a moderately warm place, and await the development of the germs in the shape of colonies.

The original glass-plate process of Koch which has just been explained has within a few years conquered the world, so to speak, and found its way into every laboratory. In fact, its advantages over the former methods are so evident and its manipulation is so simple that in this respect hardly anything seems left to improve upon. And yet it has its disadvantages. The most serious one is, perhaps, that it can only be conducted in a laboratory, since it requires a certain amount of space and certain special apparatus.

It is true we may do without the latter in case of need; the test-tubes and glass plates can be sterilized in an open flame, the plates can be laid on the level surface of a table, and the gelatin poured over them there, etc., etc. But then there remains the protection of the finished preparations from drying up and from being polluted. This requires space, transportation is almost impossible, and investigation during a journey, a campaign, etc., is extremely

7

difficult. Efforts have been made to overcome these difficulties, two different ways having been suggested.

In one of the modifications of Koch's method, flat-covered dishes, as recommended in different forms by Babes, Sayka, Petri, and others, take the place of the ordinary glass plates. The pattern introduced by the last-named investigator is the one generally employed, in which the lower dish is just 10 cm. in diameter.

A number of such dishes are sterilized in the hot-air chamber and then prepared for use. When the dilutions have been made in the usual manner, the cover of a dish is removed, the liquid gelatin poured from the test-tube into the dish and distributed over the bottom of it by a suitably tipping or slanting of the dish, which is covered again, and the gelatin left to harden. The dishes thus prepared are piled one above another, and in due time the development of germs and the growth of colonies occur.

The advantages of this system are great. The manipulation is much simpler and more convenient. The difficulties which a beginner usually has to contend with are, for the most part, avoided; the gelatin spreads, almost of itself, in an even layer over the surface, and hardens there quickly and perfectly, without the aid of the levelling apparatus; the often troublesome after-pollution of the glass plates by adventitious germs falling upon them is here prevented almost to a certainty by the protecting cover; the contact of the fingers with the layer of gelatin, which is apt to occur in removing and examining the plates, cannot take place, and the extensive liquefaction of the gelatin, the "running over" of the plates, is also impossible.

Most of the apparatus employed in Koch's original system becomes unnecessary and superfluous. Glass plates, plate-box, levelling apparatus, glass bridges, and the large bell-glass can be dispensed with, and when it is further considered how easily the "dish-plates" may be removed and transported, we cannot but acknowledge the superiority of this process in many respects.

For some purposes, however, a second modification of Koch's glass-plate process, introduced by E. von Esmarch, has still greater advantages.

The liquid gelatin is inoculated in the usual way, but then, instead of being poured out of the test-tube, it is spread over the inner walls of the test-tube itself and there left to harden. The interior surface of the tube becomes covered with a thin, even layer of gelatin, which possesses about the same extent of surface as that obtained in the case of the ordinary glass plates. The germs develop precisely in the same time and manner as on the plates; it is, we may say, nothing else than a plate bent into a cylindrical form. When a test-tube has been inoculated, we first endeavor, by slanting it gently up and down, to produce a thorough distribution of the germs; then an India-rubber cap is drawn over the cotton plug, and the test-tube is laid horizontally in a vessel full of iced water. While holding the neck of the test-tube with the left hand, we endeavor to turn it quickly round on its own axis with the right hand, not allowing one end to dip down lower than the other, as the liquid in that case flows toward the lower end and causes irregularities. In a few moments the gelatin is hard, the tube is removed, the India-rubber cap taken off, and if the operation has succeeded, the thin transparent covering of the walls can scarcely be seen. The wider the test-tube and the greater the surface over which the gelatin can spread the better, as a rule, is the result.

This process has its great advantages. One is the great saving of space, the possibility of working with precision on journeys, in a campaign, and wherever ease of transportation is a desideratum. Another is the fact that all the germs which were introduced by inoculation must necessarily develop, while if the contents of the tube be poured out upon a plate or dish, some parts of the gelatin always remain sticking to the walls of the tube, and consequently some of the germs escape observation and examination.

On the other hand, this process has also its disadvantages, some of which, however, may be easily avoided.

It sometimes happens that in one such tube no growth takes place, while it develops luxuriantly in the others. On examination it is found that the gelatin covered the lower end of the cotton plug with a thick layer, and prevented the access of air to the interior. Aërobic bacteria, therefore, cannot thrive in it, but if the plug be taken out and the crust pierced with a sterilized platinum needle, the growth of colonies will soon take place.

Again, sometimes such numbers of air-bubbles pass from the cotton-wool into the interior of the tube during the consolidation of the gelatin that the latter is completely filled with them. To avoid this, allow the gelatin to cool almost to a semi-solid state before putting it into the iced water. The tubes cannot be prevented from soon becoming useless when they contain a large number of liquefying bacteria. The dissolved gelatin runs down the sides of the glass and presently forms an opaque mass, with which nothing can be effected, while on a flat plate the horizontal position insures a much longer period of usefulness.

The employment of agar-agar in the glass-plate process is attended with much greater difficulties than in the case of gelatin

As we know, nutritive agar does not liquefy under about 90° C., and it returns to its solid state at about 38° C. As we cannot put germs into a solution of 40° or above without killing them, we are obliged, after letting the agar melt completely in boiling water, to allow it to cool down again gradually to 40° C. exactly. Then the tubes are inoculated and the dilutions made in the manner described. Yet this must be done as quickly as possible, for if the temperature sinks but a trifle lower the agar becomes solid and the labor is all in vain, for we cannot, for the above-explained reason, melt the inoculated food agar over again as we could the gelatin, without running the risk of destroying the material inoculated.

The best way is to pour the contents of the test-tube speedily into sterilized Petri dishes; in a few moments the agar is hard, and if care has been taken to tilt the dishes properly, it covers the bottom in an even, uninterrupted layer. All other processes, including the original one of Koch and the rolling process of Esmarch, are either difficult or impossible with agar instead of gelatin.

Troublesome as is the manipulation of agar compared with that of gelatin, yet agar is indispensable for many purposes, and especially in the case of organisms which only thrive or which thrive best at the temperature of the incubator. Place the dishes with their contents in the incubator, and leave them there for from twenty-four to forty-eight hours.

The third solid and transparent food medium cannot, as has already been said, be employed in like manner, since it cannot be brought into a solid state by low temperature, but, on the contrary, requires a high temperature to harden it; so high a temperature, indeed, that most bacterial germs would be destroyed by it.

Serum can, at best, only be employed for stick cultures; we pour it into little dishes, in which we make it consolidate, and then inoculate its surface. Generally we are obliged to fairly rub the inoculating matter into the jelly-like mass, and there is but a very bare possibility of afterward distinguishing the separate developing germs.

VI. PLATE CULTURES—PETRI DISH CULTURES—ESMARCH ROLL-TUBE CULTURES.

Sooner or later, according to the temperature of the place in which the plates are kept, the colonies develop in the gelatin. Generally three or four days elapse before they have acquired a tolerable size and are visible to the naked eye. At the same time it must be remarked that the energy of the growth is very different in different species—while some grow very quickly and luxuriantly, others require whole weeks to reach the same degree of development. Occasionally, no doubt, this is because our food gelatin offers better conditions for the development of some micro-organisms than for others, which can only exist upon it with difficulty. Remember that for great numbers of bacteria the ordinary food media offer no suitable nourishment, and we must, therefore, not record the number of colonies that develop as an always reliable index of the number of germs inoculated.

There is also a second reason which prevents this being done. If we have inoculated the gelatin with too many germs, they will afterward disturb each other in their growth on the glass plates; many will be completely kept back by others and will thus be lost for statistical calculation.

Lastly, it may also happen that a colony, though looking just like the others, may not have owed its origin to one single germ. In the inoculated material small, firmly-united groups of bacteria may have existed; a series of cocci, a thread of bacilli, may have passed into the nutritive liquid, and thence, undisturbed, on to the glass plate, where they could only proceed conjointly to the formation of a colony

With these exceptions, however, the colonies, as a rule, agree in quantity and variety with the original number and kinds of germs inoculated, and it is an extremely valuable quality of the solid transparent food media that they are able to give us such ready and exact information on this point. If we wish to know the quality of bacteria present in any particular substance, one only needs to bring a measured quantity of that substance into gelatin, and after a few days can read off the result from the plate. For the comparative examination of different fluids, etc., this process is extremely valuable. The particulars of its applications will be given hereafter.

Much more important, however, is the certainty with which the solid food media enable us to differentiate between the separate species.

As in every case the germs develop small pure cultures of their own species only, all the peculiarities of the species are seen in the colony in a more strongly-marked and striking manner; features that, occurring separately, would scarcely be perceived by the practised eye becoming clear and obvious when seen *en masse.* This was noticed in the case of the potato, but it is seen still more strikingly in the case of the transparent gelatin or agar. Forty-eight hours after the preparation of a gelatin plate containing ½ c.cm. of water from the Spree, a number of characteristic differences in the appearance of the separate colonies can be noticed.

Some are found which liquefy the gelatin strongly. Some form cup-like circular depressions, the edge clearly marked off against the solid substance, and the whole colony looking uniformly gray; in others a thick, crumbly mass is seen, which consists of heaps of bacteria, lying at the bottom of the liquefied depression; others distinguish themselves by secreting a strong coloring matter—not only the liquid colony itself, but also for quite a distance around it the gelatin is dyed with a peculiar yellowish-green color. Then, too, bacteria are observed which liquefy the gelatin more slowly; a somewhat more careful examination is necessary to see that the centre of the colony is slightly sunk, the edges being irregularly jagged. Then a colony may be seen looking like an entanglement of roots spreading its white branches over the plate, and beside it one transparent and of a beautiful purple color.

Then, again, there are some which do not liquefy the gelatin. Some show themselves as small white dots in the interior of the transparent food medium; others stand out like thick, transparent buttons of porcelain; also there is one having a beautiful phosphorescent green coloring substance, and one that looks like a dry, dirty gray skin.

They must be seen over and over again in order to become familiar with the different appearances and to be able to recognize them wherever they occur. To this end, nothing is wanted but an attentive eye and the necessary practice. It will be seen from experience how exactly one can fix upon the origin of such a plate, and how correctly the nature of the different colonies upon it can be determined.

All this, however, is rendered much easier, and the exactness of our investigations become indisputable, as soon as we further utilize the transparency of our food medium and examine our plates directly with the microscope.

As a rule, we employ for this purpose objectives of low power—Zeiss, 16 mm.; Leitz, 3; Bausch and Lomb, ⅔, etc.—because for our particular purposes a considerable space between the objective and the plate is desirable; the want of high magnifying power we endeavor to compensate for by using strong eye-pieces. As in examining colonies on the plate we have to deal with unstained objects, a very small aperture of diaphragm, not much above the size of a pin's head, is necessary. In order to examine the whole surface of the gelatin the microscope should have a good large stage, and, on the other hand, the plates should not be too large, but about 12 by 9 cm.

Before commencing a microscopic examination, notice the appearance of the plates with the naked eye. The original will, as a rule, be so thickly covered with germs as to be almost worthless, and therefore we generally employ the first or second dilution. The plate is pushed slowly under the microscope, the focus being moved up and down by the coarse adjustment so as to bring the entire depth of the gelatin layer under observation.

The extraordinary variety of appearance of the separate pure cultures is very striking. This variety appears still greater now than it did with the naked eye, for we are now able to note the finer distinctions in the forms of the colonies. Some are curiously granulated, others show concentric rings, many spread out quite evenly, some with ribs, like leaves; some are curled, rolled up like a ball of thread, or provided with tendrils; the majority are of a light yellow or brownish color, and many are distinguished by some definite pigment.

Yet it would be a mistake to consider two colonies as being of different species merely because they present a different appearance. A certain degree of caution is necessary, and no definite judgment can be formed without repeated experiments.

It is sometimes especially difficult to recognize colonies lying in the mass of the gelatin as being identical with others of very different appearance on the surface of the food medium. For instance, examine a plate containing colonies of typhus bacilli alone, and it would hardly be supposed that the different colonies were composed of the same micro-organism. In one place small, somewhat elliptical or whetstone-shaped, dark-brown, slightly-granulated colonies are seen, and near them may be some yellowish-white, almost transparent discs, spreading out like leaves and marked with ribs, which have scarcely a trace of resemblance with the others, but which, nevertheless, proceed from the same species of bacteria.

Conviction of this fact will be gained by noting several transition forms between the two sorts of colonies; then, too, it is seen, by prolonged observation, that one kind gradually merges into the other; further, it will be found that a microscopic examination shows everywhere the same rod-cells; and lastly, it will be found that whether new plates are prepared with one or the other kind of colony, both forms will always be developed over again.

This difference in the forms of the colonies may be easily explained.

Within the mass of the gelatin the colony, in forming, has to contend against the considerable resistance of its solid surroundings; it must conquer the ground step by step; besides which it often has to suffer from want of sufficient oxygen to enable it to thrive vigorously. On the surface the case is different: there no resistance checks its spread, and the conditions are doubtless more favorable to its development.

Thus many of the most distinguishing marks by which the different species of bacteria are recognized, in particular the liquefaction of gelatin and the formation of pigment, are only seen to their full extent in the surface colonies of a plate, and these are, consequently, of special value to us in forming definite conclusions. When more positive information is necessary as to the construction and composition of a colony, there are two methods of proceeding. Lay a cover-glass over the colony in question, press it down a little on the gelatin, then remove and mount in water on a slide. Place a drop of oil on the cover-glass and examine at once with the immersion lens. As it is an unstained preparation, employ the diaphragm, which, for the high magnifying power required, should have about the same aperture as was recommended for the examination of objects in the hollowed slide. At the edge of the colony in particular we may often make very instructive observations; the bacteria here lie somewhat freer—one can distinguish individual cells and can clearly mark how they multiply and spread further and further over the solid food medium.

In consequence of the small focal length possessed by our immersion systems, this kind of observation is hardly possible with any but surface colonies. These latter, alone and exclusively, can be examined by another somewhat similar process, but which requires the aid of staining to show the finer structure of the colonies.

Take a thin cover-glass, lay it on the gelatin plate, press it lightly upon the surface colony, and then lift it carefully up again with the forceps. An exact print of the colony sticking to the cover-glass will be obtained, which can be stained in the usual manner and then examined. Allow it (the print preparation) to become air-dry, pass it three times slowly through the flame, give it a drop of fuchsin, or gentian-violet, decolorize with water, and submit it to microscopic examination.

If the colonies are not too far advanced in their development, and the gelatin is only just beginning to liquefy, we obtain on plates of twenty-four—or at the most thirty-six—hours' standing prints of remarkable delicacy, yet of perfect sharpness. Even with a low power the colonies can be seen in their characteristic size and form on the cover-glass, but it is not till the high power with immersion is employed that the advantages of this treatment are seen at their best. What previously seemed to be the uniform mass of the colony resolves itself into a closely-crowded mass of separate bacteria, which stand beside each other, shoulder to shoulder, or lie together without order, showing in the clearest manner how such a colony is built up.

The ease with which the process is conducted and its faultless results have in a short time made it an indispensable part of our method of investigation. When working with this form of cultures, scarcely ever lay by a plate without taking one or more impressions of it.

Be on guard against a certain error in judging the plates. It is scarcely possible to prevent germs from falling upon the gelatin surface from the air during its preparation, and still less so during the after-employment of it. Such germs naturally develop; and although it is one of the advantages of a solid food medium that they are bound down to the spot on which they fall, and so cannot spread beyond a narrow limit, yet their numbers are sometimes so great--particularly when the plates have been carelessly handled or the moist chamber has been long left open—that they become troublesome hindrances to the observer.

It is chiefly mould fungi that get in in this manner, but they are little likely to be confounded with bacteria. They are recognized by there being only one, or at most but a very few, on the whole plate, and also by their occurring exclusively on the surface. If Koch's original method has not been used, but one or other of its modifications, the manner of proceeding requires but little alteration. With Petri's dishes the examination proceeds exactly in the manner described; with Esmarch's test-tube preparations, place the tube under the microscope and examine with a low power, to become acquainted with the appearance of the different colonies. Print preparations cannot, of course, be made with them. Although this must be regarded as a defect, it is almost balanced by a very considerable advantage which is peculiar to the rolled test-tubes. The plates cannot be kept for any length of time without receiving pollutions from without. It seems that quite a number of germs develop comparatively long—perhaps weeks—after the inoculation, a circumstance which escaped notice formerly only because the ordinary plate could not be kept in good condition beyond a certain limited period of time.

The agar plates require no particular mention here. No micro-organisms are known capable of peptonizing agar; it offers no facilities for judging the colonies in connection with their liquefying power, as the gelatin. As a whole, therefore, the agar plates are less useful for studying peculiarities of growth in the different species of bacteria upon a solid food medium.

It would be almost impossible to overrate the importance of the glass-plate system for our investigations. The glass plate is the invaluable, altogether indispensable means which leads us safely through the intricate world of bacteria, discovers the finest differences of species, and is able to give an answer to the most difficult questions at all times.

The more we advance in practice and experience the more we learn to prize this method, which is distinguished at once by the certainty of its operation, the rapidity with which it accomplishes its purpose, the ease of its manipulation, and the almost unlimited extent of its applicability.

These advantages are in fact so apparent that one can scarcely comprehend how some investigators can ignore them. He who brings substances known or supposed to contain a mixture of bacteria into a solid culture medium and then leaves them to their fate, shows clearly that he does not, or perhaps will not, understand the chief advantage of a plate—i.e., the separation of germs and the consequent opening of a possibility to each one germ of attaining its fair development without being crowded or overgrown by others.

Take this case: we are seeking in some organ—say the lungs—a particular species of bacteria, and we put portions of lung tissue into a test-tube containing gelatin or agar in order to let the germs develop. Perhaps there may be only ten of them altogether, and of the ten only two belonging to the species sought, the others being indifferent to us. The plate is sure, in all cases, to present these two, and to present them in their characteristic forms. If they remain with the others in the test-tube without being poured out on a glass plate or Petri dish, the two may not be able to maintain themselves against the other eight and will be lost for our observation. We then obtain an altogether false result from our examinaton; we have indeed employed the solid food medium, but we have obtained no advantage from it.

VII. PURE CULTURES.

The next step must be to definitely separate from each other the different species which the plate process has enabled us to distinguish.

The plates remain good for a limited time only, and the liquefying colonies in particular soon spoil the solid culture medium. Foreign organisms, especially mould fungi, soon show themselves on the surface, and it is therefore desirable to place in safety such bacteria as we particularly want without delay.

This is done by removing the colony in question with the platinum needle, and transplanting it into a test-tube containing solid gelatin, agar, etc.; here it can develop quietly and continue the pure culture. Certain precautions must, however, be adopted. Above all, this rule must be followed: never remove a colony from the plate without the direct aid of the microscope.

The distinguishing peculiarities of these smallest pure cultures are not clearly seen till the microscope is brought into requisition. Besides this, we can never know by the unassisted eye whether we have on our needle only one colony which we wish to transplant or whether we have at the same time touched a number of others. The colonies sometimes lie so closely together in the mass of the gelatin that a mistake may easily happen, and the greatest caution is necessary.

There is a certain definite series of manipulations which has been found in practice to be the best means of "fishing" for colonies.

We first seek a colony suitable to be transplanted—of course using a low power objective, strong eye-piece, and the smallest diaphragm aperture. If we have the choice of several colonies of the same species, we should give the preference to such as are most isolated, have already attained a certain size, and reached the surface of the gelatin. The last point in particular very greatly facilitates the operation of fishing.

Then bring a short, not too thick, previously-heated platinum wire close under the lens, and endeavor to get the end of it just over the centre of the colony. This is the most difficult part of the whole operation. The best way of proceeding is to lay the lower half of the little finger of the right hand on the stage of the microscope, then hold the wire, as horizontally as possible, close under the objective. When this is done, look through the microscope and the wire will be seen, especially if it be moved slightly backward and forward, as a faint shadow. The more it is lowered the more distinctly it becomes visible. Direct its point into the centre of the field, and dip it into the colony by a slight movement of the hand, which still rests on the stage. It is then at once raised, and the colony generally bears plain marks of the violence done to it.

It is self-evident that a wide interval between objective and object is here desirable, that we may have the necessary room for manipulation. This fishing is, however, not very easy to learn. We must avoid touching other parts of the gelatin before and after the stroke; we must strike but one colony, and that with a sure and steady hand, and this requires an amount of dexterity that practice alone can give. The colonies which lie deep in the interior of the gelatin often tax the skill of the beginner severely, and put his patience to a hard test.

The process is, however, still more difficult when we have to deal not with simple plates, but with dish or test-tube preparations. With the former, the high edge is a decided hindrance for the motions of the needle; with the latter, the cotton plug must first be removed while the tube already lies under the microscope, then the platinum wire must be carefully introduced and advanced to the part where the colony in question is situated, after which comes the fishing. It is scarcely possible to avoid touching other portions of the food medium with some part or other of the instrument.

If one has a portion of a colony on the point of the needle, it is transplanted into solid gelatin, taking off the cotton plug from a test-tube and sticking the needle deeply and vertically into the gelatin. It is then withdrawn again quickly, the plug is replaced, and the *needle-point culture* may now be left to develop. The object here being to plant an absolutely pure culture of a certain species of bacteria, special pains should be taken to prevent all foreign pollutions. We therefore endeavor, when we open the tube, to hold the mouth of it in such a manner that no germs floating in the air may fall into it. In some cases it is even better to hold the tube inverted, with the opening directed straight downward, and press it down from above upon the point of the upright needle. Of course the cotton plugs should not, in the mean time, be laid on the table, but carefully held between two fingers of the left hand.

When the needle-point culture is planted, we return to the mother colony and employ the remainder of it for a hanging drop or a stained preparation. In this way we obtain specimens of the species which have just been transplanted, which may be afterward useful for comparison.

In about the same time which was required for the formation of distinctly visible colonies on the glass plates, a culture makes its appearance along the needle hole in the test-tube. The spread of growth is, indeed, very different with different species, and as a rule all the features which were recognized in the glass-plate culture present themselves here again, with but little alteration.

If the appearance of the micro-organisms in the needle-point culture is not quite so characteristic as on the glass plate, the reason is that the masses of bacteria in these pure cultures are usually so great that many of the finer distinctions do not display themselves. Yet the experienced eye is, nevertheless, able to distinguish whether the culture is pure or not, and of what species it consists—for the differences still remain sufficiently marked.

In a series of cultures consisting of micro-organisms which do not liquefy the gelatin, some are observed to grow equally well all along the puncture, either in thick, lumpy masses or in fine, delicate granulations. Many, on the other hand, only develop on the surface of the gelatin, while the puncture itself has remained unfruitful. This is a sign that the bacteria inoculated were such as needed a plentiful supply of oxygen.

Finally, some few kinds spread themselves throughout the entire contents of the tube like misty clouds, which only become distinctly visible as fine clouds or veils when seen against a dark background. Some color the gelatin in tints varying from a bright purple-red to dirty gray; others, again, array themselves in plain white, and only become somewhat brown in old age. Numerous bacteria liquefy the gelatin, and with them we can, as a rule, only speak of plain distinctions in the beginning of their development; afterward, when the liquefaction of the solid food medium has become far advanced, the distinctive marks are lost. But at first, as already said, we do see decided differences. Some grow and decompose the gelatin rapidly, the puncture surrounds itself with a "sock," or trouser-leg, of liquefied gelatin; others, which liquefy more slowly, sink down from the surface, as it were in a funnel; some send out from the needle hole fine, bristle-like threads and branches into the gelatin, and present quite an elegant appearance. When the liquefaction has advanced still further, the motile species proceed, as a rule, to the formation of a pellicle, a thick superficial covering; the non-motile kinds sink gradually by their own weight to the bottom, leaving above them a layer of perfectly clear gelatin.

If instead of piercing the food medium with the inoculating needle we only draw it over the surface without entering the substance, we get a so-called *needle-stroke culture*. The needle-stroke culture develops best in gelatin which has hardened in a test-tube placed obliquely. The surface at disposal is then much larger, and the appearance of such an inoculating stroke is often very characteristic. Of course this is only the case with the non-liquefying bacteria: with the others the food medium soon sinks away from the wall of the test-tube.

Agar-agar, as we know, is free from the last-named fault. It is, therefore, peculiarly suited for needle-stroke cultures on an oblique surface; in fact, its chief use is for pure cultures of this description. The point of the infected platinum needle is drawn slowly across the agar without penetrating it, and then, within twenty-four hours, we can have perfectly-developed cultures of the parasitical micro-organisms, whose development, as already mentioned, may be accelerated by the aid of increased heat.

The pure culture in the test-tube does not, of course, last for an unlimited time. The nutritive substances are consumed, and the bacteria themselves excrete substances which act as a check to their indefinite multiplication.

This takes place more quickly with some, more slowly with others. As a general rule, we may say that the cultures retain vitality for three or four months, yet there are some which perish much sooner, and it is always well to renew them about once in six weeks—i.e., to transplant them to a fresh food medium.

VIII.—CULTURE OF ANAËROBIC BACTERIA, INCUBATORS, THERMO-REGULATORS, AND SAFETY BURNERS.

It is necessary to know the ways and means by which we are enabled to separate a mixture of bacteria into its component species, and to breed them in pure cultures by the aid of transparent solid food media. In this manner success will be attained with a great number of micro-organisms, and there are only a few that form exceptions, inasmuch as they require special appliances to insure the success of their artificial cultivation. We refer to those which only thrive in the complete absence of oxygen—the anaërobia, for whose culture, therefore, separate methods have had to be invented.

It may be readily imagined that the problem is a difficult one. It is not an easy task to shut off a substance which is present in all that surrounds us; we cannot succeed without special manipulations, and often we shall require special instruments. We may, therefore, at once say farewell to that laudable simplicity which is, perhaps, the greatest merit of the usual process of culture.

It is true that the difficulties are not great as long as we have to deal with liquid media, but they increase from the moment we attempt to employ the solid media. In particular, the separation of mixtures of bacteria occasions much difficulty, and the great number of means and ways that have been and are still continually suggested is in itself proof that as yet nothing has been found to answer all the requirements of the case. Any one method cannot even be recommended as being comparatively the best; all have their defects, and it must be left to the operator to select one from among the possibilities which are about to be noticed.

First of all, for the culture of anaërobic micro-organisms it is advantageous to add to the solid media reducing substances, to consume the oxygen which may be present. Of such additions the commonest are: 1% to 2% grape sugar, 1% to 2% formate of sodium, 1% to 10% resorcin, etc.

Proceed as follows: put the material in which anaërobic germs are suspected into liquid gelatin or agar, and pour it out in the usual manner over glass plates. But before the food medium is fully hardened, cover it with a sterilized scale of mica. This will stick to the viscid surface, and prevent the entrance of oxygen. In order to keep out the oxygen still more fully, seal the open edges of the mica with melted paraffin.

But this only removes the oxygen partially. We obtain much better results when we employ a larger mass of the food medium, using the medium itself as a means to keep out the oxygen. Liborius has wrought out a systematic method of culture in deep layers of solid food média. A depth of 15 or 20 cm. of gelatin or agar in a test-tube is, as far as possible, freed from air and oxygen by thorough boiling, then cooled down to 40° C., and the material to be inoculated carefully and equally distributed throughout the liquid by the aid of a strong platinum loop. The liquid must now be quickly hardened—iced water is the best means—and in the short time but little air can re-enter, while the lower portions of the food medium are, to a certain extent, protected by the upper ones from the outside air and the penetration of oxygen.

In fact, even strictly anaërobic bacteria develop in such cultures. When care has been taken by suitable dilutions, by transferring small quantities of infected gelatin into a second and third tube, with a deep layer of food medium, to bring about a sufficient distribution of germs, the colonies arise separately and in very characteristic forms, and display the peculiarities of the different species with great clearness. This method has, indeed, one advantage over all others, which makes it particularly valuable. We know that a great number of the bacteria familiar to us (in particular all the pathogenic species) are semi-anaërobic—i.e., they can thrive when oxygen is present. In all breeding processes in which the oxygen is entirely excluded from the culture vessel there is no means of distinguishing the semi-anaërobic from the strictly aërobic species, which can thrive only when the oxygen is excluded. In the deep cultures, on the other hand, the part nearest the surface remains open to the influence of the air and its oxygen, so that they give us a good key to the different amounts of oxygen required, or borne without injury to the various species which develop in them. Colonies which appear only in the upper portion of the medium belong to strictly aërobic (or semi-anaërobic) class, and those, lastly, which are only found in the deeper portions to the strictly anaërobic class. This forms a mark of distinction extremely useful in examining matter containing both strictly anaërobic and semi-anaërobic germs, and which facilitates our labor very considerably.

Yet this method is not wholly satisfactory for several reasons, and there is abundant room for improvement. The exclusion of oxygen which it affords is neither complete nor enduring. Further, it is impossible, or next to impossible, to submit the separate colonies to a direct microscopic examination. Lastly, such a culture must be utterly destroyed before we can obtain from it the material for further inoculation. The best way to get such material is to break the test-tube in a sterilized Petri dish, and then endeavor with sterilized instruments to free the gelatin or agar from the glass with a view to further operations. This will not be accomplished without some difficulty, and the danger of pollution to the culture in the course of such manipulations is always imminent.

Another process, somewhat similar to the one already described, allows of direct microscopic examination, but it is equally unsatisfactory as to further inoculations. In the usual manner bring the matter for inoculation into liquid food gelatin which has been previously well boiled, and then spread the liquid over the inner surface of a test-tube to harden, according to Esmarch's plan. The interior of the test-tube is then filled with gelatin at about 26° C.—so far cooled as to be just short of solidification. This excludes the oxygen, and the anaërobic colonies can thrive over the whole inner surface, with the exception of those parts nearest to the top. The colonies thus formed may be immediately examined under the microscope, yet the removal of portions for further inoculations is more troublesome and difficult than before.

All the other methods endeavor to accomplish this in another way, by entirely removing the oxygen from the food medium; in which latter case they either establish a vacuum or replace the air by some gas of an indifferent nature, and so create an atmosphere without oxygen.

The first of these processes is that of Buchner. A wide test-tube is closed with an India-rubber cap, and a pyrogallic solution is shaken up in it. This solution absorbs the oxygen of the air, especially if, after some time, we add a little pure pyrogallic acid. Then the test-tube culture—Esmarch's rolled system—which contains the anaërobia to be developed is fixed in this larger tube with a small wire support, and the whole immediately closed again with the gutta-percha stopper.

Yet this method does not, by any means, remove all the oxygen, and strictly anaërobic bacteria often fail to thrive with it; besides which, it is very unpleasant to work with pyrogallic acid, and even the most enthusiastic investigators dislike it.

Gruber obtains a complete exclusion of air, and therefore of oxygen, by the aid of the air-pump. About 15 cm. from the bottom of a test-tube of easily-fusible glass he makes a narrow neck. The tube is then provided with a cotton plug, sterilized in the hot-air oven, and about 10 cm. of nutrient gelatin are poured into it and rendered germ-free in the accustomed manner. Then the gelatin is liquefied, and the inoculation takes place with the platinum needle, the cotton plug being removed only as long as is necessary. An air-tight India-rubber stopper is then put on, through which passes a glass tube bent to a right angle and open at both ends. This is placed in connection with the pump. By placing the test-tube in water of 30° C. or 35° C., the gelatin is made to boil in the rarefied air, and by pumping and boiling at the same time, the air is removed in about fifteen minutes. The neck of the glass is now melted and sealed hermetically. The gelatin slowly cools and, at the proper moment, is rolled over the inside of the glass by Esmarch's method.

This process has its decided advantages, so that it can be recommended for many purposes. A microscopic examination of the growing colonies offers no difficulty. The removal of portions for inoculation can proceed as in the ordinary Esmarch tubes, after the neck has been reopened and the removal of the oxygen is thorough and absolute. If an air-pump be not at hand, or if for any other reason we wish to employ gas instead of a vacuum, we can only take hydrogen. It has been found that this is the only gas which is harmless for bacteria and does not work prejudicially on their development; while carbonic-acid gas, for example, which was formerly often employed, kills quite a number of different micro-organisms.

Prepare the hydrogen in Kipp's apparatus with zinc and sulphuric acid, and conduct the gas, before employing it, through two purifying-bottles, one of which contains an alkaline solution of lead, the other being filled with alkaline pyrogallic solution, to remove any sulphuretted hydrogen or small remains of oxygen.

The best way of proceeding is as follows. Take a number of gelatin test-tubes, inoculate, and make the necessary dilutions, etc., as usual, and then, instead of using a cotton plug, close each tube with a gutta-percha stopper, which should be previously sterilized in the steam generator, and whose two holes contain two glass tubes bent to a right angle. The longer of the two reaches down almost to the bottom of the test-tube, deep into the food medium; the other is cut off close below the stopper. In both the horizontal part is contracted into a narrow neck; the continuation of the longer tube contains, further, a pledget of sterilized cotton wool, and ends in a short piece of India-rubber tubing. By means of this, while the gelatin is still in a liquid state, we connect the test-tube with Kipp's apparatus and let the stream of gas enter. It drives out the air from the food medium and test-tube, which escapes by the shorter of the two glass tubes. After about half an hour melt and seal up, first, the short, and afterward the long tubes, and lastly, roll the gelatin over the walls of the test-tube.

This process may often be employed with success, and besides its other advantages, its simplicity and cheapness distinguish it favorably from another method which is, nevertheless, often employed, and must therefore be mentioned here. It is the method of Liborius, to whom also, as known, we chiefly owe the culture system in deep strata. Liborius also conducts a stream of hydrogen through the inoculated food medium, but he employs special apparatus for doing so. He uses small glasses, to the side of which a tube is firmly attached by melting. This tube is continued downward within the test-tube almost to the bottom. When the stream of gas passes through this tube it must first pass through the food medium before it can escape through the mouth of the test-tube, which has been contracted to a narrow neck. When the air has been carefully expelled, first the feeding-pipe and then the neck are melted and sealed up, the food medium remaining under an atmosphere of pure hydrogen.

It must be regarded as a disadvantage of this method that the colonies, as in the deep-stratum culture, can hardly be examined direct with the microscope, and can only be approached with difficulty for the purpose of further inoculations. It is also a disadvantage that we do not here roll out the food medium on the walls of the test-tube.

When we have attained, in either of the already-described ways, a distribution of germs, and have developed colonies of anaërobic bacteria so that they are ready for further treatment, proceed as in the usual breeding process to needle-point cultures. To this end I recommend the exclusive use of deep strata of solid food media.

With a short platinum wire transplant a colony as deep as possible into the solid gelatin or agar. If we transplant a strictly anaërobic micro-organism, only the lower part of the puncture will develop. There will be a certain frontier mark, showing how deep the influence of atmospheric oxygen extends. The development is stronger by far in the deeper portions. In somewhat older cultures the growths advance slowly upward, and at last there remains only a narrow ring of medium close to the surface, where no growth takes place. This is caused by the gaseous excretions of the anaërobic bacteria themselves, which probably consist chiefly of hydrogen, and collecting, expel the air which had penetrated downward from above. Thus the micro-organisms grow in their own gases and gradually rise toward the air.

That the anaërobia in particular distinguish themselves by generating gases is already known. With some species we perceive masses of gas collected under the solid bridge which separates the liquefied culture from the outer world—from the oxygen-containing atmosphere. If this barrier be punctured with a platinum needle and the food medium be stirred up a little, numerous bubbles of gas arise and seek to gain the open air.

Lastly, one more food medium will be mentioned in which we can also develop pure cultures of anaërobic bacteria. It is neither solid nor is it transparent, yet it is so undoubtedly useful in some cases that it deserves notice. It is the raw hen's egg, as recommended by Hueppe. The shell of the egg must be cleaned and sterilized with sublimate, alcohol, and water free from germs. Then the surface is dried with sterilized cotton wool, and an opening pierced in one end with a platinum needle. Through this opening the inoculation is performed, after which a small piece of sterilized blotting-paper is fastened over the hole with collodion. The small quantities of oxygen which were already in the egg or which entered it through the hole are insignificant, unless we have to deal with the very strictest anaërobic species, and are, besides, soon absorbed by the new gases which develop, in particular by the sulphuretted hydrogen. Some bacteria thrive very luxuriantly in this medium so full of unaltered albuminous substance, and produce their excretions in rich abundance.

Now that we have learned the special processes which are necessary for the culture of anaërobic micro-organisms, it remains to say a few words about those by which we are enabled to breed the strictly parasitical kinds, which need a high degree of heat.

As food media we can employ agar and blood-serum; plates can only be prepared with agar; they are kept continually in the incubator like the test-tube cultures.

There are different kinds of incubators in use, and we will endeavor to explain their construction and arrangement by referring to the two most usually employed.

One is a large rectangular ("four-cornered") sheet-iron case covered outside with felt; it has water between its double walls, and the depth of the water may at all times be seen by the gauge-glass attached at the side. The water is brought up to a fixed temperature, which it communicates to the space within. As this warming proceeds equally and simultaneously from all sides, there will seldom be an irregularity in the distribution of the heat, a point which is of the first importance. For when the temperature varies the cultures can only retain their vitality a short time: evaporation takes place here and consolidation there, the food media are robbed of their moisture, they dry up and become useless. In order to absolutely obviate these evils, it is desirable always to provide the test-tubes in the incubator with small India-rubber caps.

It is by no means easy to keep the temperature of the surrounding water constantly at the same point. This is managed by the continual self-acting control of the amount of gas supplied to the flame which heats the apparatus from below: if the water becomes too warm, a smaller quantity of gas is admitted to the burner, and vice versa.

The flame is controlled by a thermo-regulator, of which there are many different kinds in use. One of the commonest is the quicksilver regulator invented by Bunsen and improved by V. Meyer. A glass tube about 40 cm. long and closed at the bottom, shaped like a large test-tube, is divided in the middle by a diaphragm of glass into an upper and a lower half. Yet a connection exists between them, inasmuch as the diaphragm sinks into a funnel which narrows into an almost capillary tube and ends just above the bottom of the vessel. The lower part is almost filled with quicksilver; above this, however, and almost reaching the diaphragm, is a mixture of alcohol and ether, which, as we know, volatilizes when slightly heated. Indeed, it is only necessary to hold in the hand that part of the glass which surrounds the liquid and warm it a little in order to produce gas and drive away the quicksilver. The latter can, however, only escape through the capillary funnel tube, and the more it is warmed the more quicksilver will gradually rise above the diaphragm.

Into this latter space—through the India-rubber stopper which closes the mouth of the whole vessel—runs a tolerably wide glass tube cut off obliquely at the end. Above its oblique termination the tube has a small lateral hole, not larger than a pin's head, the use of which will shortly be explained. If the ether is volatilized more and more, the quicksilver above the diaphragm mounts higher and higher; first it reaches the glass tube, gradually shuts off the oblique end, and begins to approach the small lateral hole. If now the source of heat be removed—i.e., the hand—the ether begins to condense again, the quicksilver sinks through the capillary tube, the upper part becomes empty; and the same processes can be repeated *ad infinitum*. The thermo-regulator is first "set" to a definite temperature—for instance, to 37½° C. We "set" it in a large bath. When this has attained the desired temperature, push down the upper tube so far into the quicksilver above the diaphragm that the oblique opening is completely closed and only the small lateral hole remains open. The apparatus is now ready for use. Place it in the water between the two walls of the incubator, whose temperature is soon communicated to it. The tube is placed in connection with the gas-pipe; the gas which enters the regulator passes off through a small glass tube, which branches off sidewise near the top of the main tube, and is conducted to the burner under the incubator; in other words, the burner receives as much gas only as passes through the regulator.

If the flame is too large and the water becomes too warm, the ether volatilizes, the quicksilver rises and closes the oblique opening more and more, till at length only the small lateral hole remains open, and the entrance, as also the exit, of gas is reduced to its minimum; then the flame becomes smaller, the water cools down, and so on continually.

The small lateral hole is intended to leave the flame gas enough to keep it alive even when the need for cooling is at its greatest. Yet circumstances might, perhaps, cause the quicksilver to rise so high as to close even the small side hole, or some accident might put out the small flame fed only through it.

What are the consequences of such an event? The water continues to cool, the quicksilver sinks till the full maximum of gas streams through the regulator. The dangers which ensue need not be enumerated.

To prevent them from occurring Koch has invented the safety-flame, which in the supposed case would of itself cut off the supply of gas.

It is well known that different metals behave differently under the influence of heat; that they do not all expand equally; that their coefficients of expansion differ. If we make a band of two strips of different metal, closely united, and then heat it, it will bend toward the side of the metal which expands least.

The safety-burner is based on this fact. The flame is between two spiral springs, each of which is composed of two different metals—copper and iron—united in the manner just described. If the flame goes out and the spirals begin to cool, they unroll. But they are placed on a round, movable disc, which they draw to the one or the other side, according to their changes of form. This disc has at one point a catch, by which it holds a heavy lever running parallel with the gas feed-pipes and connected at its other end with a valve in the gas-pipe. As long as the flame burns the valve is open and the lever rests on the disc. If the flame goes out the spirals begin to operate, the disc is turned somewhat round, the lever-catch goes with it, the lever loses its point of support, falls, closes the valve, and shuts off the gas.

Besides incubators thus regulated, there are a number of others arranged on different principles.

The most important are those with a membrane regulator, such as the well-known thermostat of d'Arsonval. It is a double-walled copper vessel, in which an equable temperature is likewise maintained by means of regulating the amount of gas supplied to the flame which warms the water between its walls. Before reaching the flame the gas flows into a small chamber, one side of which consists of a thin sheet of India-rubber. This lies with the other side against the water between the two walls. If the water gets warmer it expands and presses the gutta-percha membrane into the chamber. The gas supply is at once considerably retarded, the flame becomes smaller, the water cools, contracts, etc.

Lastly must be mentioned a very useful and reliable kind of incubator recently constructed by Lautenschläger, of Berlin, in which the regulation is managed by electricity, by means of a contact-thermometer and a magnetic burner. A precise description of this ingenious apparatus would lead too far, however.

Thus far the ways and means which enable us to obtain a somewhat exact view of the vital processes of the bacteria have been considered.

Clearly as we may and must feel that we have hardly got beyond the beginnings of a systematic investigation, let us, nevertheless, gratefully acknowledge that the introduction of new and excellent methods has already, and in a very short space of time, revealed to us a rich abundance of previously unknown and altogether unexpected facts.

It may be hoped that the intelligent use of these valuable means of investigation will enable us to advance with rejoicing, and to maintain bacteriology in that degree of importance by means of which it has already become a centre of general interest.

CHAPTER IV.

Methods of Transmission; Special Qualities of the Pathogenic Bacteria; Powers of Resistance of the Organism; Natural and Acquired Immunity; Metschnikoff's Phagocytic Theory; Koch's Rules for the Determination of Pathogenic Bacteria; Inoculation of Animals; Methods of Infection.

I. METHODS OF TRANSMISSION AND THE SPECIAL QUALITIES OF THE PATHOGENIC BACTERIA.

We have already considered the general qualities of the bacteria and the means adopted to obtain further knowledge of their peculiarities. How far have these means enabled us to advance in our acquaintance with definite, special micro-organisms?

We shall take them in a certain succession, and the question is whether we can from any standpoint form a definite classification of the bacteria.

Attempts to form a system on the basis of their natural history or after the manner of their development are still too much in their infancy to be able to serve for guidance. Nor will we rely on purely outward circumstances or the form (the morphological behavior of the bacteria), since the difference between a bacillus and a micrococcus, for instance, is certainly not a matter of such importance that it should make a wide distinction between the different species. The bacteria interest us chiefly in an etiological point of view, because we have recognized many of them as dangerous parasites of animal organisms—that of man included—and as the exciters of a whole series of diseases, and it will be simplest and best to base the arrangement on these considerations.

On the one side, then, stand all those species which are able to exercise a noxious, disease-exciting agency; on the other, those which have not this capacity and cannot become pernicious to man. The line which is thus drawn between the *pathogenic* and *non-pathogenic* bacteria is by no means so clear as might at first be thought. A not inconsiderable number of micro-organisms are already known which generally show quite a harmless character, but which under some circumstances may become pathogenic; and many pathogenic species can be induced, on certain terms, to lay aside their noxious qualities and pass over into the ranks of the innoxious bacteria.

This apparently very striking fact will become somewhat more comprehensible if the causes of the pathogenic action of any given micro-organisms are investigated.

Here several possibilities must be taken into account.

Examine with the microscope a stained section from the kidney of a Guinea-pig which has died of anthrax. Look all through the section, and everywhere innumerable rods will be seen. The capillary vessels, and even the somewhat larger vessels, are filled with bacteria which seem to have choked up the tissues. It will then be understood why it is considered possible for the mere presence of such masses to cause seriously deleterious results by their mechanical effect alone.

In fact, the presence of so many foreign organisms can hardly fail to injure the functions of the invaded organ, and when that organ is so important as the liver, for instance, or the kidney, the whole body is placed in the greatest danger.

Yet this explanation is only possible for a small number of cases. Frequently the mass of the bacteria is so insignificant that such effects are quite out of the question. Almost always, too, we cannot adduce clear proofs of the mechanical nature of such conditions within the tissues. We may now and then find a ruptured glomerulus which was unable to withstand the pressure of the micro-organisms and was torn asunder by them. But all the other effects which we might expect as the result of such an extensive stoppage of numerous vascular regions are wanting. There is generally no appearance of hemorrhagic infarction or necrosis of tissue such as we usually see after the formation of thrombi and after embolic processes.

We must, therefore, seek other causes for the peculiar action of the pathogenic bacteria.

The micro-organisms are living beings which require a definite quantity of nutriment for their support. If they are parasitic they take this nutriment from the organism which harbors them, and which may be severely injured thereby. If the tribute exacted by the parasite exceeds the amount which the victim can pay, if he loses more in this way than he can make up in other ways, he must, if no help comes, perish sooner or later.

This tribute may be of different kinds. The chief kind, however, consists of the albuminous substances, the material of which cells are built, and which, as we already know, are a favorite food of the bacteria. With these must be mentioned oxygen, which the micro-organisms consume and which they also take from the living tissues.

Of far greater moment than the mechanical action and the consumption of alimentary matter is another factor almost always at work in the case of pathogenic bacteria.

The micro-organisms excrete, as mentioned, certain substances of peculiar composition. In most fermentations definite, specific products arise. Thus in the decomposition of albuminous matter. which, indeed, is entirely caused by the action of bacteria, particular substances are formed, a more exact knowledge of which we owe specially to the important researches of Brieger. He ascertained very exactly the chemical constitution of a number of those substances in which we see clear and tangible traces of bacterial life, and found them to be bases—alkaloids—belonging principally to the fatty compounds.

Some of these substances, which from their origin have been called by the general name of *ptomaines* (corpse-alkaloids), possess extremely poisonous qualities, so that even small quantities of these toxines sufficed to kill the larger animals in a very short time. But Brieger did not rest satisfied with these first results; he soon extended them very largely. The processes which lead to putrefaction can only be observed with precision to a certain extent, since they owe their origin to a great number of different and unknown species of bacteria. It could not but seem desirable, therefore, to examine, in one and the same manner, certain micro-organisms in pure cultures, and particularly to investigate the excretions of the principal pathogenic species. And Brieger did, in fact, succeed in making important discoveries by this method. He found, for instance, that the cholera bacteria, the typhus and tetanus bacilli, under suitable circumstances excrete specific substances out of their aliment—which prove to be genuine toxines, and are able, when inoculated into animals, to produce some of those phenomena which would be caused by the bacteria themselves.

The importance of the excretions in the pathogenic agency of micro-organisms was thereby placed beyond doubt, and efforts were now made to discover a similar agency on the part of other species besides those already mentioned.

With some—for instance, with the anthrax and the diphtheria bacillus, the vibrio Metschnikoff, etc.—a certain degree of success has been achieved, though without attaining that full degree of certainty which distinguishes Brieger's investigations. The reason is, on the one hand, the difficulty found in submitting these substances—which we have no right to suppose belong, all of them, to the bases or ptomaines like those investigated by Brieger—to chemical examinations; and, on the other hand, the fact that the different bacteria do not, as a rule, form only one definite substance, but always produce several, which by their common joint effects produce the pathological complex of symptoms which we regard the micro-organism in question as answerable for. But the greater the number of these competing poisonous substances, the more difficult is it to separate them and test their several qualities.

Although our present knowledge of this subject may still be very deficient, and though it may call loudly for completion, yet from what we do know we may safely lay down the proposition: the action of the pathogenic bacteria is chiefly to be explained by their producing specific, extremely poisonous substances which seriously injure the organism, influencing it in a definite manner and thereby causing definite independent forms of diseases.

Just as the small portion of poison injected by the sting of a bee or the tooth of a serpent suffices to cause local derangements of considerable extent, to cause disturbance in the whole organism or even to kill it, so also the bacteria, by means of their toxine, are under some circumstances able to affect parts with which they come into no direct communication. In this way we must explain the cases in which a general derangement of the bodily functions reveals a violent disease, and yet the most careful search only finds a very limited number of micro-organisms, or only finds them limited to one particular portion of the body, which for some reason or other they adhere to exclusively. In the latter case they have excreted their poison in their chosen quarters, it has been carried far and wide by the blood and other juices, and its noxious effects are seen wherever it has penetrated.

If it is in reality the excretions of the bacteria which are the great factors in their pathogenic agency, it will readily be comprehended that this agency cannot be fixed and definite in its amount, but must be subject to considerable variations.

The mere quantity of poisonous substance which, in different cases, comes into the system must be taken into account. If the disease-producing dose is attained or surpassed, pathological changes present themselves which would otherwise have had no existence.

The nature and composition of the food on which the bacteria are nourished and out of which their excretions are formed also exercises an unmistakable influence.

Here poisonous substances are composed or decomposed, while

there the necessary material is wanting; in one case pathogenic effects are possible, in another they do not occur. The albuminous substances seem to be necessary for the formation of most of the toxines, and therefore there will be only a chance of success in studying these peculiar products if we breed our bacteria on media rich in albumin.

Another circumstance deserves mention here. Under some circumstances, the excretions of different micro-organisms, which separately are harmless, may unite and develop poisonous qualities. Thus, for instance, we know from the researches of Roger that so harmless a bacterium as Micrococcus prodigiosus, which alone is almost entirely innocuous, can, in conjunction with another bacterium, which is also non-pathogenic for the animal in question, become a cause of disease. A number of other observations all point the same way, although in most of them other factors also operate. All the facts adduced thus far and the reasons for variations in the agency of the bacteria apply, in general, only to those kinds which excrete their poisonous products outside the body exclusively, which are unable to grow and thrive in the body, and which, therefore, must bring with them the poisonous products required to produce disease, without reckoning on a further increase of the same within the body. These bacteria need not be themselves transmitted: it suffices to employ the substances which they have formed, and we may deprive such a culture of its living inmates without diminishing its pathogenic strength.

It is true a certain precaution is necessary. If the germs be destroyed by submitting the food medium to heat and thoroughly sterilizing it, the very delicate products of excretion are generally also destroyed; they separate into their component parts and lose their peculiar properties. It is therefore better, in all cases in which it is desired to exclude the bacteria and study the soluble substances alone, to separate the two by way of filtration through clay cells. This is a process first employed by Pasteur.

The latter, in conjunction with Chamberland, has constructed tube-shaped filters of burnt china-clay, or kaolin, which is sure to retain all micro-organisms. At first, as Sirotinin has shown, part of the dissolved substances are retained by the walls of the filter. But after a short time this ceases, and for our purposes—the isolation of the products of excretion—these filters are extremely useful. With the last-mentioned kind of bacteria it can be proved that it makes no difference whether we employ cultures with or without germs, and this gives us a most reliable insight into the nature of the operations of these so-called toxic micro-organisms.

Under natural conditions, their importance, especially for mankind, is not great. It seldom occurs that large quantities of such substances formed by bacteria outside the body are taken into it on any one occasion. Such a possibility comes seriously in question only in exceptional cases—for instance, in cases of meat-poisoning—and it is chiefly from a therapeutical point of view that the purely toxic micro-organisms are interesting to us.

The great majority of all known bacteria, particularly the parasitic kinds, can in this manner, as experiments on animals have proved, be made to exercise a pathogenic action. But taken in its more restricted, proper sense, the word "pathogenic" is applied only to a comparatively small definite class of micro-organisms, which form a very striking contrast to those just treated of.

They possess, in fact, the capacity to multiply indefinitely within the organism which they invade. From the first original cell fresh members are continually proceeding, and in this manner are produced such enormous accumulations of bacteria in the living organs as in the case of the anthrax preparation alluded to. Here the excretions, the bacterial poisons, originate within the body, and in ever-increasing proportions, so that the quantity of germs which invaded that body at its first infection is a matter of indifference.

The difference which separates these infectious species from the toxic ones is very essential. The first are transmissible in the smallest quantities—i.e., they are always able to multiply within a susceptible organism. With the others that is not the case. If they find entrance into a body in such large quantities as to poison it, they probably find their way through the circulating blood into its different organs, and may be discovered there. But it would be a great error to confound such a case, in which the individual bacteria only play a passive part, being merely floated away and stranded again, with the other case in which the infectious species penetrate actively into the tissues and grow within the body.

To what must we refer the capability of certain micro-organisms to live in warm-blooded organisms, to multiply and produce their poisonous excretions in them? Is this a constant, invariable property of certain bacteria, or is this particular expression of pathogenic agency subject to variations which influence or obliterate the distinction between them and the harmless species?

We are still far from a satisfactory solution of these questions, but the last few years have yielded us so many valuable facts bearing upon them that we already have—or believe we have—firm ground under our feet as regards some few points.

We will allow the facts to speak for themselves. The very first experiments with pathogenic bacteria had shown that for the ascertaining of their peculiar qualities it is by no means unimportant into what kind of animal they are inoculated. One and the same micro-organism may display a decidedly injurious action on one kind of animal, while it produces no effect on another, and we will find that in all experiments we must carefully take this circumstance into account. It is only mentioned here in passing, because it stands in connection with certain observations bearing on the point in question.

The bacilli of glanders are, as we know, particularly virulent in their action on field mice, while white mice, the animals most frequently experimented upon, are quite insusceptible to them. H. Leo has, nevertheless, recently succeeded in making them susceptible to glanders by feeding them for a length of time on phloridzine, thereby putting them artificially into a diabetic condition and saturating their tissues with secreted sugar.

In a similar way Bujevid had previously remarked that the Staphylococcus aureus, which is almost without any effect when applied to the subcutaneous cellular tissue of rats and rabbits, produced a strong suppuration if applied in conjunction with a solution of sugar.

Arloing and his coadjutors proved that the bacilli of symptomatic anthrax ("black-leg") will affect animals otherwise not susceptible to them if mixed with 20% lactic acid, and in other cases a previous treatment of the tissue with sublimate, carbolic acid, or pyrogallic acid has been found to produce the same result.

It is true that the matter is not so clear and simple here as in the case of Leo's experiments. In them a change was produced in the bodily condition of an animal, enabling bacteria to multiply within it, which they would, under ordinary circumstances, be unable to do. In the cases which were next mentioned there were probably other factors also in activity, which we have already met with in the toxic micro-organisms—i.e., the excretions of the bacteria introduced were placed in a condition to act deleteriously by being united with other chemical substances.

However that may be, all these attempts to extend the sphere of activity of the pathogenic bacteria and to increase their virulence have one point in common: as soon as the special conditions under which we have placed the micro-organisms cease to act, no further change is noticeable in them. The bacilli of glanders, for instance, which are taken from the tissues of white mice have not attained the capacity to infect animals of the same kind without the previous preparation: they behave just as usual. In other words, we have not yet succeeded in producing an enduring increase of the natural virulence of bacteria, in changing toxic into infectious kinds, or in enabling the latter to infect animals which were proof against them, or to display a quicker or stronger action than was previously observed in them.

On the contrary, the opposite phenomenon, an enduring decrease—nay, even the complete irrecoverable loss—of virulence has been observed in very many cases, and in the most different species of pathogenic bacteria. In the year 1880 Pasteur surprised the scientific world by the discovery that under certain circumstances the micro-organisms of chicken cholera lose their poisonous power, to a greater or less extent, without showing any change in their appearance, way of growth, etc.

Toussaint and Pasteur found that the anthrax bacilli could be deprived of their virulence in like manner, and the same fact has since been proved with regard to the bacilli of swine erysipelas and the symptomatic anthrax, the pneumonia bacteria of A. Fraenkel, and some others.

This diminution of virulence is the result of two essentially different causes. The one, which we may call the natural cause, has recently been more fully elucidated, particularly by Flügge. It is a gradual diminution of infectious power in bacteria which we compel to vegetate for a long time separated from their natural conditions of growth, on our artificial food media, and under the atmospheric conditions of our laboratories. By a gradual adaptation to the altered, saprophytic way of life, or by a progressive selection of such cells as are naturally more capable of adopting this altered way of life, the original capacity for development within a foreign body diminishes more and more. As an outward sign of the change which has taken place, we observe that the culture now shows a much more luxuriant and rapid growth on the lifeless food-medium than was at first the case, when the conditions were yet new and strange. Not all the pathogenic bacteria possess this power of "cutting their coat according to their cloth" and adapting themselves to outward circumstances. Some cling with marked tenacity to their proper character, which they do not leave even when compelled to exist for whole years outside the body. Others, as the bacilli of glanders, the streptococci of erysipelas, Fraenkel's pneumonia bacteria, the diphtheria bacilli, etc., lose their virulence very quickly.

A similar phenomenon may also be occasionally observed in the case of the saprophytic bacteria. Hueppe and his pupils have shown, for instance, that the sour-milk bacillus and the blue-milk bacillus when bred continually and uninterruptedly on our artificial foods lose the capacity of effecting the changes from which they take their names, and Hueppe even speaks of this as a "loss of virulence" in these micro-organisms.

It lies in the nature of things that this diminution of virulence takes place gradually, proceeding step by step, and not with one great leap. Therefore we are often able to interrupt the process at a certain stage, or even to retrograde and undo the work already done. The best, and indeed the only, means to this end is to give back to the partially-weakened cells their natural conditions, and endeavor to reaccustom them once more to their former way of life.

In the case of pathogenic species, we first attempt to inoculate them into the animals most susceptible to them. If this fails, we can have recourse to the already-mentioned way of increasing the natural susceptibility of an animal artificially. Should this prove successful and the micro-organism once more establish itself on its natural soil, one may reckon with some degree of certainty on its recovering its former powers; but it is evident that this is not an increase of the virulence originally given to it by nature.

In direct contrast to the phenomena hitherto discussed is a second mode of diminishing the virulence of micro-organisms which leads to the same final results. That which brings about the diminution is, here, not the long-continued influence of altered (but not necessarily worse) conditions of life; it is the action, for a short time, of influence directly prejudicial to the bacterial protoplasm. In fact, all the means employed to produce an artificial diminution of virulence in pathogenic bacteria are such as, if applied in a slightly stronger form, would cause the destruction of the cells and would kill their contents.

Thus we breed bacteria on media to which a certain quantity of antiseptic or disinfecting substance has been added, but which just allows the microbes to exist and grow upon them. Such, for instance, is the process of Roux and Chamberland for diminishing the virulence of the anthrax bacillus, by cultivating it in a bouillon with the addition of bichromate of potassium (1:5,000 to 1:2,000), and that of Toussaint by mixing about 1% of carbolic acid with blood containing anthrax bacilli.

In a similar manner—i.e., as a disadvantageous form of nutrition—the organism of certain animals is found to act; namely, those animals which are insusceptible, or but little susceptible, to the particular kind of bacteria which we desire to weaken. Thus the bacilli of swine erysipelas lose their virulence to a certain extent when passed several times through the bodies of rabbits, as has been proved by Pasteur and Kitt.

Chauveau robbed the anthrax bacilli of their poisonous property by breeding them under a pressure of eight atmospheres, and Arloing found that sunlight is capable of weakening these same bacilli, and even their spores.

By far the surest and most commonly-used procedure is the exposure of the micro-organisms to the influence of high temperatures. Toussaint kept blood containing anthrax bacilli for ten minutes at 55° C. The bacilli were by no means killed by the heat, yet they were rendered harmless by it. Pasteur, for his experiments on a large scale, employed a considerably lower degree of heat, but he has not given full particulars of his method, so that no definite judgment can be formed with regard to it. We therefore owe our thanks to Koch and his coadjutors, who once more approached this question in a methodical, strictly scientific course of experiments, with a view to ascertain the effects of heat in diminishing bacterial virulence by studying the anthrax bacillus—the best known and most suitable species for this purpose.

Koch, Gaffky, and Löffler found that at 42° C. and 63° C. a diminution of virulence was perceptible in the cultures. They further discovered the important fact that the lower the temperature is by which a diminution takes place, the longer it takes for such diminution, but at the same time the more permanent are its effects.

Even variations of a fraction of a degree are here important. While anthrax bacilli can be rendered perfectly harmless in nine days with 43° C., it requires twenty-four days if we only employ 42.6° C., but in this latter case the new quality of the bacteria has become so thoroughly a second nature to them that they cannot throw it aside again. They not only keep it throughout their own life, but they even transmit it to their progeny. In fact, we can breed from them as many generations as we will of fully harmless bacteria.

If we endeavor to diminish virulence under the influence of higher temperatures more quickly—in a few days—the bacteria regain it with proportional rapidity.

But their virulence cannot be destroyed with one blow. Before the micro-organisms part with it completely they pass through a number of intermediate stages, each of which, with the amount of virulence still remaining, is sufficient to affect certain animals, the efficacy of the poison remaining longest for those most susceptible to it. Bacilli of twenty days at 42.6° C., for instance, will still kill mice, those of twelve days will still kill Guinea-pigs, those of ten days rabbits, those of six days sheep, etc., and this degree of partially-diminished virulence can also be preserved throughout generations of cultures.

And even if the above statements may not, perhaps, always be borne out in practice with perfect exactitude in all cases, that does not change the incontestably proven scientific fact that bacteria of a high degree of virulence may lose this quality for a shorter or longer time, or even permanently, and to any extent, up to its complete extinction.

Something similar has been observed in the saprophytic species also. Many pigment bacteria, under the influence of high temperature or of culture in media little suitable to them, lose the faculty of forming coloring matter, and sometimes do not regain it under normal circumstances till after a considerable lapse of time.

How is this extremely striking phenomenon to be explained? What distinguishes bacteria in their natural state from those artificially debilitated? Why can the former grow and multiply in the bodies of susceptible animals, and the latter not?

The circumstance that all the influences which rob the bacteria of this their (for us) most important capacity are such as are injurious and hostile to them, would lead to the supposition that an extensive degeneration of the cell protoplasm took place, which would show itself in other places also. Yet this is found to be the case to a very limited extent only. The harmless anthrax has the same appearance and the same form as the normal anthrax; its separate members show the same formation, the contents are as clear as crystal and homogeneous, the rods are motionless, they divide and produce spores as before. On the gelatin-plate and in the needle-point culture we observe the same sort of growth—in short, there are no really striking differences. It is true that on closer examination some slight traces of degeneration may be perceived. While the virulent anthrax bacillus multiplies so rapidly in the bodies of susceptible animals that the newly-formed members at once diverge and proceed immediately to further division, the anthrax of diminished virulence—for example, the mouse anthrax—frequently grows out to peculiar long threads in the organs, a sign of disturbed vital energy. The observations of Smirnow lead to similar conclusions. He found that the artificially-debilitated bacteria show in their cultures a slower, less luxuriant growth than the virulent ones, and also that they yield more readily than the latter to the influence of disinfectants. But these differences do not present themselves by any means regularly, and even if they did, they would not advance us far toward a veritable explanation of the phenomenon of diminished virulence. The question would then merely have to be put in another form: Why is the healthy protoplasm able to grow and thrive in the bodies of susceptible animals, and the degenerated protoplasm not?

We know a few facts, however, which may, perhaps, give a glimpse into these matters. We have characterized the pathogenic bacteria in general as those which produce substances poisonous to our own bodies or to those of animals. Virulent and attenuated anthrax bacilli stand in the same relation to each other as pathogenic and the non-pathogenic species. Do not the excretions here also play the chief rôle?

The investigations of Behring have yielded at least the beginning of a confirmation of his supposition. Behring found that virulent bacilli anthracis formed considerably more acid than attenuated ones, but that the latter possessed a far more decided power of reduction. These certainly appear but slight differences, but it is probable that these differences, which are perceptible on a rough examination, are but the outward expressions of finer, deeper phenomena; that the greater or less quantity of alkali produced only stands as the indicator of more complicated processes which are, as yet, beyond the limits of our perception. Not until our knowledge of the products excreted by the bacteria in general has become more complete—when, for instance, we are able to give a definite answer to the question, What chemical substances are produced by the vital functions of the virulent anthrax bacilli?—not till then can we expect final results.

The observations already made enable us, at any rate, to form an idea of these matters which may, perhaps, not be in accordance with reality, but which at least points to a possible explanation of the existing difficulties.

Let us say the relation between the bacteria on the one side and an animal organism on the other is characterized in the main by the latter opposing certain hindrances to the penetration and multiplication of the parasites which the latter have to overcome. As will presently be seen, this is by no means a mere supposition, but a fact grounded on reliable observations. It is only as to the nature of these hindrances that we are in the dark.

But let us suppose them to be of a purely chemical nature. White rats are, as a rule, insusceptible to anthrax, or in other words, the anthrax bacilli are not able to thrive in the bodies of these animals. The reason is, according to the investigations of Behring, that the blood and the tissue-juices of the rats possess an extremely high degree of alkalescence, which renders it impossible for the bacteria to live in them. Considering this with the previously-mentioned observations of the same investigator, we gain a valuable hint for the explanation of these things. Although the blood of susceptible animals may oppose to the micro-organisms a resistance equal in quality, but different in amount, an alkalescence small, perhaps, but yet injurious in its action, we may well conceive that virulent anthrax bacilli, with their decidedly acid excretions, are able to conquer this difficulty. They lessen the alkaline character of the juices of the body, and acquire the capacity of thriving in it—they take an infectious character.

Attenuated bacilli, on the other hand, are not able to make their way in like manner. If they form alkali instead of acid, they will even add to the difficulties which they found at first, and will be wrecked on this first rock, even though they may find all the other conditions for their parasitic existence.

In this way the greater or less mass of acid production would explain all the different degrees of virulence, and incline us to regard them all as purely chemical phenomena. We must, it is true, be on our guard against thinking the matter altogether so simple. The composition of blood is not always uniform, and we cannot credit it with a definite chemical formula. It depends on the state of the tissue-cells, and is only the special expression for the condition of these cells. Therefore we are only justified to a certain extent in regarding the influence of the acid anthrax bacilli on the alkaline blood as if the one neutralized the other, as one chemical neutralizes another in a test-tube. We must rather suppose that the cells of the body also come into play, since they are subject to the influence of the bacterial excretion, and in consequence of this influence they modify their own excretions, which now first become advantageous to the micro-organisms. That in reality these relations are more intricate than we might suppose at the first glance has been very clearly shown by the most recent investigations. It would seem that the powers of resistance offered by the bodies of animals against bacteria are by no means of a universal character, but possess a clearly special—one might almost say specific—character.

If we introduce bacteria, of any species whatever, into the blood-vessels of an animal, we notice that they disappear thence in a short time. Where do they remain then? It was at first supposed that they must leave the body along with its secretions, and that they would be found in the urine, the bile, etc. Yet careful investigations, among which I may particularly note those of Wyssokowitsch, have shown that the filtering membranes in their uninjured state are impervious to bacteria. In the urine, for example, micro-organisms are only to be found when a way has been opened to them by some injury to the urinary passages, some rupture of a vessel.

Wyssokowitsch found that the micro-organisms are deposited at their chief points of development by the blood-currents, namely, in the spleen, the liver, and the spinal marrow, and he believed that here they either perished or multiplied—showed themselves non-pathogenic or pathogenic.

That in reality a destruction of bacteria does take place in the bodies of living animals was proved by Fodor. Petruschky discovered that the blood of animals insusceptible to anthrax, such as frogs, killed anthrax bacilli even when all cellular portions were carefully excluded. Behring showed that the serum of white rats, even when separated from the body, possessed this capability. Nuttall was able to prove the same with regard to the aqueous humor, the ascites fluid, and other juices of the body, and this bacteria-destroying power of the blood-serum has obtained a more universal importance from the independent, though simultaneous, labors of H. Buchner and Nissen. The results of their investigations may be summarized in a few words, as follows: Germ-free serum kept for several days at a low temperature has the power of killing bacteria-germs in a very short time. It is true that this capacity has its limits. If one inoculates more than a certain quantity of micro-organisms some will be killed, but for the survivors the serum, instead of remaining a hostile element, becomes a source of aliment, and the bacteria begin to increase in it.

The different species show considerable differences of behavior in this respect. While some are particularly sensitive and sure to perish if brought into contact with serum, others are not in the least affected by it; and between these two groups is another, in which at first a slight check to development is observed, but which the bacteria soon get over and then proceed to grow and increase.

The germ-killing, disinfecting power lies exclusively in the plasma; the cellular parts of the blood, the red and white corpuscles even, counteract and paralyze it. Under the influence of high temperatures it quickly disappears, as already noted; it is also diminished by the blood being left standing for a length of time, but repeated freezings and thawings do not affect it.

Both investigators see a connection between this peculiar property of the serum and the processes which contribute principally to the coagulation of blood. This, however, does not tell us much about the exact nature of the active principle, but a subsequent series of investigations by Buchner found that the germ-killing power of serum depends on the salt which it contains: a diminution of its saltness is accompanied with a diminution of its power to kill bacteria. One might, for a moment, be tempted to suppose that it is the mineral ingredients which, directly and immediately, serve as the basis of disinfecting power in the serum. Yet Buchner shows clearly that such cannot be the case, and that the salts are only of importance because they stand in intimate relation to the albuminoid matter of the blood, the quantity and quality of which is decisive, and turns the scale one way or other. The salts serve as solvents, or agglutinants, to the albuminates, and an altogether special, peculiar, and as yet unexplained condition of the serum albuminoids is the active cause of the bacteria-killing agency.

Many celebrated investigators are of the opinion that dead albumin. such as we have hitherto almost exclusively employed in our researches, is widely different from the living albumin of the body in its chemical composition and also in its behavior. Thus it perhaps possesses qualities hostile to bacterian life which have hitherto escaped us, and it will be the task of future investigators to throw more light on all these matters.

There is no doubt that our knowledge of the means for resisting bacteria which an organism possesses has been very considerably increased by the observations whose results we have just considered. In one point, however, they are not quite satisfactory. Buchner and Nissen experimented almost exclusively with the blood of dogs and rabbits; both apply the results obtained in a general manner, and speak simply of "blood" and "blood-serum," without distinguishing the animals whence it was obtained. But the experiments of Petruschky, and still more those of Behring, have shown that the different species of animals show marked differences, and that the blood of the frog and the rat, for instance, show qualities not to be met with in that of other animals. This would lead to the supposition that the differing susceptibility of the animals might be an important factor in this question, and might deserve more attention than it has hitherto received.

However that may be, what has been written can only strengthen our belief that the processes by which the body resists the bacteria are chiefly of a chemical nature.

But it would be unfair to mention this view of the case and no other. A large number of investigators whose opinion is worth listening to, do not believe that the organism makes use of such means to attain its object, but that the relation between the two hostile elements expresses itself in an immediate struggle between the cells of the body and their foreign invaders.

The most enthusiastic and ingenious supporter of this view is Metschnikoff, who has developed it into a definite theory, which he has endeavored to prove by direct experiments. Metschnikoff saw that in certain daphniaceæ a yeast fungus (blastomyces) which had been received along with food occasionally brought about a general infection of the whole body. The vegetable parasite, boring through the alimentary canal, found its way into the tissues and multiplied within them. Frequently, however, the attack does not end fatally, and in such cases Metschnikoff invariably observed a remarkable phenomenon.

If we examine a preparation of an infected water-flea strongly magnified, we will usually notice without difficulty that a portion of the sharp fungus-spore penetrates the intestinal wall. This part looks somewhat degenerated, as if it had been gnawed; and further, it will be seen to be surrounded by a throng of white blood-corpuscles, whose numbers will increase every moment and which will attack the micro-organisms on all sides.

It is they which, according to Metschnikoff, play the most important part in the whole process. As descendants of the medial germinal layer, or mesoderm, they are closely related with the elements of which the digestive apparatus of all the higher animals consists. They consequently possess the capacity of receiving and devouring foreign bodies, and in the struggle between cells and bacteria they make free and full use of their power. If they succeed in overpowering and destroying the yeast-fungus germ, the organism is victorious and the animal is saved. If they fail to destroy the germ, the foreign invaders are victorious and the scene ends with the death of the animal.

Generalizing his observations and their consequences and applying them to the human body, Metschnikoff came gradually to the conviction that in every case of infection it is the white elements of the blood that, as scavenger-cells (*phagocytes*), have to save the organism if they can. If bacteria attack any part of the body, these cells, favored by their mobility, at once appear at the place of danger and rush upon the invaders. If they are able to make the latter innocuous no infection takes place; if, however, these defenders of the organism struggle ineffectually and yield, the enemies begin to multiply and spread themselves over the unprotected domain. The latter is always the case when the attacking bacteria are virulent to the animal attacked. If, for instance, we inoculate a mouse or a Guinea-pig with full-strength anthrax bacilli, there is no appearance of a reception of bacilli on the part of the blood-cells. Yet such a reception does take place when we inoculate an animal susceptible to full-strength anthrax with attenuated anthrax bacilli. These attenuated bacilli have lost their invulnerability, have been robbed of their dangerous character, and this time they become the prey of the phagocytes. The phagocytic theory cannot explain why, in one case, the bacteria always conquer, in another always the cells; why the virulent micro-organisms alone are able to conquer the opposition which the tissue elements offer to them. It contents itself with stating the fact, and does not attempt an explanation. Serious objections have been raised by very high authorities against Metschnikoff's views. Flügge, Baumgarten, and Weigert, in particular, have opposed them, stating that a reception of bacteria by the cells of the body only takes place when the former have been already killed, or at least been deprived of much of their vital energy by other influences. The phagocytes, they maintained, did not form an active and dangerous weapon of defence for the organism, did not stand in the foremost rows in the battle against the invading parasite, but were the open graves behind the line of battle, destined to receive the fallen enemies or any other lifeless bodies or substances. Nothing, they said, compelled us to believe that these cells possessed a peculiar devouring and digesting power: they were, on the contrary, nothing but buriers of the dead, removers of decaying matter. They maintained, further, that whenever healthy, vigorous bacteria entered the cells these latter always fell a sacrifice to them and were irretrievably lost.

In fact, the opinion that bacteria are destroyed by influences lying outside the cells has received a strong confirmation in the recent observations of the germ-killing power of blood-serum separated from cells.

It cannot be denied that Metschnikoff's theory agrees ill with the revelations of physical science in general, which exhibits life of all kinds as depending on processes, physical and chemical, which obey the simplest laws. It would, as already said, be more comprehensible if we could see the cells in a less immediate strife with the bacteria, and if both were represented to us as the special exciters of definite processes, chiefly of a chemical nature, which then in their turn began to act upon each other. Yet the phagocytic theory, perhaps on account of its palpable character, has, nevertheless, gained numerous adherents, who hold fast to it and defend it against all attacks, so that it cannot be left out of account in discussing the matters with which we are dealing.

Whether the obstacles which the organism opposes to the invasion of bacteria be of a chemical or cellular nature is not the main point of the question. We may conveniently summarize our opinions by saying: The *pathogenic bacteria are those which, by their vital action, produce excretions injurious to the bodies of men and animals;* in the infectious species the excretions possess, further, the capability of overcoming the resistance offered to them in the organism of susceptible animals, thereby enabling the bacteria to multiply within the tissues. This capability may be lost if the excretions alter their composition, which happens either where the bacteria adopt a saprophytic mode of life or under the influence of injurious agents affecting the cell-protoplasm.

What conclusions may be drawn from these premises? Let us consider a few of them.

In the course of this work we have often considered the differences of susceptibility in the same species of animals as regards one and the same micro-organism. This explains itself now as a differing amount of resisting power in the body of the species in question. Reference is here directed to the view taken when considering the attenuated anthrax bacilli, and it will be conceivable that an insusceptible species of animal may be made susceptible by a chemical influence affecting the tissue fluids.

We owe to Behring a convincing proof of the correctness of this assertion. Rats, as a rule, are not susceptible to anthrax infection, probably because their blood possesses a particularly high degree of alkalescence. If fed exclusively with vegetable nutriment, or if acid phosphate of lime be added to their food (both being measures by which the formation of acid in the organism is promoted), this immunity disappears: the animals die when inoculated—i.e., anthrax bacilli have become infectious for rats so prepared for them.

Leo's discoveries will now be comprehended, and for many varieties of susceptibility within one and the same species otherwise inexplicable, we have here, it would seem, found the right explanation. Thus it will be understood how, especially for the less susceptible species, the quantity of infectious matter introduced may be of importance. Let us adhere to our often-employed example of the anthrax bacilli and suppose that some species of animal which is insusceptible to them owes its immunity to a high degree of alkalescence in its juices. A small number of anthrax bacilli are not able to reduce the alkalescence sufficiently. If they are introduced in increased numbers some indeed perish, but these dead cells pour out their "acid" contents into the surrounding fluids and open a possibility of life to the still living cells near them. Thus at some one spot a field arises offering the conditions for bacterial development, and when the stone once begins to roll there is no stopping it—the insusceptible individual has become susceptible.

In other cases this change is brought about, not by an alteration in the state of the body, but by the degree of virulence in the bacteria. We have seen that the excretions are the chief agents of infection, and that even the slight variations in their composition can essentially influence their poisonous power.

In our eyes and for our interests the virulence of the bacteria is doubtless their most important property, while for the bacteria themselves it is not of such paramount importance. In many species of bacteria their pathogenic action is the most variable thing about them, being subject to numerous external influences, standing on tottering foundations, and depending on slight modifications of condition. If a small increase of alkali can turn the scale and out of perfectly innocuous anthrax bacilli, for instance, produce a modified breed which kills mice, then we must say that almost everything is possible in this department.

If we further consider that for general reasons, depending on their natural history, every pathogenic species of bacteria must once have been non-pathogenic; that a micro-organism which is now only capable of living as a parasite must once have led a saprophytic life; that most of them at the present time still possess the faculty of existing in both ways; that the pathogenic agency has only proceeded from an adaptation to special conditions and a particular nourishment, we shall be able to see in the loss of this property nothing but a return to former habits and customs.

It will also be seen that the readiness with which different species lay aside their virulence is very variable—that some hold more tenaciously to the acquired faculty than others. In all of them, however, *it is an acquired faculty*, an accessory, or, if you like, a secondary faculty, which can be changed in any direction.

Therefore differences of virulence must not induce us to separate into different species bacteria which are otherwise identical.

Virulent and attenuated anthrax bacilli, as already said, stand to each other in the same relation as pathogenic and non-pathogenic species. How would it be, then, if chance had made us first acquainted with the harmless variety and afterward brought the poisonous ones under our notice? Should we not commit a great error if we regarded them as two altogether different things? How much less, then, ought we to insist on far slighter variations, such as the difference of effect on different animals.

For practical use such a way of proceeding—particularly when the differences are of a somewhat enduring nature and of regular occurrence—may perhaps be convenient and therefore permissible. Yet we must always keep in mind that by adopting such a course we may easily lose the solid ground under our feet, and find ourselves all adrift with our artificially-built-up system of differential diagnostic criteria.

As will hereafter be recorded, we are acquainted with quite a number of micro-organisms, evidently closely allied to each other, which slight indications only enable us to distinguish from each other and to arrange in definite groups. Wherever the attempt has been made to employ differences of virulence or infective power as grounds of separation within these definite groups, it has always been found, sooner or later, that this proceeding yields no reliable results, and that it is useless in the long run.

When it is understood that the pathogenic qualities of the bacteria are their least constant qualities, that from clearly intelligible causes they form the most variable item in the picture of micro-organic appearances, we shall be very little inclined to cite variations in this particular as arguments against the law of constancy of species before alluded to. In fact, it has hitherto, in every case, been possible to recognize permanently as a separate species those micro-organisms which have been once recognized as such. It is true that, under some circumstances, a very careful and precise consideration of all their peculiarities is necessary: yet where this condition has been fulfilled one species has never been found to merge into another and display its characteristic marks.

In conclusion, one more point will be considered, which grows naturally out of what has just been written.

That the virulence of a given micro-organism may, under natural conditions, be sometimes greater, sometimes less, we have already seen. This fact has been, with great probability, set down as one of the reasons why the same infectious disease often occurs with different degrees of malignancy, a phenomenon which would otherwise be very striking. It is known that the higher plants show something similar; that, for instance, the fox-glove from a given locality will yield in one year a very potent poison, in the next year a very much weaker one. Now, if diphtheria bacilli of a peculiarly virulent kind infect a human being and pass over from individual to individual, at length causing an extensive epidemic, will not, or cannot, its character depend on the original qualities of the disease-exciting micro-organism?

These considerations might be spun out to a considerable length in various directions. But we will not exhaust ourselves in theories nor run on in advance of our real knowledge. It will not have escaped notice that many of the deductions already laid down stand on a weak foundation; that a firm footing exists only here and there, our knowledge resembling a patchwork and being greatly in want of additions and completions.

II. THEORIES OF IMMUNITY.

Further information in the matter just treated of, especially the question of the peculiar action of the pathogenic bacteria, is very desirable. It is immediately connected with another matter, darker perhaps and more mysterious, but certainly far more important for us, because it touches the great problem of the cure of infectious diseases. Certain powers of resistance which the body employs against the parasites must be conquered by the latter if they are to multiply and make use of their powers. We have also seen that we can influence the relations of the two hostile powers and turn the scale in favor of the one or the other: in favor of the parasites if we diminish the body's power of resistance, for example, by introducing injurious substances such as sublimate, pyrogallic acid, the excretions of other micro-organisms, or, as in the experiments of Leo and Behring, by a change in the entire process of assimilation; in favor of the body if we alter the nature of the bacteria and weaken their special products. But two further possibilities are here present. The scale will turn in favor of the micro-organisms if we succeed in sufficiently strengthening the bacteria; an experiment which, as already mentioned, has not yet been successful. On the contrary, the body gets an advantage if we can succeed in increasing its natural power of resistance, in heightening its bacteria-killing force, and in this particular field recent investigations have engaged with remarkable energy and, as we may perhaps say, with remarkable success.

If particular animals are by nature safe from infection by particular micro-organisms, the reason is that they possess a high power of resistance in the constitution of their tissue-juices or in the behavior of their cells. Besides this kind of non-susceptibility, this inborn immunity, there is also an *acquired immunity.* We know a whole series of diseases, particularly the exanthematic affections, such as small-pox, measles, scarlet fever, etc., which usually attack the organism but once and fortify it against after-attacks—i.e., they place it in a like position with that of individuals previously rendered non-inoculable.

The immunity thus brought about by nature may also be produced artificially. There are proofs, for instance, that in China more than 3,000 years ago men endeavored to combat that terrible disease, small-pox, by inoculating persons, after a special preparation, with genuine small-pox virus, and that they were able, by suitable measures, to attenuate the effects of the infection, and by these means insure those so treated against the full force of a natural attack of this dreaded malady.

This process was successful, it is true, but it was nevertheless dangerous and defective. In lieu of it came, as is well known to all, toward the end of the last century, the practice of vaccination with the so-called cow-pox virus, as introduced by Jenner. Whether the cow-pox virus is identical with or nearly or distantly related to the genuine small-pox virus is still an unsolved question. Suffice it to say that the disease called cow-pox gives immunity from the variola vera, and this fact has led to very important discoveries, which we owe more particularly to the epoch-making experiments of Pasteur.

Pasteur started from the view that cow-pox virus is an attenuated form of small-pox virus, and proceeded to apply the experiences attained through vaccination against small-pox to all cases in which he was in possession of attenuated bacteria virus. Success crowned his experiments and showed the correctness of his *a priori* reasoning.

He and his scholars succeeded, in the case of chicken-cholera, and later in the case of anthrax, swine erysipelas, symptomatic anthrax (quarter evil, black-leg), and, as we know, also in the case of an affection called canine madness (which is, perhaps, a bacterial disease), by the cautious inoculation of micro-organisms in various degrees of attenuation, to give immunity to animals against inoculations of material which would otherwise have been infallibly fatal to them.

Artificial immunity is usually the more perfect and durable the more cautiously the protective inoculation is performed. One must not employ the unattenuated virus directly after the virus in its fully-attenuated form, but must proceed gradually to the former, employing gradually-increased strengths. Pasteur employs two "vaccines" of different degrees of virulence: the weaker of the two, the "premier vaccin," generally makes the animals moderately ill; the stronger, the "deuxième vaccin," may, under some circumstances, have fatal effects. Yet this occurs only in rare cases, and generally speaking the animal gets over it without injury, and is then able to receive and to support the unattenuated poison.

It may easily be conceived that such a surprising discovery at once led to a host of the boldest hopes and plans, and that the practical worth of the fact was regarded as extremely great.

Yet how necessary is reserve of opinion in this matter, has been shown by the experiments of Koch and his collaborators, Gaffky and Löffler, with regard to anthrax. They also found it possible, by subcutaneous inoculation with attenuated bacilli, to make animals capable of withstanding an after-inoculation with cultures of the highest degree of virulence. But they remarked, at the same time, that this experiment did not by any means succeed equally with every kind of animal, and, further, that even in the most successful case a full security against all attacks of anthrax disease was not obtained. As we shall yet see, the animals, when infected under natural conditions, usually received the virus along with their food and in the alimentary passags. Material containing spores goes through the stomach without the destruction of the spores, into the intestines, and thence the bacteria spread to other parts. Against this kind of infection the inoculation grants no unconditional and reliable protection, so that the animals sometimes die of anthrax in spite of the inoculation.

If this be placed along with the already-mentioned fact that the vaccination itself, particularly with the stronger second virus, sometimes ends fatally, it is natural that opinions are divided as to the practical value of protective inoculation. This is a question of utility which can only be decided by means of statistics, and statistics show that for anthrax the use of protective inoculation is desirable in countries or districts in which this disease occurs regularly and spreads widely. Experience has also shown that this inoculation yields better results with cattle than with sheep—a fact which deserves due attention.

In the case of swine-erysipelas and chicken-cholera, the success has not as yet been very satisfactory, and qualified advisers, therefore, pronounce against the use of inoculation as a protection against these complaints. In the case of symptomatic anthrax, cattle-doctors pronounce with one voice in favor of inoculation, which may be regarded as a useful and effectual means of protection.

But these purely practical considerations do not touch the heart of the question, and do not invalidate the very important scientific fact that under some circumstances the most virulent matter may be rendered ineffectual by inoculation with attenuated virus.

It may be readily supposed that for a phenomenon so extremely striking and important, causes and explanations have been sought. Yet in spite of the most persevering efforts which of late have almost exclusively occupied the attention of investigators, no certain result has hitherto been attained, and to the question as to how the acquired immunity is brought about, we are still unable to give a decided answer.

Before discussing the various, and often contradictory, views held by different authorities, let us state the simple facts, such as they are. As just mentioned, it is possible to fortify animals, by inoculating them with attenuated anthrax bacilli, against the virulent bacteria. Bitter has shown that the thus inoculated bacilli develop only in the immediate neighborhood of the spot where the inoculation took place, and are not carried into other organs by the circulating blood. Hueppe and Wood found that a species of bacteria, clearly distinct from the anthrax bacillus, apparently innocuous and strictly saprophytic, was able to secure even very susceptible animals, such as mice and Guinea-pigs, against anthrax. Roux and Chamberland found that cultures of virulent bacteria freed from all micro-organisms and the tissue-juices of killed animals yielded protection against anthrax and a number of other diseases. Before and after these investigators, Salmon and Smith, Beumer and Peiper, as well as several others, have observed similar facts in regard to swine-fever, typhus, etc. Foà and Bonome successfully employed upon the proteus a definite chemical substance which they supposed to be excreted in large quantities by these micro-organisms, instead of using the real excretions, and Wooldridge was even able to obtain immunity from anthrax by means of a substance which has no connection with the vital process of the bacteria, viz., by means of albumin from the tissues, peculiarly changed and prepared.

How can all these so apparently contradictory facts be harmonized with each other? Only, it would seem, by drawing the conclusion that the immunity is brought about, not by the micro-organisms themselves, but by certain chemical substances which are, for the most part, bacterial products. These substances are produced within the animal, at the place of inoculation, by the attenuated bacteria in continually-increasing quantities, and spread throughout the body. They are also contained in the cultures of virulent bacteria and in the serous fluids, which—for example, in malignant œdema—develop in the subcutaneous cellular tissue of the infected animals. If we inoculate with such matter, but without special precautions, we obtain no success with infectious species, because live germs begin to grow in the new individual, and bring death, instead of protection, from disease. But if we remove the micro-organisms by careful filtration or cautious application of heat, their excretions remain, which do not multiply in the animal, but, on the contrary, are diluted by mixture with the blood and the tissue-juices, and their effects being thus diminished, they are able to produce immunity.

The real efficacious substances are sometimes formed not only by one definite species of bacteria, but by several in the same manner. Thus the results obtained by Hueppe and Wood may be explained as also those of Roux, Chamberland, and others, who by inoculating with one kind of bacteria produced immunity from several diseases, and who observed the frequent occurrence of reciprocal inoculative protection by which one micro-organism secured immunity against another, and vice versa. Some of the substances recognized as bacterial products are chemical bodies well known and exactly defined as to their composition.

As neurin is one of them, we can understand how Foà and Bonome were enabled to employ it with success. The results obtained by Wooldridge may also be explained in a similar manner. At all events, we have, on the one hand, in protective inoculation certain chemical substances, excretions of bacteria, as an active principle, and on the other hand the acquired immunity—i.e., a condition which resembles that of the individuals to which nature has granted immunity. How can such a cause produce such an effect?

Of the many attempts at explanation which have been suggested, we can only enumerate a few of the more important ones. There is one supposition which we can clearly refute by the light of what has already been written, and which we only mention on account of its historical interest. It is the so-called theory of exhaustion set up by Pasteur and Klebs, which supposes that on the "first invasion" a number of substances are consumed in the body, which form a necessary nutriment for the invading species of bacteria. As these substances are not afterward renewed, a second attack becomes impossible; the exhausted soil has become unfruitful. Being thus incompatible with the already-developed opinion that it is not the bacteria themselves, but the chemical substances that decide the question, this theory could no longer be maintained after it had been discovered that soluble substances freed from all living germs, which certainly could not produce such an exhaustion, were, nevertheless, able to produce immunity. The observation of Bitter, too, that the growth of the inoculated bacteria takes place within a narrowly-limited portion of the organism, goes to disprove this explanation, which supposes a general diffusion of micro-organisms throughout the body and a complete exhaustion of the available nutriment.

The second explanation, the so-called hypothesis of retention, forms a direct contrast to the one just mentioned. Its principal champion is Chauveau. It supposes that the excretions of the bacteria remain in the body after the first invasion and prevent the return of the same species. This view was based chiefly on the already-named circumstance that in the fermentation of lactic acid, butyric acid, and urine, as also in the putrefying decomposition of the contents of the intestines, substances arise which at last put a stop to the further development of the micro-organisms, the originators of these processes, some of which even possess antiseptic properties and are hostile to bacterial life. A similar process was supposed to take place in the body, the accumulated products offering an enduring resistance to after-attacks from the same kind of bacteria.

The hypothesis of retention agrees with our views, in that it regards the action of chemical substances as the chief factor. In fact, Chauveau was led to the formulation of his theory by the same considerations which we have been pursuing in our discussions. He saw that the blood of animals suffering from anthrax was able to grant immunity after having passed through a filter impervious to bacteria. This filter was, indeed, no artificially-constructed instrument of plaster-of-Paris or porcelain, but a natural one supplied by the body itself, the placenta. When he inoculated ewes during the latter part of their pregnancy, first with attenuated, then with full-strength anthrax bacilli, the lambs, when born, were not susceptible to anthrax infection.

From the formerly-made investigations of Brauell, and more especially after those of Davaine, it had been considered as an ascertained fact that the placenta formed an impenetrable partition, a "barrière infranchissable," for micro-organisms of every kind. It seemed, therefore, that the inoculative protection could only be caused by soluble substances which passed from the maternal to the fœtal organism through the blood.

But more recent experiments (I will only name those of Malvoz and Jacquet) have shown that the placenta is by no means such an impenetrable barrier for the progress of bacteria. The various species of animals show differences in this respect, and the numerous kinds of micro-organisms also. But it is beyond doubt that frequently, especially when lacerations—even the smallest—take place in the placenta, bacteria find their way into the circulation of the embryo, and that this occurs frequently in cases of anthrax. For this reason I purposely abstained from referring to this fact (immunity of lambs whose mothers had been inoculated during their pregnancy) while endeavoring to prove the importance of the excretions in the question under discussion. We were able to cite so many other convincing reasons in favor of our view that by this alone the hypothesis of Chauveau is not disturbed.

But there are more serious objections to it. Flügge and Sirotinin have discovered that pathogenic bacteria in their artificial cultures do, in fact, produce substances which, sooner or later, put an end to their further increase. But in such cases we have nothing but an excess of accumulated alkali or acid, and Flügge is certainly right when he says such substances have a very poor chance of lasting for any length of time in a living organism and of granting permanent immunity.

However, one must not regard the results of test-tube experiments as necessarily applicable also to living animals. Nothing can disprove the supposition that the same bacteria may produce very different substances in two such very different cases, and that we have been unable hitherto to find out the difference between the substances in consequence of their intricate composition or delicate nature.

But if we grant this, the second part of Flügge's objection loses most of its force, although it may be difficult, with our present knowledge of such matters, to conceive how chemical substances can remain for an unlimited time in the body. On the other hand, there exist no convincing proofs that such cannot be the case.

We may suppose that we have here to deal with substances very difficult to dissolve or to diffuse. Such, for instance, are the albuminoids, which therefore can only be excreted slowly and imperfectly and do not disappear from the organism for some time.

The presence and action of such substances is further rendered probable by the fact that the immunity does not follow immediately upon the inoculation, but that days or even weeks must elapse before the former has been developed to its full extent; that is, the time required for the solution of a requisite quantity of the chemical substances introduced. That this process is not accomplished without difficulty is evident from the considerable increase in the temperature of the body—the "inoculation fever"—which generally occurs, and the importance of which for the production of immunity has recently been insisted on by Gamaleïa. Another portion of these substances which remains for a time unaffected is gradually brought to solution and absorbed or excreted, thus producing an enduring immunity.

The hypothesis of retention is able to explain all the phenomena hitherto observed, possibly not in its original form, but yet unaltered in its main features. Until a better theory is found to replace it, it may continue serviceable in many cases and may reckon on our support.

Many persons, however, are not satisfied with such a simple explanation of the matter, and the fact that the acquired immunity extends over years, and even over decades, drives many to the conviction that there must be a permanent alteration in the condition of the whole body which could only be accomplished by the influence of the active tissue cells.

These tissue cells form the basis and centre of a third hypothesis, which, unlike those hitherto discussed, is founded on direct observations and incontestable experiments. We already know that Metschnikoff ascribes the chief rôle in the relations between the body and the bacteria to the white elements of the blood, as phagocytes. He found that virulent anthrax bacilli were received into the white blood-corpuscles of insusceptible animals, attenuated ones into those of susceptible animals, and, as he believed, there devoured and digested, while a similar process was not discoverable in the case of susceptible animals and virulent bacilli. From these facts Metschnikoff drew the following conclusions:

The presence or absence of immunity depends on the ability or inability of the cells of the body to devour and kill off the bacteria. This ability may be natural or acquired. In the latter case, the cells where they have once had an opportunity of devouring attenuated micro-organisms with a milder poison which nature enables them to withstand are so far accustomed to it that they can bear a stronger dose without injury, and at last are able to devour the most virulent material with impunity. This can be effected both by gradual functional adaptation and also by a kind of selection in which only the strongest and most vigorous cells remain and transmit the acquired faculty to their descendants. The leucocytes are but short-lived formations. A permanent resistance of the organism to a disease which it has once passed through or against which it has been protected by inoculation is, therefore, only conceivable if we grant to the cells the power of transmitting an acquired property unaltered to their children and their children's children.

This hypothesis, as must have been seen, presupposes an extraordinary docility in the protoplasm of the white blood-corpuscles, to which it attributes something like feeling, thinking, and acting—a sort of mental perception.

But even if we raise no objection to this, there remain plenty of reasons for combating the phagocytic theory. In our opinion, the fact that it is essentially the excretions of the bacteria which produce or are able to produce immunity is difficult to harmonize with Metschnikoff's hypothesis; for if no living micro-organisms are present none can be devoured, to accustom the cells to the poison and prepare the way for resisting more virulent successors. To overcome this difficulty, it would be necessary to suppose that the reception of attenuated germs acts upon the cells only as a specific stimulant, to which they answer by a functional reaction, and that this stimulating power exists in the same degree and works in the same manner also in the bacterial products.

Another objection which we have already alluded to is perhaps still more serious. We know that a number of very celebrated investigators are of opinion that the leucocytes are only able to receive bacteria which have been killed, or at least weakened, by some influence or other, and that they cannot master a healthy living enemy. In fact, an agency outside the cells whose existence was formerly only a conjecture has lately become known to us in the power of the serum of the blood to destroy bacteria. How far this fact may be able to explain the phenomenon of acquired immunity cannot yet be stated with certainty. Lubarsch justly remarks that the experiments hitherto made have shown no difference between the power of the blood in susceptible animals and in those artificially rendered insusceptible, whereas we might expect such a difference if we regard this power in the blood as the chief agent in removing the micro-organisms from bodies protected by inoculation.

It is possible that the future may throw more light on this dark question. But even if we do place peculiarities of the serum, instead of direct cellular influence, in the foreground of the picture, we only touch those diseases whose exciters increase and exert their influence within the vascular system, as the strictly septicæmic diseases, such as inoculated anthrax, chicken-cholera, and swine-erysipelas. In other diseases the blood holds a secondary place. For the toxic species of bacteria, for instance, we should be compelled to admit a possibility which lies entirely outside the region to which we have hitherto confined our observation; we should have to admit that in this case the body gradually accustoms itself to the poison. The adaptability of our organism to bear the action of several kinds of poison, provided they be administered in slowly-increasing doses, is great, and the influence is therefore justifiable that a similar adaptability may here also play an important part. One indubitable fact may certainly be gathered from this mass of contradictory views and observations, namely, that artificial immunity is acquired not by one regular process, but sometimes in this manner and sometimes in that. It is possible that chemical substances retained in the body (as supposed by the hypothesis of retention) are often the cause of the phenomenon; that in another case, as Metschnikoff supposes, the cells are the active agents, or that chemical properties of the blood or of the tissue juices participate; or, lastly, it is possible and even probable that causes are in operation which are as yet unknown, and which future investigators are destined to discover.

In the second and third cases there would be an artificial increase of the power given to the body by nature for resisting the attacks of micro-organisms. With the phagocytes there would be an augmentation of their peculiar function in the blood, an increase of bacteria-killing power. The latter, too, would indeed have to be referred also to a special action on the part of the cells. A reactive change in the fixed tissue element under the influence of definite bacterial products, which shows itself in a change of the composition of the blood, would then have to be regarded as the cause of the phenomena; the time elapsing between the protective inoculation and the acquired immunity would be the time required by the organism for the development of its mysterious powers, and the strengthening of that degree of resisting power which has been given to it by nature.

I have considered these matters at such length, notwithstanding their uncertainty and vagueness, because, as I have already said, they are closely connected with the important question of the healing of infectious diseases. It is true that we cannot speak so much of a healing as of a means of *protection*, but we see before us a possibility of mastering the most dangerous enemies of human life and suppressing their destructive effects. A healing in the strict sense of the term would indeed not be effected without our being able to check and artificially remove the affection when it was already in progress. This has also been attempted, and the present is perhaps the best opportunity to consider the progress hitherto made in this direction. The most natural way of proceeding is, of course, to combat the further increase of the bacteria already in the body, by the use of means known to possess qualities hostile to bacterial life and calculated to kill bacterial germs. The natural powers of the organism are in this case not called into requisition; but an artificial aid, to which they stood in no relation previously, is offered to them.

Unfortunately, the results obtained by this method have by no means fulfilled the hopes with which it was first welcomed. All the substances which, outside the body in the test-tube, have a decided influence on the micro-organisms are powerless in the living tissues unless they are employed in doses which would be directly deleterious to the body, and would destroy it even more quickly than the bacteria would have done. Yet it would be unwise to give up in despair on account of the want of success hitherto secured. The valuable results obtained by the purely empirical treatment of malaria with quinine and of syphilis with mercury encourage us to persevere and warn us against a too early abandonment of our efforts.

Although we have not yet succeeded in discovering with complete certainty the exciting causes of these two diseases, yet there is every reason to believe that they belong to the class of diseases occasioned by parasites of the lowest forms, and if chance has enabled us to find specific remedies for these evils, we may hope that strictly methodical experiments and systematic investigation will also lead to like success in other cases.

That it is in reality possible to paralyze and get the mastery over micro-organisms, even of the most dangerous kind, in the bodies of susceptible animals, has been proved by the interesting observations of Pawlowsky, of Bouchard, and more particularly of Emmerich. These investigators succeeded in saving from almost certain death rabbits inoculated with virulent anthrax bacilli, by bringing a considerable quantity of erysipelas micrococci, of green pus bacilli, or of Micrococcus prodigiosus into the circulation either before or after the anthrax inoculation. The animals continued to live, but without immunity against a later inoculation with anthrax, to which they regularly succumbed.

Here, then, there was no acquired permanent immunity, but a real cure of a parasitical affection, attained by the agency of a second species of bacteria, and we have only to inquire how this counter-effect is to be explained.

That there exists a decided antagonism between certain micro-organisms has been shown by the experiments of Soyka, Garré, Freudenreich, and others. If sterilized gelatin be allowed to harden on a glass plate and the surface of this solid culture medium be planted with various kinds of bacteria by means of needle strokes at short intervals from each other, after a few days it will be seen that the expected growth here and there fails to make its appearance.

Further examination will show that the excretions of certain bacteria act as a decided check to the development of certain others, and sometimes even exercise a destroying influence and kill the neighboring growth.

He who would simply apply these facts to the living body would find himself sometimes unpleasantly disappointed, for the living tissue is no plate of gelatin and no test-tube. But for the explanation of our case of cured rabbits these facts may, we think, be utilized—with a certain degree of caution, of course. It might be supposed that the soluble products of the injected erysipelas cocci render a further growth of the anthrax bacilli impossible, or that they possibly destroy them altogether. Also that the one decomposes the products of the others already formed and neutralizes their influence; and if such results are not observable in experiments made in our laboratories, the reason may be that the excretions of the streptococci do not attain their active properties outside the living organism.

Lastly, one can suppose, with Emmerich, that the tissue cells exercise a temporary reaction; the power of resistance to anthrax bacilli which nature has given to the cells being increased for a short time and thus enabling them to gain the desired ascendency.

III. EXAMINATION OF INFECTED ANIMALS—METHODS OF INOCULATION.

The pathogenic bacteria are by far the most interesting to us. Wherever and whenever we find micro-organisms, in the human body or in animals, the question always arises whether we have this variety of micro-organisms or not. The answer is often difficult and must always be according to fixed and definite rules which were first laid down with exactness and precision by Koch many years ago.

These rules require that a micro-organism, to be recognized as a specific agent in the production of pathological alterations, should fulfil three conditions:

First, it must be proved to be present in all cases of the disease in question.

This is a matter of course, and scarcely requires a justification; for if the disease can occur without the bacterium, the latter is not unconditionally necessary and cannot be regarded as specific.

Second, it must further be present in this disease and in no other, since otherwise it could not produce a special definite action. We often meet with apparent violations of this condition and find bacteria which we regard as specific, in the sense already explained, and whose occurrence is, nevertheless, not confined to one disease. Such exceptions are, however, explainable by the differences which various kinds of bacteria show according to the place of entrance, the organ they have attacked, the degree of susceptibility in the organism attacked, the degree of virulence which they possess, etc., etc. A careful consideration of all these different factors will suffice to guard against errors and false conclusions.

Third, a specific micro-organism must occur in such quantities and so distributed within the tissues that all the symptoms of the disease may be clearly attributable to it. On this point also the peculiar manner in which, under certain circumstances, the bacteria produce their effects demands careful attention. The answer to the question whether a particular bacterium fulfils these conditions or not is afforded by microscopic examination.

This is conducted according to ordinary well-known rules and requires no special explanation.

The results obtained with the microscope are efficiently aided and protected by the collateral use of artificial breeding. Here we often find an obstacle in the fact that our artificial food media, as we know, do not offer a suitable habitat for all species of bacteria. Yet many an apparently insuperable impediment may be removed.

Such was the case with the tubercle bacilli in breeding which Koch achieved a brilliant success, though perhaps such ingenuity and acuteness as his may rarely be met with. In many cases the culture to be successful must be conducted throughout with all imaginable precautions. The following illustrates this: Given a rabbit which has, within a few hours, died of a disease the nature of which, let us suppose, we are as yet in ignorance of. By making simple cover-glass preparations of the blood or tissue fluids, it may at once be seen that bacteria are very probably connected with the death of the animal. Supposing these bacteria present in the form of short rods, our next task is to develop these suspicious-looking forms artificially, in order to learn further details regarding them. As it is, of course, very important to keep the original material free from pollution by foreign matter, the greatest care should be taken during the following procedure.

The body of the animal is laid on its back and fastened to a board, the pelt being washed with $\frac{1}{10}\%$ solution of sublimate, to free it from all dirt, before an incision is made. We must sterilize beforehand, in a flame, a number of knives, scissors, and forceps, and place them under a bell-glass for protection. Now take the scissors, cut open the skin along the median line, without injuring the abdominal walls, and push it as far back as possible on either side —i.e., skin the animal as far as necessary.

Fresh instruments are then taken in order to avoid pollution by stray germs. The abdominal walls are cut through, and now, with fresh instruments as often as possible, the organs of the abdomen are taken out, and after the breast-bone has been removed, those also of the thoracic cavity. Each organ is placed in a sterilized glass dish. The order usually observed in removing the organs is as follows: First the spleen, then the liver, then the kidneys, next the heart, and lastly the lungs. As a matter of course one is not strictly bound to this order of proceeding; other organs than those just named will sometimes require attention, and also different species of animals will occasionally require special methods of dissecting suited to their peculiarities.

The chief point is never to lose sight of the extreme importance of exactitude and cleanliness. Nevertheless the effects of decomposition, which (in the case of small animals, such as mice) generally begins very soon after death, are apt to interpose and render the determination of the true cause of disease much more difficult.

When the dissection is completed, proceed to further investigations, as follows: Small quantities of blood from different parts, portions of tissue from organs which frequently harbor large quantities of bacteria, such as the spleen, the liver, and the lungs, are placed in our liquefied culture media, the usual dilutions are made, and we ascertain, by means of the glass-plate process, whether there are bacteria in this original material, and if so to what species they belong.

If the presence of strictly parasitical micro-organisms, which can only thrive at the temperature of the living body, be suspected, we must prepare agar plates and keep them in the incubator. If we have reason to expect anaërobic bacteria, we must adopt suitable measures for their examination.

When the colonies begin to appear the plates are subjected to a close scrutiny. We must find out whether only one kind of bacteria or a number of different kinds have developed, and in the latter case, whether one kind distinguishes itself by its numerical superiority or by any peculiar qualities. Attention will, of course, be specially directed to such, and we endeavor to find out with the aid of the microscope whether in form and appearance they are identical with or simulate the bacteria already observed in the cover-glass preparation.

If we are able to examine many cases in which the disease is apparently one and the same, it becomes much easier to form a judgment based on the indications of the plates, since we may expect that the particular micro-organism will present itself everywhere in the same manner.

If such a bacterium which (firstly) occurs in all cases of the disease in question and (secondly) occurs there in large quantities also presents marked peculiarities of growth or form, enabling us to distinguish it from other species and also to prove (thirdly) that this one definite species is found only and exclusively in connection with this one particular disease, we may fairly conclude that it stands in a particular intimate relation to this disease, and the probability of its being also the cause of this disease is so strong that it approaches very near to certainty.

The last link in the chain of evidence is of course supplied only by a successful transmission, before the overwhelming force of which all opposition must yield. Until that is obtained the objection is always possible that the bacteria may be a regular sequel or accompanying symptom of the disease, in consequence of the fact that certain morbid metamorphoses offer peculiar facilities for the development of certain bacteria. This view of the case has, it is true, an immense weight of probability against it, but it cannot be finally disproved without separating the bacteria from all their natural surroundings and experimenting with the pure culture, to find out whether the specific qualities attributed to them are still there or not.

As long as we continue to transmit virus obtained directly from a diseased organism, there is a possibility that other substances are inoculated along with the bacteria, and that these other substances contain the disease-causing matter. If, however, we take as our point of departure a culture which has been extended through several generations, the above objection collapses, and by the successful transmission—the reproduction of an affection like the original disease—the specific character of a given bacterium is indisputably proved.

Unfortunately, we cannot yet fulfil this requirement in all cases. We have already often referred to the very different degrees of susceptibility possessed by the different species of animals as regards the attacks of the pathogenic micro-organisms, and when we reflect on the obstinate tenacity with which more highly-organized parasites often attach themselves exclusively to one species of animal, which they never leave under any circumstances, we will no longer find it astonishing that the lower parasites cannot always be transmitted from one species of animal to another.

For this reason, from this consideration we may at once draw the conclusion that in our attempts to reproduce a bacterial disease we should, as much as possible, experiment with those animals which are susceptible to it under natural conditions. Now it is the infectious diseases of man which possess by far the most interest for us, and we know that many of them attack human beings exclusively. We must not here require impossibilities from our method. Even if the attempt to transmit the disease to animals failed, that is no reason for doubting the specific agency of the bacteria in question. In the same way, when a particular micro-organism affects an animal, we must not expect to see precisely the samo phenomena produced which we were accustomed to observe in human beings. Occasionally chance comes to our aid and brings about an unintentional transmission to a human being, or some particularly zealous investigator goes to the length of experimenting upon himself. As a rule, however, we must direct our attention chiefly to the animals more nearly related to mankind, and the example of relapsing fever, which can be transmitted to human beings and monkeys alone, would seem to indicate that we are on the right road.

Monkeys, however, are expensive animals and therefore not numerous in our laboratories. As a rule, we keep mice, Guinea-pigs, rabbits, and occasionally a few varieties of poultry (pigeons and fowls). These are the ordinary materials from which we seldom depart, and it is astonishing that so much success has, nevertheless, been obtained with these restricted means. The reason is probably that the animals just named are, generally speaking, easily susceptible to infection, whereas the dog, which was formerly almost exclusively employed, proved far less satisfactory, being insusceptible to the influence of most organized poisons.

Another cause of failure in numerous attempts at transmission has been the defective manner of conducting them.

It is by no means immaterial in what manner we apply the inoculating matter: we must here, too, endeavor to imitate the operations of nature so far as they are known to us. There are three ways in which micro-organisms usually penetrate into our bodies:

First, from the surface of the skin, generally after it has been injured in some way—i.e., through a wound, from whence the bacteria find their way into the blood or juices. It does not, indeed, always require such a special door of entry. The investigations of Garré, Schimmelbusch, Roth, Braunschweig, and others, have proved that the uninjured skin or mucous membrane is penetrable to infectious germs, and therefore it does not offer the unconditional protection often attributed to it.

Second, the digestive canal, into which the bacteria pass along with the food. Many, it is true, cannot pass through the stomach in their usual form, being destroyed by the action of its acid contents. Other kinds are less sensitive, and when spores are present, or when disease has altered the character of the digestive fluid and weakened its bacteria-killing power, there is no further obstacle to the passage of the parasites.

Third, the respiratory organs can afford entrance to the bacteria. Although the body has arrangements for excluding and repelling foreign matters, especially in the upper part of the respiratory tract, yet they are only able to perform this office within certain limits.

Numerous experiments, especially those of H. Buchner, leave no room for doubt that bacterial germs may be received by the organs of respiration, may be inhaled, and settling on the uninjured surface of the mucous membrane of the lung, may there produce a general infection of the organism.

Our artificial so-called methods of infection have been formed after these natural models.

The first of them is simple inoculation. By this we understand a slight lesion of the cutis, to which the virus is applied, and whence it is distributed over the body, chiefly by the circulating fluids. It is rather difficult to perform this in the case of mice; in the ear, however, it is sometimes possible to effect so slight a cut into the outside skin that this alone is affected without injuring the tissue beneath it.

More frequent is the use of subcutaneous application. Here the subcutaneous cellular tissue is the place of deposit for the micro-organisms, and the currents of the blood aid in their further distribution. With mice an incision is made above the root of the tail, the skin being carefully separated from the subcutaneous tissue by means of a scalpel or lancet-shaped needle, and thus a small cavity formed, into which the inoculating material may be introduced with a platinum wire. Or a portion of the skin on the back is undermined with the scissors or forceps, and in this cavity the silk thread charged with bacteria, the portion of tissue from another animal, etc., is placed.

With Guinea-pigs the abdominal region is more commonly selected. The hair is removed at the spot chosen, a portion of skin is nipped and raised with the forceps, a transverse cut is made with the scissors, and a pouch is formed with the blunt-pointed scissor-blade. Care must be taken to cut through the muscular layer, which lies immediately beneath the skin, since otherwise the matter may fail to reach the inner tissues. With rabbits one proceeds similarly, but any method is applicable and suitable by which an entry is effected to the subcutaneous tissues without serious injury to the animal. The operation must always be performed so carefully that but little blood flows from the wound, otherwise there is danger of the inoculating material being washed away and rendered inefficacious.

Similar to the subcutaneous inoculation is a special kind of inoculation, the introduction of the material into the anterior chamber of the eye. It was first employed by Cohnheim and Salomonsen. It is extremely well calculated to yield information as to the special symptoms in the course of a disease, and is therefore of great value. It is performed by treating the eyeball with cocaine, pressing it forward out of its socket, and then making an incision from above, at the junction of the cornea and sclera. Some of the aqueous humor flows out, and when this has taken place the virus is introduced through the incision.

Essentially different from simple inoculation and from subcutaneous application is the process by which bacteria are brought directly into the blood-vessels, and thus at once enabled to spread through the entire organism. For this purpose we endeavor to open one of the large veins and introduce the substance for inoculation into the circulating stream, by means of a syringe. We either expose the jugularis communis or externa, or, what is much easier, we take, in the case of larger animals such as rabbits, one of the veins of the ear, the best being the one that runs along the exterior edge of the ear. With the canula of the syringe we pierce the blood-vessel and inject the virus into it, either directly through the skin or, if necessary, after cutting through the surrounding tissue, and after having compressed the vein below the point selected for puncture, so as to make it swell as much as possible. The virus must of course be applied in the form of a liquid solution. It requires some practice. Beginners find that the elastic walls of the veins are very apt to slip under the hand, in which case the inoculating fluid is injected into the subcutaneous cellular tissue, as may be plainly seen by the appearance of a thick lump near the vein, which never occurs after a successful injection.

If we desire to give the virus a particularly wide distribution, we inject directly into the large cavities of the body. We pierce the thoracic or peritoneal cavity with the canula and inject the material. The danger of injuring the intestines or one of the larger vessels is not great, as these flexible parts usually slip out of the way of the syringe.

Yet we must never forget how serious a matter such an operation is, even in the best of cases. We know well, from human pathology, that the serous linings of the thoracic and abdominal cavities are exceedingly sensitive to injuries of all kinds, and therefore we cannot be too cautious in drawing conclusions from results obtained in this manner. Only in a restricted sense can we here speak of infectious processes, properly so called, since in the great majority of cases it will be a question of undoubted and immediate phenomena of poisoning. And what degree of value can be attached to experiments in which such tiny creatures as mice receive several degrees of fluid from a Pravaz syringe injected into the thorax? No wonder that they perish after it. The diminutiveness of all the proportions makes it almost impossible to inject the material into the pleural cavity alone. One penetrates also into the lungs, and indeed it is somewhat similar to injecting three or four litres of some liquid into the human respiratory organs by means of a fire-engine. It is high time that these mouse experiments should be confined within the limits of true usefulness.

If bacteria are to be brought into the organism by way of the digestive organs, they can be administered along with the food, or they may be brought direct into the stomach by the œsophageal probang, or into the intestines per anum. With rabbits the former process is easy; the catheter is put into the lateral gap between the teeth and cautiously pushed further; with Guinea-pigs the two rows of teeth must be held apart by a perforated wooden gag, and one must be particularly careful not to employ too much force, as otherwise the epiglottis is easily pushed aside and one penetrates into the lungs instead of the stomach.

If we wish to cause absorption of virus through the lungs, we should employ the inhalation method. The best process is that described by Buchner, which approaches more nearly the natural course of things. The material to be inoculated is diluted with sterilized water or bouillon poured into a spray apparatus and thence dispersed into the air. The spray thus produced is, however, so dense that the animals, if directly exposed to it, become dripping wet and are apt to swallow a considerable portion of the virus. Buchner, therefore, places the entire spray apparatus in a vessel of considerable size—for instance, in a Woulff's flask or a wide test-tube with a doubly-perforated India-rubber stopper, and conducts the exceedingly fine vapor which comes out, by means of a tube, into the closed box containing the animals to be experimented upon. The list of means at our disposal for conducting experiments of transmission has by no means been exhausted, but they will be found to depend, one and all, on the principles enumerated, and we may therefore be excused from going into all the possible details.

Of course the most careful observation of all those precautions of bacteriological manipulation which should in no case be neglected is doubly necessary here.

At the place where the animals are to be inoculated hairs

should be removed and the skin should be freed from accidental impurities with a $\frac{1}{10}$% solution of corrosive sublimate. The instruments must be reliably sterilized and should be cleaned again immediately after being used.

This is difficult only in the case of the syringes employed. Much ingenuity has been expended in the attempt to devise an instrument in which it shall be possible to sterilize the piston. Koch has cut the Gordian knot by dispensing with the piston altogether. His syringe is of glass and is, together with its needle, heated in the hot-air oven each time before being used. The best way is to put it into a wide test-tube. The pressure is exercised by a small India-rubber ball, which we fix upon the syringe, and which, as it does not come into direct contact with the inoculating fluid, requires no sterilizing.

Another instrument very suitable for our purposes is that recently invented by Stroschein. Two short tubes, resembling test-tubes, are placed one within the other in such a manner that the inner one, which is somewhat shorter and narrower, is held in connection with the outer one by a stiff, broad India-rubber ring. The inside tube has at one end a continuation upon which the needle fits, and at the other end a small aperture connecting it with the space between the inner and outer tube. If, with a twisting motion, the two tubes are drawn asunder as far as the India-rubber ring will allow, a partial vacuum is produced, and if during this time the needle be dipped into the injecting fluid, the latter will be sucked up into the inner tube. A pressure with the thumb now causes one tube to ride over the other. This little apparatus is easily constructed, cheap, easily sterilized, and very convenient, especially as it can be worked with one hand only.

CHAPTER V.

SPECIAL PART.

I. NON-PATHOGENIC BACTERIA.

Micrococcus Prodigiosus, Bacillus Indicus, Sarcinæ, Bacillus Megaterium, Potato Bacillus, Bacillus Subtilis, Root Bacillus; Micro-organisms in Milk, Bacillus Cyanogenus, Bacillus Violaceus, Bacillus Fluorescens, Phosphorescent Bacteria, Bacterium Termo, Bacillus Spinosus, Spirilla, Spirillum Rubrum, Spirillum Concentricum.

I. HAVING now made some acquaintance with the general characteristics of the bacteria as a class, let us pass over to the treatment of their particular species.

First of all, we will briefly consider some non-pathogenic micro-organisms; for though they possess less interest for us, there are some among them which deserve our attention, both on account of their wide distribution or frequent occurrence and on account of their special peculiarities.

If within this restricted circle we further confine ourselves to those bacteria which have been sufficiently studied and have become sufficiently well known to deserve special mention, our selection from among the great number of non-pathogenic micro-organisms which have been described of late will not be particularly extensive.

The mode of our investigations will always remain according to modern methods. In the first place, the hanging drop shows us the form and appearance of the micro-organism under examination; the same examination also shows us whether spontaneous movement is perceptible or not, and gives us information on the question of spore-formation. The cover-glass preparation perfects the first part of the examination, and settles the question of behavior under the influence of staining matters. The breeding begins from the glass plate; the peculiarities of the colony, which is the starting-point of the pure culture, already betray to us some of the relations of the bacterium under examination to the solid food media, specially to gelatin. Next comes the test-tube culture on different media—gelatin, agar-agar, and blood-serum; lastly, the potato culture will often be found useful. Let me here remark that in future, whenever we speak of gelatin, etc., we always mean a 10% meat-peptone gelatin and 1½% meat-peptone agar, unless some other proportions are specially mentioned.

The appearance of the colonies on the glass plate to the maximum of their development, as described, takes place in two or three days of twenty-four hours, at an ordinary room-temperature of 14° to 20° C.

Lastly, we shall consider any particular qualities that may happen to distinguish the bacterium under treatment, and thus collect all the particulars which contribute to give a full and exact picture of each individual species.

Micrococcus Prodigiosus.

The prodigiosus is found in nature sometimes on substances containing starch, on moist bread, boiled potatoes, and fresh-starched linen, sometimes on hard-boiled white of egg, on meat, in milk, etc.

Since its growth is accompanied by the development of a bright-red coloring matter, we cannot be surprised that it excited attention in early times, and that it is one of the first bacteria that was recognized as such. In times when people neither had nor could have any idea of the existence of such things as micro-organisms, the appearance of the prodigiosus led to the wildest suppositions. All the cases of miraculous, blood-covered bread, weeping hosts, etc., which are reported, may be safely referred to this bacterium, as may also those in which the reddening of bread was supposed to result from diseased corn, and the reddening of milk from a special disease of the cows. To determine the cause of the latter special commissioners were appointed, but it was some time before the real nature of the phenomenon was recognized. It was Ehrenberg who first gave a good description of the "Monas prodigiosus," which has since retained this specific name.

In consequence of its striking peculiarities, which make it recognizable at all times and without difficulty, the prodigiosus is a favorite for experimental purposes. In all laboratories where bacterial labors are carried on it is, therefore, to be found, and with it the beginner often makes his first modest attempt at bacterium culture.

The prodigiosus is not, properly speaking, a globular bacterium, but a short rod, for one diameter of its cells considerably exceeds the other. Chains of ten or more links are also occasionally observed, particularly when the process of segmentation takes place slowly, when slight obstacles to growth prevent a too speedy disintegration of the individual cells into new-formed elements. Thus in slightly acid food media which are not quite favorable to the development of the prodigiosus, Wasserzug and Kübler saw such strings, or chains, arise very frequently—a fact which, of course, removes the last doubt as to the fact that this micro-organism is a bacillus, and not a micrococcus.

The Micrococcus prodigiosus, in consequence of the facts just mentioned, has been robbed of the name it had so long borne and under which it had become known to science, and has been described as "bacillus" prodigiosus. It possesses the faculty of spontaneous movement. It is true that generally its movement is but little noticeable. It is best seen under the unfavorable conditions of growth just mentioned—for example, in certain substrata or in strongly-diluted fluid media. Schottelius explains this fact by saying that, as a rule, the cells are surrounded by a close, glassy mucus, which hinders locomotion and cements the individual organisms to each other. Only when the formation of this membrane is checked is spontaneous motion visible.

The existence of spores has not yet been observed. Nevertheless it is a remarkable fact that the prodigiosus retains its vitality for a long time in a dry state. A certain amount of moisture is doubtless necessary for its growth; yet if we transfer small portions of a potato culture to a silk thread, or lay them between sheets of blotting-paper, we can from this seed provoke a rich growth on a fresh food medium, even after several months.

The Micrococcus prodigiosus does not thrive as well in the incubator as at ordinary room-temperature. It is so little sensitive to a want of oxygen that it may be reckoned among the semi-anaërobic species.

Differences in the formation and composition of the cells are not, as a rule, visible; yet occasionally, when the cover-glass preparations are treated with anilin stains, one may notice in the middle of the cells small shining gaps, which at the first glance might be taken for spores. They are, however, connected with the process of segmentation, and show the spot where fission is commencing.

On the gelatin plate the colonies of the M. prodigiosus have a different appearance, according as they lie in the interior of the mass or reach the surface of it. The former appear to the naked eye as small white points, while the microscope shows them as greenish-brown roundish masses, irregularly fringed at their edges. The surface colonies show the two chief peculiarities which the prodigiosus develops when it is in contact with the oxygen of the air; it liquefies the gelatin and, somewhat later, generates a special pigment which is at first pink, but afterward of a deep blood-red.

With the naked eye one sees, at first, only pale, saucer-like hollows in the culture medium, the bottom of which are occupied by whitish central masses of bacterial growth. Colonies which are further advanced show clearly the full red color, and the microscope also displays the central mass as granular and deep red, while the color appears paler or dark brown toward the borders. The solid gelatin does not everywhere present an equally definite border dividing it from the liquefied part; it often shows a marked waviness and has a collar-like edge. The great rapidity of growth which the prodigiosus develops on gelatin, as well as on all other artificial media, is particularly noticeable. In test-tube cultures, too, a very quick and equable liquefaction of the gelatin all along the puncture early makes its appearance; it soon becomes so extensive that it reaches the walls of the tube. The color develops at first only on the surface, but gradually sinks to the bottom in crumb-like portions and granules. As it is continually produced in the upper part, the whole culture at last appears saturated with color throughout its entire depth.

The pigment develops particularly well on agar-agar, and on the obliquely-hardened surface a massive deep-red covering forms, the color of which does not penetrate into the medium.

Blood-serum is liquefied by the prodigiosus, though less quickly than gelatin, and also shows the development of color. That it grows very rapidly on the potato, forming large blood-red blotches, is already known. Older cultures have a peculiar play of metallic color which strongly reminds us of the appearance of crystalline undissolved fuchsin.

The chief peculiarities of the prodigiosus are, therefore, the liquefaction of gelatin and the production of pigment. The former is caused by the action of a peculiar ferment, which dissolves glue and fibrin, which can be separated from the bacteria, and the qualities of which have recently been experimented upon by Fermi. As a rule, the softening of the stiff gelatin by the prodigiosus is, as already noted, very considerable. From this circumstance we may take the hint to prepare at least three or more dilutions, instead of the two ordinarily made, in order to get plates with well-distanced colonies.

Under some circumstances, the peptonizing power of the prodigiosus may be lost to a certain extent. Thus if it be cultivated for a long time in the acid solutions already mentioned and transplanted to ordinary food gelatin, the liquefaction at first displays itself within very narrow limits, but a return to the normal conditions of growth soon fully restores the interrupted power.

The formation of coloring matter is also capable of artificial attenuation. The M. prodigiosus, as we have said, can only thrive with difficulty at high temperatures, chiefly on account of its saprophytic nature. If it is compelled to grow for many generations at the temperature of the incubator, the red pigment disappears more and more, and it is even possible, as Schottelius has shown, to get perfectly colorless white cultures. Sour bouillon produces similar effects, but the loss of developing power is in both cases only temporary. Two or three transplantations to potato or agar-agar at ordinary temperature suffice to restore strength and beauty to the pigment and counteract the previous attenuation. The pigment is formed in the cells as a chromogenic substance, a white body, so that it cannot be seen in them. It is only outside the bacterium and in contact with the oxygen of the air that the color develops, appearing in clearly-visible granules and extending to its immediate neighborhood. Its formation, therefore, ceases as soon as the access of oxygen is diminished or made impossible, and in our colonies it is only developed at the surface.

The absence of light is without influence on the production of the pigment.

Nothing is known as to the precise chemical nature of the pigment. The fuchsin film already mentioned reminds one of the behavior of the anilin colors.

The coloring matter is insoluble in water, soluble in alcohol and ether; when treated with acids it becomes paler and light red; if it is then mixed with strong alkalies, such as ammonia, for instance, it regains its former appearance.

It is further to be noted that both in its growth on gelatin and on the potato the prodigiosus emits that penetrating and unmistakable odor of trimethylamin which is so characteristic of herring-brine. In milk the prodigiosus gradually causes the separation of casein, and gives to this food medium a deep red color, without causing further decomposition. It causes a solution of sugar to ferment, forming alcohol and carbonic acid.

The soluble products of the prodigiosus produce some effect on the bodies of animals. Grawitz and DeBary have shown that large quantities of prodigiosus culture are able to produce inflammatory symptoms; Roger discovered the interesting fact that animals which are otherwise not susceptible to malignant œdema may be infected by the bacteria of this disease if, at the same time, 1 or 2 c.cm. of a prodigiosus culture be injected; Pawlowsky, lastly, showed that rabbits can outlive an inoculation with anthrax if they are afterward treated with prodigiosus. In the one case the excretions of two different micro-organisms unite in common action; in the other they counteract each other and become harmless.

Bacillus Indicus.

A kind of bacterium which has many points of resemblance in common with the Micrococcus prodigiosus, which is, therefore, usually described immediately after it, is the Bacillus indicus. It derives its name from its having been found by Koch in the intestines of an Indian monkey when he was in India seeking the cause of cholera. He thought it worthy, on account of its striking peculiarities, to be introduced into the cultures of our laboratories.

The Bacillus indicus is small, slender, extremely lively in its movements, and has slightly-rounded corners. The presence of spores has not yet been clearly established, yet the indicus, like the prodigiosus, has the faculty of resisting for a long time various deleterious influences—especially that of desiccation—without having any recognizable permanent forms.

For instance, a small quantity of potato culture grown in India was laid between blotting-paper, and in this state sent to Germany in a letter. This letter on its way was subjected to all the measures employed by the sanitary police of the different countries through which it passed for disinfecting the mails coming from cholera districts. It was perforated and fumigated with chlorine and sulphur, according to the postal regulations, but the first experiments made at the Imperial Health Office in Berlin at once showed that the vitality of the bacteria had suffered no harm whatever from all these operations.

As might be expected from its origin and its character as a parasitic bacterium, the indicus thrives without difficulty at body temperature; yet it is also able to grow at lower temperature, and even develops some of its peculiarities better outside the incubator.

It is an optional anaërobe, capable of living without air, but thrives more luxuriantly when it has access to the atmosphere.

It takes the anilin stains readily.

On the gelatin plate it is distinguished, like the prodigiosus, by a remarkably rapid growth and a very vigorous liquefaction of the culture medium. In order to get isolated colonies suitable for close investigation, we must prepare four or even five different plates. On the last dilution small white specks below the surface, can be seen with the naked eye, while the surface colonies show liquefaction and appear as round depressions with gray-colored contents and well-defined edge. If we examine the plate under the microscope we see, in the former case, irregularly-formed greenish-brown granular masses, in the latter case (or surface colonies) grayish-yellow dense masses, finely and evenly granulated. The edge is fringed with short fibres; attentive observation shows a backward and forward motion, a lively swaying to and fro in the colonies, even with a low magnifying power.

When the development proceeds still further the gelatin becomes slightly red, which gradually deepens and at last becomes brick-red.

Similar appearances are seen in test-tube cultures. All along the puncture rapid liquefaction occurs; also accumulation of dense, flaky, grayish-white masses in the deeper layers and a delicately-folded, strongly-reddened film on the surface. Obliquely-hardened agar is covered over—in twenty-four hours in the incubator, in a few days at room temperature—with a shining crust, which after a time, as a rule, shows the brick-red color over its entire surface.

The best medium for the pigment development is the potato. Here a thick, greasy growth quickly appears and soon takes the faint red color.

Yet the pigment formation is a very capricious quality of the indicus, and is still more dependent on exterior influences than in the case of the prodigiosus. Even under ordinary circumstances the coloring sometimes fails to appear, or it appears in full strength only on one part of the culture, the rest remaining pale. The edges of potato or agar growths in particular are almost always white; in the incubator no pigment is developed, and even the immediate progeny of deep-red bacteria often show no tendency to color.

Like most of the parasitic species, the indicus possesses the quality of acting toxically on animals when administered in considerable quantity. Guinea-pigs and rabbits are killed by the injection of 20 c.cm. of a fresh bouillon culture into the peritoneal cavity. It is worthy of remark that death also takes place when the same quantity is introduced into the circulation by injection into a vein, but that in this case symptoms of intense inflammation of the intestinal mucous membrane are generally observed, which are sometimes even accompanied by the formation of ulcers.

SARCINA.

The bacteriological examination of the air has made us acquainted with a number of different micro-organisms, few of which, however, possess special interest, and it is hardly worth our while to consider them here, although many have been exactly described and their peculiarities carefully investigated.

We will only mention a few of the sarcinæ which occur in the air, because they display the peculiar cell-division characteristic of this class of bacteria, in which the segmentation proceeds equally in all directions.

Yellow Sarcina.

The yellow sarcina derives its name from the beautiful sulphur or lemon coloring matter which its cultures display.

The separate cells are colorless. They are somewhat large, globular, or slightly-flattened cells, always occurring in the well-known bale-like or packet-like arrangement. They stain readily, yet in the colored preparations the peculiar form of the bundle often becomes indistinguishable.

On the gelatin plate the colonies of the yellow sarcina grow but slowly. They appear under the microscope as roundish, slightly-granulated sulphur-colored masses.

In the test-tube cultures the yellow sarcina grows freely only at the surface, where it forms a moderate-sized yellowish accumulation, which is continued a short distance down into the puncture in the shape of clearly-isolated granules. A little lower down their size and frequency greatly diminish, and lower still no growth at all takes place. In older cultures a very slow and slight liquefaction of the culture medium is generally perceptible.

On oblique agar the yellow sarcina soon produces a thickish coating of a canary-yellow color.

On potatoes it grows slowly and very gradually produces small yellow spots and grains.

The yellow sarcina is a strictly aërobic bacterium; it also thrives in the incubator.

The white sarcina differs from the yellow one only in the absence of the coloring matter; no other differences have been remarked.

The orange sarcina is distinguished by its forming a golden-yellow pigment, as well as by its somewhat intense liquefaction of gelatin. Further, the separate colorless cells are perceptibly smaller than those of the yellow sarcina.

On the glass plate it grows in the form of round granular colonies with clearly-defined edges and orange-yellow color, which liquefy the gelatin if they are on the surface.

In the test-tube the gelatin is softened throughout the whole depth of the puncture, but principally in the upper portion, in which also the pigment is developed. In advanced cultures the great mass of bacteria has sunk and lies at the bottom, while the upper portion of the culture has become quite clear.

On agar the orange sarcina produces a very beautiful golden-yellow, shining crust; on potatoes it grows slowly, producing, however, its characteristic pigment.

It is also strictly aërobic, and does not thrive, or but very imperfectly, in the incubator.

Lastly, a red sarcina is, perhaps, worthy of mention, since it has been found by Menge that it can under some circumstances be the cause of a reddening of milk under natural conditions.

The sarcina in question is a rather large one, which on the gelatin plate slowly forms colonies of moderate extent. These gradually liquefy the culture medium to a very slight degree. They form an intense rose-colored pigment, which appears in scratch cultures on oblique agar, on potatoes with alkaline reaction, and particularly in sterilized milk. This latter becomes, at length, so strongly colored that one might fairly call it "red milk." In non-sterilized milk, which usually falls a speedy prey to lactic-acid fermentation, the red sarcina is unable to thrive.

It is a strictly aërobic micro-organism and ceases to grow almost entirely at the temperature of the incubator.

Bacillus Megaterium.

Bacillus megaterium, so named by De Bary, who first described it, and which is of interest for us because De Bary employed it in his important investigations concerning spore-formation and the sprouting of spores, is a species of bacteria.

Megaterium was first discovered, by pure accident, on the leaves of boiled cabbage, but it develops without difficulty on our ordinary food media.

It is clearly a rod, about three times as long as it is broad, of clumsy appearance, with strongly-rounded corners, frequently somewhat bent, so that it has also been called the "large comma bacillus."

Peculiar to it is the granulation of the cell-contents, which, unlike those of most other bacteria, do not appear evenly transparent and homogeneous, but covered with little granules and dark spots, showing differences in the state of contraction of the protoplasm for which we can give no explanation.

The Bacillus megaterium has a strong inclination to produce involution forms, to degenerate; it seems almost as if a continued nutrition with our usual food media did not agree with this micro-organism. The cells, originally so distinctly rod-shaped, swell and become deformed and assume irregularly-rounded forms; the cell-divisions, so clearly visible in the healthy forms, disappear, the contents become quite turbid, and one might be tempted to suppose that a new variety had arisen, were it not always easy to breed normal cells again from these monstrous and crippled forms by employing a more suitable culture medium.

This micro-organism possesses a marked inclination to form groups of cells; as a rule, two or more are found connected, but long threads are not unusual.

Its locomotion is not very lively; it usually moves in a peculiar crawling manner, which reminds one of the amœboids.

It often bears spores, and the alteration in the appearance of the cells before mentioned, the granulated arrangement, has been referred to the process of sporulation. Yet we must state that Ernst's method for displaying the sporogenic granules within the bodies of certain bacteria does not bear out this view. Either, therefore, Ernst's method must be defective, or these appearances have nothing to do with sporulation. It is certain, however, that when a cell begins to sporulate, this is indicated by a special arrangement and separation of its contents. A portion of these contents contracts and becomes denser, flows to a definite spot, increases its power of refraction, takes a definitely-circumscribed shape, becomes covered with a capsule of its own, and thus becomes a fully-developed spore. This spore is, in the case of Bacillus megaterium, about as long as the spore-bearing cell, but much narrower, and the latter does not alter its form during the whole process.

Later the ripe spore becomes free. When it, in turn, prepares for germination, the spore-membrane breaks at the middle, and the young bacillus at first drags about with it the remains of the ruptured membrane, in the form of a cap at either end.

Bacillus megaterium thrives best at a temperature of about 20° C., but it can also support the temperature of the incubator. It is strictly aërobic, and is quite dependent on oxygen for its existence. It takes the ordinary anilin stains well, yet the granulation of its cells often remains visible in the stained preparation, the separate granules appearing now darker, now lighter, than the rest of the cell-contents. The fully-formed spores are easily made conspicuous by the ordinary process of spore-staining.

On the gelatin plate Bacillus megaterium grows with moderate rapidity and requires a certain time for its full development.

At first the colonies in the mass of the gelatin appear to the naked eye as whitish dots. The microscope shows them as yellowish, somewhat irregular lumps, without any particular character. The surface colonies which have access to the oxygen of the air slowly liquefy the gelatin. As a rule, they then take a very remarkable typical appearance. The colonies have the shape of a kidney or crescent, and are peculiarly grained, looking like shagreen leather.

In the test-tube a liquefaction of gelatin is observable all along the puncture, but it is generally far greater in the upper portion, and extends but gradually to the lower part. The mass of bacterial growth sinks gradually to the bottom, and only a slight turgidity of the upper part indicates that remnants of the culture are still present there. A decided crust is never formed on the surface. Even in large masses the culture remains quite colorless. On obliquely-solidified agar Bacillus megaterium forms a dull white or light-gray covering, which can be easily separated from the medium below it.

On potatoes it grows as a thick, greasy, whitish-gray film, which is usually very rich in spores and involution forms.

Potato Bacillus.

In the preparation of potatoes I recommended a particularly careful cleansing and sterilization. I advised, too, that whenever this precaution is neglected a pollution of the food medium is always observed, proceeding from one particular and constantly-returning species of bacteria, whose germs are distinguished by a high degree of resisting power, and which, on account of its special relation to the potato, has been named the potato bacillus. There are, in fact, several micro-organisms comprised under this collective name and distinguishable from each other by slight differences; yet we are not called upon here to explain these differences, and shall only devote a few words of description to the commonest species.

It is found in the uppermost layers of cultivated land, from which it gets on to the surface of the potato. It is also found in the fæces of men and animals, in putrefying fluids from various sources, river-water, etc.

Its separate cells are small rods, with rounded corners, often united in twos, but seldom forming long threads. It has a lively power of locomotion.

On our usual culture media it thrives extremely well and generally produces spores. These are endogenous forms which, in the shape of shining oval bodies, almost fill the entire cell and display an exceedingly high degree of resisting power against exterior influences of all kinds. The most tenacious vitality as yet known in micro-organisms is met with in the class of potato bacilli. Globig found that the spores of one species were able to survive an exposure of more than five hours to a jet of steam at 100° C. The potato bacilli, as already stated, occurring frequently in substances containing albumin, and liable to putrefaction, we must be particularly careful when sterilizing, or when employing such substances for bacteriological purposes. A neglect of this precaution may lead to unpleasant results. Thus, for instance, the unfortunate and much-talked-about cancer bacillus proved to be nothing but a harmless species of potato bacillus, whose germs had not been exterminated in the blood-serum used as food medium.

The potato bacillus thrives at incubator temperature and belongs to the aërobia.

The bacilli readily take anilin stains; only when they are on the point of sporulating, some portions of the cells show themselves less sensible to the dyes. The fully-formed spores may be stained separately.

On the gelatin plate we see, first, yellowish-white roundish spots, slightly granulated and with somewhat irregular edges. When further advanced they liquefy the gelatin quickly and strongly. The colonies then appear as gray circular depressions. The microscope shows them as disc-like plates, with contents resembling a ball of string, and with a tender, bright, white, finely-patterned edge. Later on these distinguishing marks disappear and the entire colony lies then as an impenetrable solid mass, with fibrous outlines, in the midst of the liquefied gelatin surrounding it.

In the test-tube also the gelatin softens rather quickly, and much more energetically in the upper part of the puncture than in the lower. The liquefied gelatin remains turbid from the granulated, crumbly masses of the culture. On the surface a thin scum forms, which appears folded and has a dull gleam.

On agar the potato bacillus produces a thick, wrinkled crust, of a dull white color, which can be easily raised and removed from the surface on which it grows.

On potatoes the peculiar manner of growth is seen still more plainly. Here the bacillus speedily covers a whole slice with a film which at first appears white, then grayish, and lastly brown, and which displays numberless elegant foldings and windings, and often appears as if strewn over with white powder. If we endeavor to remove a portion of this moist layer with the platinum needle, we find that it is held together by the agglutination of the intimately-joined bacteria. Thus one can draw out threads a foot long, which are only held together by the shining, swollen envelopes of the separate rods.

It is remarkable that the surface of the potato itself, under the influence of the bacterial mass growing upon it, often takes a slightly-red, sometimes a decided red, color, which extends deep into the substance.

BACILLUS SUBTILIS.

The hay bacillus (Bac. subtilis—Ehrenberg) is one of the most widely distributed and frequently occurring of all bacteria. As its cells appear in the form of very large and clearly-recognizable rods, it of course early attracted attention and became a subject of study. F. Cohn observed the formation of spores in it, and a whole series of facts, which were afterward found applicable to the bacteria in general, were first noticed in connection with this particular species.

The spores of Bacillus subtilis are found in the air and in the water; the upper layers of the soil, the dust of dwelling-rooms, the fæces of men and animals, putrescent fluids, etc., all contain them in rich quantities. It has been called "hay bacillus," from its being regularly found in hay and in vegetable infusions of all kinds. If we cut some dry grass into small portions, put them into a flask, add a moderate quantity of distilled water, close the vessel with a pledget of cotton-wool, and boil for about a quarter of an hour, the greater part of the germs contained in it are destroyed. Only those of Bacillus subtilis, in consequence of their high power of resistance, remain alive and capable of development, and after two or three days a whitish covering forms on the surface of the fluid. This covering or film consists of a luxuriant growth of hay bacilli.

If we examine a small portion of it under the microscope, we see large rod-like cells, somewhat slender, about three times as long as they are broad, with slightly-rounded corners and perfectly homogeneous, bright, translucent contents. The hay bacillus has a strong proclivity to form groups. Isolated cells are seldom observed, and long threads crossing the whole field of the microscope are by no means rare. This is the immediate consequence of its energetic manner of growth. Attentive observers assert that a cell can, under favorable circumstances, divide and become two new ones by transverse segmentation within half an hour, and that this rate of increase can continue unabated till checked by exhaustion of the nutritient medium.

It has a considerable motile power, which is shown in a very peculiar way. The rods do not glide elegantly and smoothly through the fluid, but throw themselves, as it were, from side to side, and "waddle" across the field of the microscope. The Bacillus subtilis is one of the species in which the organs of motion have been distinctly seen as flagella at either end of the rods.

Under some circumstances, as yet but imperfectly understood, the Bacillus subtilis proceeds to sporulation. The appearance of the spore-bearing cell usually undergoes no change, yet the fully-formed spores, though considerably shorter than the mother cell, are often somewhat broader and thicker. They are formed in the middle of the rods. They are egg-shaped bodies, which have a very brightly-gleaming appearance and a high degree of resisting power. On silk threads they retain their vitality unimpaired for years. They outlive an exposure for above an hour to dry heat of 120° C., and are equally insensible to the influence of chemical substances. The process of sporulation is very peculiar in the Bacillus subtilis. The strong envelope of the spore bursts transversely at the middle, but instead of being completely severed, it remains undivided at one point. The young cell then escapes, as Prazmowski has observed, from the gaping aperture at right angles to the longitudinal axis of the spore. According to De Bary, the budding cell, when it has attained a certain length, makes an evolution of 90°, and protrudes at right angles to the rupture in the membrane. The expanding new cell, by the hindrance which the tough spore membrane opposes to its extension, is forced to turn toward the ruptured opening in the middle and to seek exit there.

The hay bacillus belongs to the strictly aërobic species: it can thrive between great extremes of temperature, from 10° to 45° C.; its optimum is about 30°, which is also the best temperature for sporulation, while the germination of the spores takes place between 30° and 40° C. The rods take the anilin stain, and its spores are particularly well adapted for double staining.

If we grow Bacillus subtilis on the glass plate, small white dots at first appear, which, under the microscope, appear as irregularly-rounded, green-shining, slightly-granulated masses. Yet the growth of this micro-organism is so energetic that this first stage is of short duration. Very soon the colonies expand rapidly and reach the surface of the gelatin, which they quickly and extensively liquefy, thus offering the true characteristic forms of hay-bacillus colonies. With the naked eye one sees the small pure cultures as saucer-like gray, translucent depressions in the gelatin; the larger ones also present a similar appearance. As the growth, beginning with the one original germ, proceeds with perfect regularity in all directions, the colonies always form an exact circle, and look as if stamped into the gelatin with a round punch. They are of a delicate grayish-white color, but in the middle at the deepest point they show a white spot, consisting of the first beginnings of the colony which have sunk to the bottom. The principal mass, consisting of gray, crumbly flakes, fills the rest of the space up to the sharp white edge which separates the solid from the liquefied gelatin. Frequently, too, a curious radiated "star-fish-like" arrangement of the mass is observable.

A much more striking picture is seen under the microscope. In the centre lies a small, grayish-yellow, dense mass. This is surrounded by an irregular entanglement of very thin fibres, which, by careful examination with the ordinary magnifying power employed for examining colonies (for example, Leitz 3, Eyepiece 2, or Bausch & Lomb ⅔", Ocular A), is seen to consist of separate rods whose spontaneous movement may be recognized and watched. The colony is surrounded, as it were, by a "corona radiata." The bacilli which occupy the foremost rank at the extreme edge always bore vertically into the still solid gelatin, and stand like an army with lances stretched out on every side.

Thus the colonies of hay bacillus have such a characteristic appearance that they may at once be recognized, and cannot be mistaken for those of any other species.

The same may be said of the hay bacillus test-tuoe cultures. In the gelatin the strong liquefaction of course strikes the attention. It takes place equally throughout the entire puncture. Very soon the chief mass of bacteria sinks down in whitish flakes, the portion of liquefied gelatin above them, which was for a time cloudy and turbid, becomes clear, but on the surface there forms a dense, dry, and brittle pellicle, looking as if composed of separate scales held together by rods which have become immobile and have coalesced into a zoöglœa.

On oblique agar the hay bacillus spreaas out as a wrinkled, whitish covering, arranged in regular folds and easily lifted from the food medium. In its appearance it has a strong resemblance to joints of a tape-worm. Blood-serum is quickly liquefied, also forming a pellicle with folds. On potatoes the hay bacillus thrives excellently; forming a white, cream-like covering, which, especially in older cultures, contains great numbers of spores and also yields involution forms of the rods.

This bacillus has no pathogenic qualities, and even large quantities of it may be introduced into the system with impunity. If spores of Bac. subtilis are introduced into the blood of an animal they are soon expelled again, being deposited in the liver and spleen, as has been proved by Wyssokowitsch. Here they may remain for months without exercising any influence on their surroundings and without themselves being altered—i.e., killed.

This negative behavior is worthy of remark, inasmuch as formerly, when it was not yet possible to distinguish the different species of bacteria by discriminative marks and signs, as we do at the present day, the hay bacillus was confounded with the anthrax bacillus, on account of a certain superficial resemblance in the form of their cells. It was intended to change anthrax bacilli into harmless hay bacilli, and, vice versa, to breed from the latter virulent anthrax, but nothing has been heard of the success of this magnificent intention.

Root-form Bacillus.

The root-shaped bacillus is a species which has taken its name from the appearance of its colonies on the gelatin plate. It is pretty common in river-water and well-water, and occurs almost everywhere in the upper layers of the soil in gardens and fields.

Its rods are about as long as those of Bac. subtilis, but thicker. The corners are very little rounded; the cell contents are homogeneous. The separate members have a tendency to remain together after segmentation, and thus we often find very long chains and threads. The root bacillus possesses a very slight power of spontaneous motion, and it requires very careful and repeated observation to be convinced of the change of place which the cells effect. Spores occur as large, egg-shaped, glittering bodies (growing in the middle of the cell), and, like those of Bac. megaterium, are peculiarly suited for double staining. The bacillus belongs to the strictly aërobic species, and thrives at ordinary temperatures and also in the incubator. It may be stained in the usual manner.

The forms of the colonies on the glass plate are very peculiar. At first they appear as whitish, cloudy spots, which, however, soon reach the surface, liquefy the gelatin, and then develop further in a peculiar manner. From the centre of the whitish-gray colony extending to a distance are the twisted and branched outshoots, which at the very first glance remind one of the tangled roots of a tree. From the chief diverging lines others proceed which cross one another at numerous points, and so produce an elegant and wide-spread network. The effect under the microscope is similar. A close tangle of yellowish-brown threads radiates from the centre outward on all sides; some larger threads may be easily distinguished, from which smaller ones proceed. Toward the edge of the colony some of these threads are often so thin and transparent, so curiously turned up and apparently ramified, that one might confound them with the mycelium of a mould fungus.

In gelatin culture the root bacillus offers a very peculiar appearance during the first few days. The puncture is surrounded by large numbers of those processes—those ramified tender offshoots, so that the whole looks like an inverted fir-tree. In the mean time liquefaction is going on at the surface; a dense, moist, gleaming white skin is formed, and under it a cloudy turbidness shows that here too there are large quantities of bacteria. Later on these sink to the bottom, and the culture resembles the appearance described in Bac. subtilis; at the top the crust, then the clear liquefied gelatin, and at the bottom the whitish, crumbly flakes. Yet the pellicle is distinctly different in appearance from that of Bac. subtilis.

On oblique agar the root bacillus produces—beginning from the inoculation scratch—a grayish-white moist layer, which quickly covers the whole surface. At first this covering also looks like roots, but later on there is nothing of this appearance to be seen, except at the edges, while the middle portion is occupied by a thick, even film.

On potatoes it grows as a greasy white coating. It is without pathogenic action, even when administered in large quantities.

Micro-organisms in Milk.

Unboiled milk is extremely rich in saprophytic micro-organisms of different kinds. The investigations of Lister and Meissner have shown that, like most animal secretions, it is indeed germ-free at the moment it leaves the body, but as soon as it is removed from the mammary gland and comes in contact with the air, with the human hand, or with non-sterilized vessels, bacteria find their way into it.

Milk being an excellent food for most of the known micro-organisms, we cannot be surprised that they soon begin to multiply very vigorously in it. If we put a few drops of milk, obtained some hours before without special measures of precaution, into gelatin, and spread the latter out on glass plates, a development of numerous and various forms soon takes place. Among the rest, there is always one of the hyphomycetes—Oidium lactis—whose colonies appear like white stars, or Chinese asters, dotted over the gelatin. Although this fungus permeates the milk in great masses and covers the cream on the surface with a tightly-cohering, velvety, dense coat, yet it seems to take no essential share in the important changes which the milk so quickly undergoes.

It is universally known that milk turns sour and curdles if it stands for a certain time. The slightly alkaline or amphoteric reaction of fresh unboiled milk disappears, and the albumin of the milk—the casein—is separated from it. This change, which is usually called lactic-acid fermentation, is the work of certain micro-organisms. Not one single species, but a whole series of different bacteria—as, for example, the potato bacillus, already described—possess the power of causing this decomposition of milk.

In the great majority of cases, however, the change is due to one and the same, regularly-occurring micro-organism—a bacillus carefully studied and precisely described by Hueppe, to which its discoverer has given the name of bacillus of lactic acid fermentation—Bacillus acidi lactici.

The rods are quite short and thick, scarcely longer than they are broad, generally united by twos, rarely in long chains. They are non-motile, but in consequence of their small size generally show the Brownian molecular movement very clearly. Sporulation has also been observed in them—small globular bodies with strong refractive power at the ends of cells and able to withstand high degrees of heat, thereby showing their importance as enduring forms.

The Bac. acid. lact. thrives at temperatures between 10° and 45° C.; it belongs to the semi-anaërobic species and is little sensitive to the want of oxygen.

The cells are stained with the usual anilin colors.

On the gelatin plate at first appear small white dots, afterward grayish-white, shining blotches like glazed porcelain, with transparent edges, which do not liquefy the gelatin. Under the microscope, one sees in the deeper-lying colonies small yellow masses, without any particular characteristics; the surface colonies, however, look like flat, extended leaves, with irregularly-jagged, very tender edges. In the middle they are yellowish, but the color fades toward the margin and shows rather more distinctly an elegant folded pattern.

In the test-tube the gelatin even of old cultures is not liquefied. At first the growth proceeds equally throughout the whole length of the puncture, forming a delicate layer composed of small isolated granules. Later the development at the surface becomes particularly luxuriant, displaying a moderately thick, dry, brittle covering, of a shining grayish-white, and which often breaks up into separate flakes. Elegant salt-crystals are nearly always separated from the gelatin, and hang down from the under surface of the bacterial coating, like little roots. This is a result of a change of reaction, brought about in the gelatin by the action of the Bac. acid. lact., causing the previously alkaline food medium to become distinctly acid.

On agar-agar the growth of Bac. acid. lact. offers nothing remarkable. On potatoes it grows as a brownish-yellow greasy covering.

The peculiar action of the bacillus is best seen if we place a small quantity of a pure culture in absolutely sterilized milk. The latter, it is true, is by no means easy to obtain. Milk often contains germs of very tenacious vitality—for example, those of the potato bacillus, which one can only destroy by adopting very thorough measures. One must therefore either keep the milk for several hours at 100° C. in Koch's steam generator, or employ fractional sterilization, warming it up to 60° C. on five or six successive days. The first process is the more reliable, and the changes which it produces in the composition of the milk, though sometimes important, are of no consequence in this case.

If we inoculate milk thus sterilized with the Bac. acid. lact., we soon perceive that the sugar of milk turns into lactic acid and carbonic acid, thereby causing an acid reaction.

This result then leads to the separation of the casein, which is here brought about by the lactic acid, just as it might be by any other acid; for example, acetic acid. At a suitable temperature, the best being from 35° to 40° C., the whole process is completed in eight or ten hours, and a homogeneous gelatinous coagulum, which we call "curds," has been formed.

While the bacillus itself, as already stated, belongs to those which can, under some circumstances, exist without air, it seems to require oxygen for the exercise of its decomposing powers, which it can only employ when air is obtainable.

According to the investigations of Grotenfelt, the sour-milk bacteria belong to those saprophytic species which, like many pathogenic micro-organisms when kept for a long time apart from the conditions of their natural existence, lose part of their specific power. Thus on our gelatin which contains no sugar they gradually lose more and more of their capacity for causing lactic acid fermentation, they become "attenuated," and this process can only be counteracted by returning them repeatedly to their natural food medium.

A chemical change, very different from the one just described, is brought about in milk by a special kind of bacterium, which has also been carefully studied by Hueppe.

Its rods are rather large, slender, and elegant, with rounded corners; they often occur in groups of two, but rarely form long chains. They have a very lively power of locomotion, and at a somewhat high temperature, say about 30° C., they produce spores in the middle of the rods. These are gleaming oval-shaped bodies which can be stained so as to distinguish them from the cell contents. The non-sporulating cells take the anilin colors without difficulty.

On the glass plate at first appear small white points which quickly gain the surface, liquefy the gelatin very energetically, and soon make the observation of individual colonies impossible. Under the microscope the deeper-lying colonies show yellow, lumpy masses. When the liquefaction of the gelatin begins, the edge of the little swarm of bacteria "unravels," and the colony soon has the appearance of a grayish-brown mass, evenly granulated.

In the test-tube the gelatin is quickly and extensively liquefied, the liquefaction proceeding as a rule pretty generally through the entire depth of the puncture, the gelatin being colored with a slight yellow tinge. On the surface a thin whitish-gray skin, with delicate wrinkles, is formed, but the chief mass of bacterial growth remains suspended in the liquefied gelatin as a thick opaque cloud.

On oblique agar the micro-organism grows as a light yellow, greasy coating.

If a portion of pure culture is put into sterilized milk, a series of chemical changes take place which are much favored by the temperature of the incubator and free access of oxygen.

First comes, without any perceptible change in the amphoteric reaction of the milk, a general coagulation of the casein. It sinks to the bottom in dense lumpy masses, and then begins at the end of about eight days to decompose. The separated albumin is changed into peptone and other products of decomposition, among which ammonia holds a prominent place. At the same time the milk becomes decidedly bitter to the taste.

From this fact Hueppe drew the conclusion that the rod-cells just described must be identical with the much-talked-of bacilli of butyric fermentation, to which the same or similar influences have been attributed and which are generally acknowledged to possess such powers.

There can, however, be no doubt that Hueppe's bacteria are not identical with those exciters of butyric fermentation which have become well known to us through the investigations of Pasteur, Fitz, van Tieghem, and particularly of Prazmowski. Although the observations of these men were made at a time when the advantages of a solid food medium were still unknown and when, consequently, some important means for testing and proving the results elicited were wanting, yet we know sufficient about the micro-organisms in question to be able to form a pretty comprehensive idea of them.

We know that they are large, broad rods with clearly-rounded corners and often growing out into long chains. They have a lively power of locomotion. They frequently, proceed to sporulate, and the spore-bearing cell always changes its previous form. It becomes inflated at the place where the spore forms, and the rods thus swollen take a spindle-shaped appearance; they take the form known as a clostridium. When the spores begin to germinate, the membrane bursts at one end of the oblong spore, and the germ becomes visible; the empty envelope of the young cell often remains hanging for a considerable time.

This Bacillus butyricus (also called Clostridium butyricum) belongs to the strictly anaërobic bacteria. It thrives only when completely cut off from oxygen and ceases to perform its functions the moment air gains access to it. This is also the reason why the attempts made to breed it artificially have not yet led to satisfactory results.

The Bac. butyricus has, under certain circumstances, the quality that parts of its cell-body become of a deep indigo-blue or blackish-violet color when they come in contact with an aqueous solution of iodine. This is particularly noticeable when the bacilli have grown on a medium rich in starch; young cells then become fully blue, while older ones only change their color at certain of their cross stripes. As this peculiar reaction resembles that of granulose, van Tieghem has taken occasion to describe the Bacillus butyricus as Bacillus amylo-bacter.

In solutions of starch, sugar, and lactic salts, Bac. butyricus produces large quantities of butyric acid, developing at the same time carbonic acid and hydrogen. It is the same in old milk, but the sugar of milk must have been previously fermented into lactic acid, for the butyric-acid bacilli cannot bring about this last change with their own unassisted powers. They are, however, able to slowly dissolve coagulated casein, and the importance of the rôle which they are thereby called to perform in the organic world seems to be even still more extensive. According to some observations, they are said to cause the putrid decomposition of vegetable substances when kept damp (for instance, in the case of the wet-rot of potatoes), and even to separate cellulose into its component parts.

The manifold action of the butyric-acid bacilli occurs at 40° C., and only with restricted admission of oxygen.

It hardly needs to be added that this species of bacterium has an immensely wide diffusion in nature. It is worthy of mention that traces of its existence have been found as far back as the coal period—at least van Tieghem has been able to recognize bacteria in thin sections of the roots of conifera from that age, which from their form he thought himself justified in pronouncing to be cells of Clostridium butyricum.

More exact investigations with regard to this so important species are certainly very desirable. Such investigations will, perhaps, show that just as the Bacillus acidi lactici is in most cases the exciter of lactic fermentation, so also Clostridium butyricum generally causes the formation of butyric acid, while here also quite a number of other micro-organisms possess the same power.

BACILLUS CYANOGENUS.

We have already seen that a special disease of cows was for a long time supposed to be the cause of that peculiar red color in the milk which is produced under some circumstances by Micrococcus prodigiosus, as also by Sarcina rosea, and other micro-organisms. A similar cause was formerly supposed to bring about that blueness of the milk which, especially in the summer months, is no rarity in our North German dairies. Bad food, damp meadows, etc., were often blamed as the more or less direct cause of its appearance.

The investigations of Fuchs and Neelsen proved, however, that this discoloration of the milk is due to the action of bacteria, and indeed of one definite species, which has therefore been called the blue-milk bacillus (Bac. cyanogenus).

They are rather slender rods, about twice or three times as long as they are broad, with slightly-rounded corners, which are often united in twos, but are scarcely ever found in more numerous groups. They have an extremely lively spontaneous movement, which continues for hours in a hanging drop. The formation of spores has been observed as early as the third day in milk, in gelatin, and still more favorably in a slimy decoction of marsh-mallow roots. Small, gleaming bodies are seen to arise at one end of the rod.

On the glass plate the naked eye perceives, first of all, small white dots in the mass, while at the surface thick bluish-white knobs, which shine like porcelain, swell up. Under the microscope both kinds appear as dark-brown solid discs, with a smooth, sharp edge. Toward the margin the strong coloring fades a little, but there is nowhere to be seen any delicate pattern-work in the coloring, nor even a distinct granulation of its contents.

In the test-tube cultures the blue-milk bacilli show a decided tendency to surface-growth. Thus the development generally fails entirely in the lower part of the puncture; higher up a thin white thread is formed, while the chief mass of bacterial growth spreads over the surface of the gelatin as a dirty-white or light-gray covering.

The gelatin is never liquefied, but it gradually takes a peculiar discoloration which proceeds from the culture, and by no means always shows itself in the same manner. The differences which are observed depend on the reaction of the gelatin. The less alkaline it is the more clearly is the blue color seen, and in slightly-acid gelatin the formation of coloring matter reaches its maximum.

Agar-agar also shows this discoloration, while the mass of bacteria itself is seen as a dirty-gray, moist, thickish covering in the immediate neighborhood of the puncture.

The growth on potatoes is very characteristic. The whole surface of the slice is quickly covered by a thick, fatty mass, which saturates the culture medium with pigment. Thus the whole potato appears discolored up to its very edge, varying from blackish-blue to yellowish-brown, according to its reaction.

If we transfer a small portion of such a pure culture into ordinary unboiled milk (skimmed milk being, as Heim has shown, the best for the purpose) some peculiar changes take place. First of all, a few blue spots appear on the surface; these coalesce, the milk becomes covered with a colored film, at the same time the casein coagulates and the fluid becomes distinctly acid. It is otherwise if sterilized milk be inoculated in the same way. Then the precipitation of casein does not take place, the alkaline or amphoteric reaction is not altered, and the discoloration is by no means so striking as in the first case, and instead of being "sky-blue," the milk appears slate-gray or dirty-violet. This shows clearly enough that the blue-milk bacilli are only able to produce the pigment without being concerned in any of the other changes which take place in the milk, and which are all caused by other organisms, two of which have already been noticed.

The coloring matter is formed outside the rod-cells at the expense of the food medium on which they thrive, and out of which they take what they require to form the pigment. The appearance of the latter is, therefore, dependent on the chemical qualities of the substratum, which explains its variable behavior. In milk it takes its origin in the casein and more closely resembles pure blue in its appearance the more distinctly acid the reaction of the liquid becomes, in consequence of the lactic-acid fermentation.

According to the investigations of Scholl, the Bacillus cyanogenus loses in long-continued, uninterrupted culture on our ordinary gelatin more or less of its power to form coloring matter. It becomes attenuated, like Bac. acid. lact., and then produces no pigment in sour milk, even under the most favorable circumstances.

Bacteria of Drinking-water.

Every one knows with what care, at the present day, we watch over the water which we drink, and how we endeavor by periodically-repeated examination to know at all times what it contains. These examinations are both chemical and bacteriological. The latter are intended to inform us as to the micro-organisms present in the water. The particulars as to the manner of conducting such examinations will be considered later. We will here only remark that they have made us acquainted with a number of bacteria which occur with a certain degree of regularity, and some of which are distinguished by striking peculiarities.

Although they are without any special importance, yet we will dwell briefly on some of the best-known species.

Bacillus Violaceus.

In river-water we sometimes find a bacterium which produces a beautiful violet pigment and which liquefies gelatin rather quickly.

This is Bacillus violaceus, a slender rod-cell about three times as long as broad, which is often found in moderately-long strings. It has a very lively spontaneous motion and forms spores in the middle of its cells.

On the glass plate its colonies at first appear to the naked eye like little air-bubbles inclosed in the gelatin. On closer examination it will be found that these bubbles are nothing but an even liquefaction of the gelatin, which proceeds also in a downward direction, and at the bottom of which the whitish culture lies. Under the microscope the colonies appear as small irregular heaps, with confused, fibrous, loose edges. The larger colonies which have already reached the surface show a circular, sharp border, with strong refractive power, which is, in fact, the line of demarcation between the liquefied and still solid gelatin. At the bottom of the former lies the granulated mass of the colony, which already shows the bluish-violet coloring matter. The further the development proceeds the more closely the pigment becomes visible to the naked eye.

In the test-tube this bacillus liquefies the gelatin in the shape of a funnel, and throughout the whole depth of the puncture. On the surface there is generally formed a bubble-like contraction, while the chief mass of the culture sinks to the bottom of the funnel-shaped liquefaction in small coiled-up bluish-white masses. On oblique agar a deep bluish-black coating forms, which shines as if lacquered. On potatoes the bacillus grows at a moderate rate and forms a bluish-black covering.

There is also a red-water bacillus. It is extremely mobile, and shoots hastily across the microscopic field in long threads. Its separate cells are about the same size as those of the violet bacillus.

On the glass plate the red-water bacillus appears in small, yellow, lumpy colonies, which soon become surrounded with a tender, transparent, collar-like border. Then the liquefaction of the gelatin commences, and the colony becomes an evenly-granulated mass. In the test-tube the secretion of a yellowish-red coloring matter occurs, together with the liquefaction of the gelatin. On the surface a thin, somewhat crumpled skin is formed, under this is a layer but slightly colored, and at the bottom the yellowish central mass of the culture, consisting of slimy threads. On agar a thin covering spreads out, which is clearly yellow in the middle, while the irregular edges are paler.

The best field for the development of the pigment peculiar to this species is the potato. Here the whole surface is quickly covered with a rusty-red or orange-yellow coating, which is scarcely to be found elsewhere in such perfection.

Fluorescent Bacteria.

Several of the bacteria which occur frequently in water produce on gelatin a green pigment, which under some circumstances is beautifully fluorescent, and the color of which reaches far into the culture medium. Two of them, differing in the forms of their colonies on the glass plate and in the appearance of the cultures in test-tubes, liquefy the gelatin, and two of them develop without liquefying it.

One of these latter is a small, fine bacillus without any power of spontaneous movement.

On the glass plate it forms large, iridescent colonies, with jagged edges, which, under the microscope, appear as light yellow, delicate disc-like plates, and display very fine, elegant, leaf-like markings. In the centre one often sees a darker speck from which the colony began. In the test-tube it grows almost entirely at the surface; the puncture remains sterile. A tender, circumscribed growth takes place with irregular edges, which permeates the gelatin for some distance with a beautifully gleaming, iridescent pigment.

The other non-liquefying bacterium is also a bacillus, but with the power of spontaneous movement, much larger than the one just treated of, and characterized by its forming large central spores which possess a peculiar, strongly-red gleam and glitter. This may, under some circumstances, be so striking that one is led to suspect at first sight that it has been stained with fuchsin. It has, therefore, been described as Bacillus erythrosporus.

On the glass plate and in the test-tube it grows like the previous species, only that its colonies do not show the same beautiful markings, and that some growth occurs in the puncture, at least in the upper portion of it. The production of the iridescent coloring matter proceeds just as in the last-mentioned species.

All these water-bacteria have certain qualities in common: they are all aërobic species, which are distinguished by particular sensitiveness to any want of oxygen; they will not thrive at high temperatures, which, of course, prevents them from developing any pathogenic qualities; lastly, they find in ordinary water, without any addition of nutriment, all the requisite conditions for their growth and propagation, and thus their unpretending nature may even, under some circumstances, enable them to live in water that has been repeatedly sterilized and distilled, and thus to subsist on a quantity of organic substance so small that it may be said to hardly exist at all if judged by our ordinary standard.

Phosphorescent Bacteria.

The water of the sea is likewise the home of a special group of micro-organisms, which have become known to us chiefly through the investigations of B. Fischer. They all possess the capacity of shining in the dark, they are phosphorescent, and in artificial cultures this peculiar phenomenon is readily observed.

We know at the present time three different species of bacteria which glow in the dark. One, called Bacillus phosphorescens by its discoverer, was found by Fischer in West Indian waters, and isolated by means of the usual glass-plate process. It is a rod-cell of medium size and with a lively power of spontaneous motion, and which grows without difficulty on our usual media—gelatin, agar, bouillon, etc. The gelatin is speedily liquefied, and for a considerable distance around, the appearance of the colonies on the glass plate and in the needle-point cultures is not characteristic.

As a genuine inhabitant of the tropics, this bacillus requires a comparatively high temperature; it cannot grow at all below 15° C., and its maximum of growth is seen at about 30° C., at which temperature the phosphorescence also appears very strikingly. The best place for observing this phenomenon is, however, the surface of boiled fishes. When inoculated with a small quantity of artificial culture, it is covered in a few hours with a fatty-looking bacterial coating, which in the dark emits a beautiful bluish-white light.

Fischer discovered another bacillus in the harbor of Kiel, which has a certain resemblance to this West Indian phosphorescent bacillus, and which he calls the "indigenous glow bacillus." Here we again have a motile bacillus whose cells are generally somewhat shorter than those of the species last described. It thrives on gelatin and agar. It liquefies the former, but more slowly and to a less extent than does its West Indian relative. The colonies soften the solid food medium very gradually. If they reach the surface, circular, bubble-like hollows appear, which look as if clearly stamped out with a punch, and at the bottom of which the yellowish mass of the culture lies, and at last the appearance of such a plate is similar to the violaceus, only that here the pigment is wanting and the growth is more tardy.

The slowly-softening test-tube culture, with its narrow funnel of liquefaction in the immediate vicinity of the puncture, and the culture on agar offer nothing peculiar.

In contrast to the West Indian bacillus, the indigenous one requires low temperatures, thriving best below 15° C.; it even belongs to those micro-organisms which Fischer found able to grow at a temperature below 0° C.

Down to this point, too, the cultures display phosphorescence, which, as in the West Indian species, is of a bluish-white color, and is developed best on the surface of boiled fishes.

The third species has by far the widest distribution of all the phosphorescent bacteria, and is, therefore, quite apt to come under notice in the course of our bacteriological investigations. As the phenomenon of phosphorescence also shows itself very markedly in this species, often continuing for months in our artificial cultures, and thus admitting of extended examination, we shall devote some space to the consideration of this micro-organism.

Take a number of fresh sea-fish which have not yet become dry on the outside—haddock or herring are the best—and keep them bewcen two plates at about 15° C. As a rule, at the end of twenty-four hours some of them show shining spots which soon increase in size, and often on the second day spread over the whole body. Later on, as the decomposition begins and progresses, the phosphorescence diminishes gradually in intensity and at last disappears altogether.

In this same manner other kinds of food (bread, raw beef, fat, etc.) become phosphorescent, and as investigations thus far made have shown, always in consequence of the presence of one and the same micro-organism which has been described by Fischer as Bacterium phosphorescens.

It is a short, thick rod-cell with rounded ends which frequently forms globe-like cells, when a rapid fission of the rods takes place, thus reminding us in its outward behavior of Micrococcus prodigiosus. Like this, its cells often cohere in twos or threes, and occasionally form long threads. We must mention, too, the regular and early occurrence of involution forms, which often take a very peculiar shape. It is not motile; the existence of spores has not yet been observed. Like the other phosphorescent bacteria, it can be stained without difficulty with the ordinary anilin colors.

On the gelatin plate its colonies develop at a moderate rate. They are small, white, and have a gleam like that of mother-of-pearl, never become larger than a pin's head, and never under any circumstances liquefy the food medium. Under the microscope they appear as small, roundish, yellowish-white drops with sharply-defined but irregular edges and granular contents, often arranged in several concentric layers.

In the test-tube culture the growth proceeds all along the puncture in white globular grains, yet far more luxuriant at the surface, where a grayish-white thin coating forms, which occasionally breaks up into separate flakes. In older cultures the culture medium becomes of a yellowish-brown color in the vicinity of the bacterial coating.

Scratch-cultures, on obliquely-hardened gelatin, show a thick growth, which is confined to the immediate vicinity of the inoculating scratch; also on oblique agar and potatoes the growth does not extend far beyond the inoculated portion.

The Bacterium phosphorescens does not thrive at the temperature of the incubator, and even at 30° C. its development is but slight. It thrives best at 15° or 25° C. Yet according to the investigations of Forster, Tilanus, and Fischer, it can, like the indigenous glow bacillus, live and increase even below 0° C.

It is a semi-anaërobic micro-organism, which, in a space deprived of oxygen, in an atmosphere of hydrogen, or even of carbonic acid, continues to grow. The development of phosphorescence is dependent on the access of air—i.e., on a process of oxidation, both in Bacillus phosphorescens and also in the two other glow bacilli. Therefore this phenomenon is seen only in the upper portion of puncture cultures, and disappears lower down; its occurrence is restricted to the same degrees of temperature which were mentioned as important for the development of the bacillus itself—i.e., between 0° and 25° C. While the West Indian and the indigenous bacilli show a bluish light, the glow of Bac. phosphorescens is greenish or greenish-white. It is, under some circumstances, so considerable that by the light of a gelatin plate or of one of Esmarch's rolled tubes the hands of a watch can be recognized on the dial-plate, and Fischer has even succeeded in photographing the shining cultures by their own light.

It is worthy of mention that cultures of Bacterium phosphorescens which once show this property retain it sometimes for months. In other cases it disappears much sooner, only lasting a few days during the commencement of the development and gradually dying away. This is specially noticeable when the bacteria have been kept a long time and through many generations on our ordinary gelatin. Here they fall a prey to natural attenuation, and at last lose their luminous power altogether. Yet there exists a means by which it can be restored at any moment, and which has never yet failed; this is the addition of 2 or 3% of common salt to the artificial food medium. On gelatin prepared in this manner the growth of the bacteria is particularly luxuriant, and the phosphorescence always appears clearly and strongly even if the very last traces of it had disappeared from previous cultures. Natural or artificial sea-water and the surface of boiled fishes are also an excellent field for the display of the luminous phenomenon. Contrary to the other two glow bacilli, the phosphorescence when transmitted to boiled fishes yields a light that is confined to the point of inoculation.

How the phosphorescence is produced is not yet known with certainty. As we know, Lehmann and Tollhausen attribute it to an intra-cellular process, saying that as from the protoplasm heat, carbonic acid, etc., are developed by molecular metabolism, so also in this case light is developed; and that this development of light is nevertheless only functional, as, for example, contraction is functional in the muscles, which always produce heat and CO_2; that under certain conditions (absence of oxygen) the phosphorescence can cease without any interruption to the growth of the bacteria, will be for future investigators to determine.

None of the glow bacteria possess pathogenic qualities.

Bacterium Termo.

If we examine a putrid solution of organic matter in a stained preparation or in a hanging drop, we find a surprisingly large number of micro-organisms, and we already know that putrefaction only takes place in consequence of bacterial action. In particular we will almost always find certain very mobile rod-cells of moderate size, which certainly stand in a very intimate relation to the decomposition of substances containing albumin.

Of what species are the bacilli which seem destined to perform such an important part in so essential a process?

As long as the microscope was the only means of gaining an insight into bacterial life, all these micro-organisms were considered as belonging to one single species, to which the name Bacterium termo was given. The name was given by Dujardin and Ehrenberg, and a precise description was supplied by F. Cohn.

According to the latter, the Bac. termo occurs in the form of rod-like cells, twice or three times as long as they are broad, often united in pairs, but rarely in long rows, and with lively powers of locomotion. In Cohn's nutritive solution, the composition of which has been given, they thrive very luxuriantly, the liquid becomes turbid, and a greenish, iridescent scum forms on its surface.

This is certainly a pretty full description, which satisfied the demands of bacteriological science so completely that Cohn himself regarded this species as clearly characteristic and well defined, calling it the "ferment of putrefaction, without which the latter could neither begin nor continue," and attributing to it an office of the very highest importance in the economy of nature.

But such conclusions cannot stand before the requirements of more recent times. They want precisely that proof which at the present day every bacteriological fact must be able to show. They have not passed the ordeal of pure culture and the application of modern methods of examination.

If these methods be employed upon the material which is found in the observation of putrefaction, the single Bacterium termo resolves itself into a number of the most different forms. A simple experiment will suffice to illustrate this, superficially at least. Take a drop of any putrefying liquid, put it into gelatin, and spread it over a glass plate; it will always be found that quite a number of different colonies appear, which, of course, must be carefully examined as to their appearance and behavior.

Moreover, the process of putrefaction is attributed, by a very high authority, chiefly to the action of strictly anaërobic bacteria, while the micro-organism described as Bacterium termo was certainly not anaërobic.

Bacterium termo is for us, therefore, an abstract idea, and does not apply to any one species. It should be banished from our nomenclature until some well-defined species is found that performs a particularly important and conspicuous part in the process of putrefaction, and to which we might then justly give the name "Bacterium termo" as a title of honor.

To arrive at such a result extensive investigations would be necessary, for hitherto the important question of putrefaction has remained an almost unploughed field of bacteriological research.

Hauser, it is true, had described, under the name of proteus, some species in which he believed he had discovered important exciters of decomposition in organic substances containing albumin. But the results which he has made known are not yet so fully established as to be regarded in the light of indisputable fact.

Of the three species described by him, the one which he calls Proteus vulgaris is interesting to us on account of the peculiar formation of its colonies.

It is a small, slightly-curved rod-cell, very mobile, which often occurs in pairs, seldom in long chains. It has a remarkable tendency to deviate from its normal form, and sometimes produces cells like those of the cocci, sometimes twisted threads which have hardly a trace of resemblance to the original shape—a phenomenon which justifies the denomination "proteus."

On the glass plate an extensive liquefaction of the gelatin soon appears, and with this remarkable disposition of the colonies: they lie as yellowish-brown, bristly heaps in the centre of the liquefied portion and appear to have bunches of hair along their margin. The liquefaction of the gelatin spreads over the medium in such curious intricate twistings and arabesques, and often produces such peculiar figures and patterns, that this micro-organism has been called the figure-forming bacillus. These intricate markings are pale and colorless, and look as if cut or scratched into the gelatin. They are caused by the dissolving power of the bacterial cells. If we make a cover-glass preparation of such a colony, we see that all these branches and outgrowings of the little culture come out as colored markings. A stronger magnification with oil immersion will show that the ramifications consist of separate rod-cells arranged in strings, which in their rapid growth yield such strange forms and figures.

It is true that it is not easy to get preparations in which all this is clearly displayed. The liquefaction of the gelatin is exceedingly rapid and spreads widely; it requires four or five dilutions to yield a plate not too crowded, and even then it is often difficult to select the right moment for observation.

In the test-tube culture the liquefaction of the gelatin proceeds evenly from the puncture and soon the entire contents of the tube are melted. At the surface a thick layer of whitish-gray, cloudy substance collects, but does not advance to the formation of a crust. Under it is a somewhat clearer fluid, and at the bottom lies the chief mass of the culture in flakes and crumpled portions. Agar is quickly covered over with a moist, shining, grayish-white thin coating. On potatoes a dirty-colored, fatty coating appears, which has no peculiarities worthy of note.

This bacillus thrives excellently in the incubator; on media rich in albumin it forms poisonous excretions, and hence has toxic qualities. If we introduce a considerable quantity (3 to 5 cm.) of such cultures into the peritoneal cavity of a rabbit or Guinea-pig, or if we inject it into the blood of these animals, they perish in a short time. A dissection often shows symptoms of acute peritonitis or of a well-pronounced inflammation of the intestinal mucous membrane.

The Proteus vulgaris is the species against which Foà and Bonome were able to grant immunity by means of a definite chemical substance. An increased power of resistance on the part of an animal against a strictly toxic micro-organism such as the proteus, can only be brought about by gradually accustoming the animal to the poison. Foà and Bonome proceeded from the *a priori* supposition that neurin was a chief product of the proteus. Their experiment succeeded; they were enabled, by the use of this substance, to bring about immunity against the effects of proteus cultures.

The proteus species (vulgaris, mirabilis, and Zenkeri) discovered by Hauser have since been increased by others (Proteus hominis and P. capsulatus) which have been particularly studied by the Italian investigators Bordini-Uffreduzzi and Banti, and which play a part in human diseases. It would lead us too far to enter into a particular description of them here. Suffice it to say that they seem to be inhabitants of the intestinal canal which are harmless under ordinary circumstances, and only under special conditions—for example, when the power of resistance in the tissues is diminished—penetrate into the latter, multiply in the blood, and so take a pathogenic character, which is clearly perceptible also in experiments performed on animals.

Bacillus Spinosus.

It is desirable that we should consider a certain harmless, non-pathogenic member of the important class of anaërobic bacteria, by the examination of which we may be enabled to study the peculiarities of these remarkable micro-organisms. We refer to the Bacillus spinosus of Lüderitz, which displays a very high degree of sensitiveness to the effects of atmospheric oxygen.

It may be remembered that the anaërobic bacteria are far more widely diffused in nature than one might at first suppose, and that the anaërobic micro-organisms are almost always to be found in water, in putrefying liquids, and in the upper layers of earth, especially garden earth. It is from the latter source too that we obtain the Bacillus spinosus. Put a little dry earth into a pouch in the abdominal wall of a Guinea-pig, and after about two days the animal generally dies from the effects of a pathogenic, anaërobic species which will be further considered in its proper place. This species, however, is never found alone. The material used for inoculation always contains the germs of several kinds of anaërobic bacteria, which multiply in the animal's body and adopt a parasitic way of life, but which, as far as we have yet been able to perceive, do not seem able themselves to play a pathogenic part.

Among these is the spinosus. It is a very large, moderately-thick rod-cell, with brisk spontaneous movement, which generally proceeds quickly to sporulation. The spore-formation takes place in the middle of the cell, and the spore-bearing bacillus often bulges somewhat at the place where the spore lies, producing forms which remind us of those of the clostridia. The cells take the anilin stains well; the spores may be brought into prominence by special staining in the usual way.

Our artificial food media receive, as we know, for the culture of anaërobic bacteria, an addition of 1% or 2% of grape-sugar, or some other reducing agent. If the spinosus be put into gelatin thus prepared, the inoculating material distributed in a thick layer, and the colonies allowed to develop, we will find that at ordinary room-temperature the growth will become clearly visible in two or three days. First in the original tube, later in the dilutions, small white dots appear, which quickly increase in size and liqufey the medium to a considerable extent. The appearance of well-isolated colonies is then very characteristic. Balls as large as a hemp-seed are formed, which shine in varying gray tints, showing a peculiar star-like arrangement in their interior, and being generally surrounded with numerous sharp prickle-like spurs. Under the microscope, one sees even with a low power a brisk motion in the liquefied portion, and sees the irregular, fringe-like border of the colony still more distinctly.

In the interior of the liquid balls, small bubbles of gas soon make their appearance, gradually grow larger, and at last even unite with each other. Hand in hand with this goes another phenomenon. While at first the growth was strictly confined to the lower two-thirds of the gelatin, and the upper portion, to a depth of several centimetres, remained quite barren, the culture now extends upward. The solid covering at the top of the gelatin becomes thinner and thinner, till only a few millimetres remain. The cause of this is that the gaseous excretions of the bacillus expel the air more and more, and with it the oxygen contained in the upper layer of the gelatin, and the bacterium thus with its own products paves the way for the conquest of new territory.

All these things may be observed exceedingly well in the puncture cultivation in gelatin with much grape-sugar. For a few days the development takes place only in the deeper layers. The neighborhood of the puncture is rapidly liquefied, the deeper the more completely. Numerous thorn-like processes push out into the still solid portion, and the culture at this stage looks, as Lüderitz says, like a hairy caterpillar. As time passes on, the caterpillar (the culture) creeps higher up, so that at length the entire gelatin is liquefied from bottom to top, and the test-tube filled with a grayish-white viscid mass, consisting of zoöglœa of the bacilli. If we insert a stout platinum needle into the culture and stir it up, great quantities of gas-bubbles rise, and the liquor fairly sparkles.

The gases which escape have a very unpleasant odor, reminding one of rotten cheese and onions combined. Nothing is as yet known with regard to the nature of these gases.

In grape-sugar agar the growth is also rapid, the food medium is usually burst open in numerous places by the great development of gas, and occasionally it is even forced out of the test-tube along with the pledget of cotton-wool.

This bacillus thrives at ordinary temperature and in the incubator. It forms spores at ordinary room-temperature. Gelatin cultures, therefore, retain their vitality for months and can be transplanted to a fresh food medium.

As already said, the spinosus does not possess any pathogenic qualities.

SPIRILLA.

All the micro-organisms as yet treated of belong, morphologically, to the globe and rod bacteria, and we have not yet made the acquaintance of a single screw-shaped bacillus. This is not because of their rare occurrence; on the contrary, they are very widely diffused in nature. For instance, if we let the blood of oxen or sheep stand for a considerable time, we will almost always find that in the upper parts of it extremely long, briskly motile, strong, well-formed spirilla are developed. Still oftener in the saliva of the mouth, in intestinal contents, etc., one meets with short, simply bent forms, which are commonly called vibriones, and which in cover-glass preparations only appear as bent bacilli. It is not till they stretch themselves and begin to grow that one perceives how, together with the bending, there is also a turning, a leaving of the surface, and how the supposed rod in its development does not follow its longitudinal axis alone (which would change it from the semicircular to the complete circular form), but grows in a screw-like or corkscrew-like fashion.

An exact study of all these forms is attended with some difficulties, inasmuch as the spirilla or vibriones almost without exception show very little inclination to develop on our artificial food media, while, as Weibel has found, highly-diluted food media are much more to their taste than the concentrated ones which we ordinarily employ.

With these latter we have, nevertheless, succeeded in breeding some members of the class of screw-shaped bacteria, and the investigator just mentioned has isolated quite a series of species found in the mucus of the nose, the coating of the tongue, and particularly too in canal mud.

Most of these appear only as short, bent rods or vibriones.

The regular, or at least frequent, occurrence of long screws, on the other hand, has hitherto been noted in only two micro-organisms of this group, viz., in the Spirillum rubrum, discovered by E. von Esmarch, and in the Spirillum concentricum, discovered by Kitasato.

The former was accidentally found in the bacteriological examination of the putrid body of a mouse. Whether it has any direct relation to the process of decomposition cannot as yet be stated absolutely.

These bacteria are tolerably thick, quite transparent, and pellucid, and show perfectly regular screw-like windings. The length varies extremely, according to the more or less favorable conditions of development, so that one may sometimes find spirilla of only three or four windings, sometimes of more than forty. They are very motile, and the shorter ones especially shoot quickly across the field with a twisting or boring motion, while the larger ones gradually grow less active, and at last quite lose their power of locomotion. The organ of locomotion is easily seen by the aid of Löffler's process as a wavy flagellum at either end of the screws.

The spirillum multiplies by means of transverse segmentation. A thread breaks up into several more or less equal portions, which again in their turn grow out into individuals of greater length.

This micro-organism has a marked inclination to produce involution forms. Older cultures in particular contain hardly a single individual which would fully answer to the above description. One sees only quite short threads often curiously dilated, or thickened at the end, which do not recover the normal form till they are transplanted to a fresh food medium.

The investigations of v. Esmarch leave it undecided whether the spirillum forms spores or not. In the unstained preparation one often notices that a number of screws have bright, sharply-defined spots in their interior, and these when treated with anilin stains, even if the latter are warmed and allowed a long time to act, remain as unstained gaps. Such spirilla possess a very considerable power of resistance against desiccation, and when preserved on silk threads are capable of propagation after eight weeks. On the other hand, they perish from high temperatures above 50° C. at once, and as a special staining of the already-mentioned spots in the manner of double spore-staining has hitherto always failed, we must reserve our judgment as to the sporulation or non-sporulation of the spirilla until further investigators have thrown more light on the question.

The spirillum thrives best at temperatures between 16° and about 40° C., while 37° suits it best. Yet the capacity of spontaneous movement can only be permanently retained at low temperatures. If the access of oxygen is restricted, moderate growth still continues.

The growth of the spirillum is extremely slow and tardy on all our food media. On the gelatin plate, for instance, the commencement of its development is not recognizable till after five days, and weeks pass before colonies visible to the naked eye are formed. When they appear they are seen as grayish-red rounded heaps as large as a pin's head, which never liquefy the gelatin and which appear under the microscope as finely-granulated, yellowish-red discs with smooth edges.

In the test-tube culture thickly-crowded, roundish grains gradually form along the puncture, which in the deeper parts take a beautiful wine-red color, while in the upper part, near the top of the puncture, they remain colorless. This fact stands in direct contrast to what is usually observed, viz., that the pigment-forming micro-organisms need oxygen for the development of their coloring matter, which therefore appears chiefly or even exclusively at the surface.

Only one other bacterium, the Bac. lactis erythrogenes, which has been specially studied by Hueppe and Grotenfelt, behaves in this respect like Spirillum rubrum. It may be that Hueppe's view is correct that such bacteria form their pigment as a direct excretion as "color ptomaine," quite independent of other influences and conditions.

On oblique agar and blood-serum the Spirillum rubrum produces a coating with well-defined edges and a moist, transparent appearance, which does not spread far from the inoculation scratch. It is at first whitish-gray and afterward rose-red in thickish layers.

The potato, too, forms a suitable medium, yet here the growth is tardy, and the deep red colonies do not become larger than a hemp-seed.

While the spirilla generally form only short, incomplete screws on all the solid media, in fluid nutritive solutions, specially in beef-bouillon and sterilized milk, we often find those long-drawn spirals with many coils which have already been mentioned.

The Sp. rubrum seems to possess no pathogenic qualities.

Spirillum Concentricum.

The Spirillum concentricum is found in putrefying ox-blood, and is perhaps identical with the thick screw-like bacteria which are usually met with in the same.

Morphologically it behaves in just the same way as Sp. rubrum: it forms short, imperfect screws on solid media, while in fluid ones it produces long screws with many turns with brisk spontaneous movement and provided with flagella. Like rubrum, too, it is inclined to degenerate into involution forms, and is without anything that can be clearly proved to be a spore.

Its growth takes place within tolerably wide limits as to temperature, yet it thrives better at ordinary room-temperature than in the incubator.

On the gelatin plate grayish-white, round colonies of middle size develop rather quickly. These are seen under the microscope as sharp-edged, slightly-granulated discs. If they reach the surface of the food medium, they often show a peculiar concentric arrangement, narrow transparent rings, alternating with broad opaque ones, so that the whole bears some resemblance to a cockade.

The gelatin is not liquefied.

In stab-culture the growth takes place more at the surface than in the depths. It is striking that the neighborhood of the puncture is gradually occupied more and more by the culture. The spirilla bore into the solid gelatin, and push forward through it; this behavior is particularly noticeable in scratch cultures on oblique gelatin. From a tough bacterial growth on the surface a cloudy-looking whitish-gray substance extends down to the very bottom.

The development on agar offers nothing remarkable; on potatoes no growth takes place.

As far as we know, the Spirillum concentricum is as destitute of pathogenic qualities as the Sp. rubrum.

This is all that will be said of the non-pathogenic bacteria, although the consideration of them has been incomplete in two respects.

On the one hand, there are many bacteria which have been examined by the new methods of research and found to be well-defined and separate species, but which have not been described because they do not possess for us sufficient importance to merit special attention.

On the other hand, there are other bacteria which occupy an important and prominent place in books on bacteriology, especially older ones, and might therefore seem to deserve notice. But all of them—Micrococcus ureæ, Bacterium aceti, Bacillus ulna, Ascococcus Billrothii, etc.—share the fate of Bacterium termo. They are so many denominations which will remain without value for us until the modern means of investigation shall have yielded more information about them and enabled us to say precisely what we understand by them.

Until that has been done such names have only a historical importance, and the writer must be excused if they are passed over in silence.

CHAPTER VI.

The Pathogenic Bacteria; Anthrax Bacillus; Bacillus of Malignant Œdema; Tubercle Bacillus; Lepra Bacillus; Syphilis Bacillus; Bacillus of Glanders; Asiatic Cholera Bacillus; Finkler-Prior's Vibrio; Deneke's Vibrio; Vibrio Metschnikoff (Gamaleia); Emmerich's Bacillus; Bacillus Typhosus; Spirillum of Relapsing Fever; Plasmodium Malariæ; Friedländer's Pneumococcus; Fraenkel's Pneumococcus; Diphtheria Bacillus; Bacillus of Rhinoscleroma; Pyogenic Bacteria; Staphylococcus Pyogenes Aureus; Staphylococcus Pyogenes Citreus; Streptococcus Pyogenes; Bacillus Pyocyaneus; Bacillus Pyocyaneus B. (Ernst); Gonococcus; Tetanus Bacillus; Bacteria of Septicæmia Hæmorrhagica; Bacillus of Hog Erysipelas; Mice Septicæmia Bacillus; Micrococcus Tetragenus.

THE PATHOGENIC BACTERIA.

WITH regard to the strictly pathogenic infectious bacteria, a careful selection, a conscientious sifting of the very extensive mass of material which our young science pours in upon us in daily-increasing quantity, is far more requisite than with the harmless species. It has been recognized that the facts discovered by modern bacteriology are of extreme importance in understanding diseased conditions, and consequently a flood of enthusiastic, active research has poured over the newly-opened field of investigation.

The desire to make discoveries has been great, and there is now scarcely a malady which can be supposed due to parasites which has not been referred to some particular micro-organism. Most of them, it is true, will not long retain the position to which they have been raised, and this whole system of jumping at conclusions would be harmless enough did it not offer a serious danger for the further development of our science. If false ideas of the progress already made gain credit, we lose the clear perception of what still remains to be done.

Those things will be treated of which will bear a strict criticism, and only well-founded facts and reliable observation will be considered, and therefore the limits of our present knowledge may appear somewhat narrower, perhaps, than expected.

It need not be repeated that we recognize a bacterium as the undoubted exciter of morbid affection only when it is proved to be regularly present in that affection; when, further, it can be cultivated outside the organism; and when, lastly, it is able to reproduce the same pathological effects when its artificial cultures are inoculated. There are, it is true, but few bacteria that are in the happy position to fulfil all these requirements. With one species this link, with another that link, is wanting in the chain of irrefragable proof; but it has already been stated that, supported by other experiences and known facts, we at the present day are in some cases justified in calling a certainty that which, properly speaking, is only a very high degree of probability.

As to the order in which we take up the different pathogenic species, it will be an advantage if we bring together some smaller groups whose members offer certain points of resemblance and can be regarded from a common point of view. The order of the members within such a group is a matter of indifference.

I. ANTHRAX BACILLUS.

Anthrax is one of the most widely-spread and most destructive diseases which attack cattle of all kinds, and it is not infrequently communicated from them to mankind. In its appearance and progress it offers so many peculiarities that investigators long since turned their attention to it and sought to discover its exciting causes. The discovery of the anthrax bacillus was the first step into a world till then almost unknown.

In 1849 Pollender saw rod-like forms in the blood of cattle affected with anthrax, and shortly afterward Brauell, quite independent of the first-named examiner, noticed the same thing. Both recognized the vegetable nature of these forms, perceived that they were foreign to the animal body, but both failed to comprehend their true importance.

It is to Davaine, who published his celebrated observations in 1863, that the most credit is due for the elucidation of this subject. He was the first to assert definitely that the rod-cells were the cause of the disease, and if not able to prove the truth of his assertion, at least to make it extremely probable by a series of admirable investigations. He showed that the bacteria are a constant accompaniment of anthrax, and completed his discovery by a very convincing experiment, which had been tried in a similar manner by Brauell before him. If he inoculated healthy animals with the blood of sheep affected with anthrax after the bacteria had been separated from it, they bore it without any injury; but they perished without exception when the inoculated blood contained the rod-cells. This experiment has since been repeatea times without number, and always with the same issue. It may here be mentioned, too, that Pasteur, after the method of Klebs and Tiegel by filtration, arrived at similar results.

The great progress which followed in our knowledge of the anthrax poison—that is to say, of the bacillus and its peculiar qualities—we owe, in the first place and chiefly, to the investigations of R. Koch, whose brilliant career of general bacteriological research commences here. He saw the development of the cells, the spores in the interior of the cells, and thereby correctly classified anthrax bacilli. What is of still more importance, he succeeded in breeding the rods artificially outside the animal organisms, and transmitted them successfully to susceptible animals, thus establishing an irrefragable proof of their specific importance.

Since then the study of the anthrax bacillus has not rested for a moment, and has been the means of adding many useful facts to our stock of knowledge. It will therefore be apparent that the anthrax bacillus, the best known of all the bacteria whatever, which indeed serves as a sort of paradigm for all other pathogenic micro-organisms, deserves to be treated very thoroughly and carefully.

The anthrax bacillus (Bacillus anthracis, the "Bactéridie du charbon" of the French) appears, when taken from a young culture made from the blood of an animal which has perished of anthrax, as a large, evenly-pellucid rod-cell with slightly rounded ends. The separate cells are of various lengths, but are generally somewhat shorter and considerably narrower than a human red blood-corpuscle.

The cells are perfectly motionless under all circumstances. The slight trembling and waving of the rods which is sometimes observed is always the result of small currents in the surrounding fluid.

If we place such bacilli in a drop of our ordinary food bouillon and inclose the drop in a hollowed slide, we may, under the microscope, watch the entire further development of the bacteria step by step. At a somewhat high temperature the rods begin to grow by continuous transverse segmentation and the formation of new cells, and to extend out into the culture medium. After twelve to twenty-four hours the short cells have become long threads, which stretch across the entire field, and only here and there, by a slight constriction or bending of the thread, show that they really consist of short pieces joined together.

The transparent nature of the protoplasm has now disappeared. The cells look clouded and, as it were, granulated. They contain a great number of very small grains and dots, which are irregularly distributed over the bacterial body and which appear sometimes quite dark and opaque, sometimes with a gleam peculiarly their own.

Twelve hours more, and the picture is once more transformed. The number of threads has still further increased; they now fill the whole preparation in dense masses and generally show an arrangement which is characteristic of the anthrax bacillus. They twist in numerous windings round each other, and thus sometimes form very remarkable groups which might remind one of the appearance of a Chinaman's pigtail, or a ship's cable with the twisted threads that compose it.

The granulation has in the mean time become more distinct, and now at a suitable temperature usually develops into spore-formation. The small aggregations of more solid protoplasm flow gradually toward the middle of the cell and unite there to form a large, strongly-refracting body, which gleams forth out of a darker background as a bright spot with somewhat irregular boundaries. This body increases in brilliancy, its form becomes well defined, it becomes surrounded with a capsule, which may be recognized as a sharp contour, and the *spore* is now fully formed. The fruit-bearing cell has not altered its appearance during this process, and the spore lies, an egg-shaped, bright, shining thing, in the middle of the cell, than which it is considerably shorter, although about equally broad. If the sporulation extends at the same time to all the members of a thread, it yields a particularly beautiful sight. Like pearls on a string, the gleaming little balls lie at regular distances from each other.

Presently the transparent remainder of the cell-contents which was not employed to form the spore dissolves and vanishes—the spore is free.

If the latter finds its way into fresh nutritive solutions, it begins to sprout and becomes a rod-cell. In order to watch this process under the microscope the spores should be transferred into a hanging drop of nutrient gelatin, or still better, food agar. The drop quickly hardens and imbeds the separate spores which it contains so firmly that they cannot get away from the spot, and can thus be subjected to an uninterrupted examination. We then see that the spore begins to lose its gleam, stretches itself longitudinally till the tough spore membrane breaks at one pole, and the young rod-cell makes its appearance at the opening. It stretches itself in the direction of the longitudinal axis of the spore, pushes off the membrane completely, and so brings the simple course of development of an anthrax bacterium to its conclusion; from the bacillus to the spore, and from the spore round again to the bacillus.

It is true that the process is not always completed as above described. If the conditions of growth and nourishment are less favorable, the rod-cells become atrophied. The anthrax bacillus has a decided inclination to produce involution forms, which generally appear as lumpy, swollen, irregular bodies.

The healthy development of the bacilli depends on the proper nourishment, of which we shall speak presently, and also on proper temperature and atmospheric conditions. Under 16° C. this bacillus cannot live, and the upper limit is about 45° C.; the optimum of temperature is at about 37° C., yet 30° C. suffices to enable it to develop very luxuriantly and perfectly. In our artificial cultures Bacillus anthracis shows itself very sensitive to a want of oxygen. A slight diminution of this gas is sufficient to check its growth, while in the living body other conditions prevail.

The occurrence of sporulation depends on the fulfilment of special conditions.

It takes place only within certain limits of temperature, in particular not under 24° to 26° C.; therefore as a rule not in our ordinary gelatin, which, as we know, cannot be exposed to such a temperature without losing its solid character. Further, there must be an unchecked access to oxygen. This explains the fact that the anthrax bacillus forms no spores in the living bodies of animals or in their uninjured dead bodies, and that in fluids of some depth when the bacteria sink to the bottom sporulation hardly ever takes place. The best fields for the development of spores are, therefore, the surfaces of our solid media, agar and potatoes, and then shallow vessels of bouillon or highly-diluted human urine, etc., which in all parts are accessible to the air.

These facts point also to the cause of the fruit formation. The formerly often-mentioned exhaustion of the food medium has certainly very little to do with it, for there is no difficulty in bringing about an extensive and rapid sporulation in a rich food solution, which would nourish a thousand times more bacteria than it contains. It seems, on the other hand, as already mentioned, as if the occurrence of sporulation was the expression for the culminating point of growth in the anthrax bacillus as well as in other plants, and as if it showed an extremely perfect stage of development.

It is remarkable that the anthrax bacilli in some circumstances permanently and completely lose the power of producing spores. Lehmann, Heim, Buchner, and particularly Behring, have described these sporeless varieties, which differ in no other respect from the normal bacilli. The cause of this striking phenomenon is to be sought in the action of deleterious influences, which rob the protoplasm of a power which it had previously possessed. Thus, according to Roux, one can with certainty obtain permanently sporeless bacilli by breeding them for some time in a food fluid containing just so much carbolic acid—about 1:1,000—as not to completely stop the growth of the bacteria.

The anthrax spores are of great importance to us for many reasons. The bacilli themselves are comparatively tender organisms, which at high temperatures (above 60° C.) quickly perish, and which are also so little able to bear desiccation that even in tolerably large portions of tissue they can at most retain their vitality for a few weeks. The spores, on the other hand, possess all the qualities of a genuine permanent form and are able to preserve the species against all hostile influences. It is with difficulty that we can destroy them by chemical or physical means, and of the forces which nature can bring to bear against them, we know of none that could harm them with the one exception of sunlight.

In consequence of its high power of resistance, the anthrax spore has become the most favored of all test-objects for experiments in disinfection. We require of a means for disinfection that it shall with certainty destroy the source of infection which offers the greatest resistance to our destructive efforts, and that source is certainly the anthrax spore. We test a disinfecting process by exposing to its action silk threads on which spores have been dried. The process has not passed the test till the threads can be afterward placed on a fresh culture medium without causing a development of bacteria.

In all such tests we must take a phenomenon into account which has been treated of at length by E. von Esmarch, namely, that there exists a striking difference in the power of resistance of spores obtained from different sources. The considerable differences to be observed in the behavior of different bacteria of the same species as regards their external influences have already been noticed, and this is a particularly striking proof of the correctness of that assertion. It is, therefore, well to always examine spore-threads and ascertain the degree of their power of resistance before employing them as tests.

In the preparation of spore-threads, it is best to take the material from the surface of good agar cultures. In the incubator they usually form abundant spores in twenty-four to thirty-six hours. When convinced of their existence by means of a hanging drop, then scrape the coating from the culture medium with a sterilized scalpel or platinum hook, and mix it with sterilized water in a dish until a homogeneous opaque, grayish-white liquid mass is formed. A number of pieces of silk thread about ½ cm. long have previously been sterilized in a test-tube with dry or moist heat. These are now laid in the spore-solution, well stirred up with it, and then laid out in rows on a sterilized sheet of glass. A bell-glass protects them from impurities in the air, and in a few hours they are completely dry and can be taken up with the forceps and preserved in a safe place. They remain efficacious and reliable for years.

We have stated that the anthrax bacilli appear in the hanging drop as rod-cells of perfectly uniform transparency and with slightly-rounded ends. It is otherwise, however, when we employ staining solutions. As a rule, it is true, the cells take the color equally in all parts. Bacilli taken from a fresh culture have, in a stained cover-glass preparation, a strong resemblance to other large rod-cells, for example to those of the hay bacillus, only that they show decided corners and do not appear pointed at the ends like the hay bacillus.

But if the bacteria are derived from the blood or tissue fluids of animals which have died of anthrax, a very peculiar behavior is noticeable in the staining. Occasionally a narrow central zone in the interior of the cell, running parallel to its longitudinal axis, proves particularly susceptible to the stain, and stands out as a dark mass from its paler surroundings, which look like its capsule or halo. This effect is obtained specially by quick staining with Ziehl's solution or carbol-methyl-blue, which would lead to the supposition that we have in this case a bacterial nucleus with its protoplasmic body.

In most cases, however, the appearance of the rod-cells in the stroke-culture is different. They show no difference between centre and circumference, but are striking from the very curious form of their ends, which are distinctly concave. Thus it happens that where two cells come together end to end, an oval free space is left. As one must regard a single cell not as a flat body, but as a rounded or cylindrical staff, the formation of the ends must be something like that of the top end of the radius where it joins the bone of the upper arm.

When a long row of rod-cells is seen united, one is reminded of a bamboo-cane, the thickenings and constrictions occurring at regular intervals bearing a certain resemblance to the joints of the cane.

What may be the cause of this peculiarity is difficult to decide. Perhaps the membrane (which, as in other bacteria as well as the anthrax bacillus, reaches its fullest development only in the living body and not in our cultures) may have something to do with it, or it may be that the hypothetical protoplasmic body contracts in this peculiar manner under the influence of the alcoholic staining solutions. It must be noted that all staining matters are not equally inclined to produce this "bamboo form." Bismarck-brown and methyl-blue do it best; stronger agents are apt to block up the interstices and efface the characteristic appearance.

It makes a difference which process of staining is employed. If we use Gram's method, for example, which is of easy application for the anthrax bacilli, we see nothing of the things just described. From contact with the iodine in iodide of potassium solution the rods frequently even cease to look homogeneous, they become granulated, seem to be composed of separ te grains, and altogether lose their normal appearance.

The peculiar form of the ends of the cells which appears under certain definite conditions is without doubt a characteristic mark of the anthrax bacillus. It may be distinguished with certainty from other bacilli by this same peculiarity, and is, therefore, in contrast to the great majority of other micro-organisms, distinguishable by its morphological qualities, its outward form, and by microscopical examination alone.

In the case of fruit-bearing cells, the spore remains unaffected by the usual staining process. The spore double-staining is, however, possible, although the spores of the anthrax bacillus take carbolized fuchsin much less readily than those of some species already considered; for instance, the hay bacillus or the Bacillus megaterium.

If we inoculate nutrient gelatin with anthrax germs and pour it out upon glass plates, the colonies develop in a few days. They appear at first to the naked eye as small white points, which increase in size at a moderate speed, reach the surface, and there begin to decompose the food medium. They then lie as white pellicles with jagged edges in the liquefied portion.

Under the microscope we see in the body of the gelatin finely-granulated, green, shining, roundish or egg-shaped accumulations, whose color passes gradually more and more into brown.

The larger surface colonies offer a very peculiar appearance. The centre is slightly depressed, of somewhat granulated structure (as far as can be judged through the thickness of the layer), and of a yellowish color. The edge is surrounded by a close tangle of fibres, which wind about like whip-lashes or twist round each other like the snakes on the head of the Medusa.

In watching the growth of the anthrax bacilli in a hanging drop of bouillon, the decided tendency of this species to produce zoöglœa rolled up into curly bundles was remarked, and the same thing is here once more presented.

It is well worth while to make print preparations of such surface colonies by pressing the cover-glass upon the colonies as already described. Even with a low magnifying power they have an extremely elegant, characteristic appearance with their polyp-like branches and continuations; and if we call in the aid of oil immersion, we perceive that the blue or red stained, curiously-interlaced markings of the print resolve themselves into unbroken rows of separate, closely-pressed rods, lying side by side in regular order like so many bricks.

It is very similar on the agar plate. The surface of this medium becomes covered in twenty-four hours with numerous colonies, intricately interlaced with the most peculiar arabesques, which frequently unite and run into each other like a fine-woven texture. After a short time the spore-formation also begins, and the interior of the small pure cultures then takes a granulated, shagreened appearance.

In the puncture or stab cultivation the anthrax bacillus also grows very characteristically. All along the puncture, fine white threads penetrating into the gelatin are seen, which as they advance branch out or unite with each other, and in some places appear like bunches of the finest bristles or prickles. The liquefaction of the culture medium advances slowly, beginning from the free surface. Here a thick, slimy, white layer appears, consisting of bacteria which gradually sink to the bottom by their own weight as the softening of the gelatin advances. The bacilli having no locomotive power cannot rise again, and so it happens that older cultures present a very peculiar appearance. The upper portion of the gelatin is completely liquefied, but clear and free from all admixture, and without any film on its surface; lower down, where the still solid part of the gelatin forms a definite limit, the closely-felted mass of the culture lies in whitish clouds and flakes; below these may generally be seen the remainder of the puncture still preserved with its small, elegant, thorn-like projections.

On the surface of the obliquely-hardened agar the anthrax bacillus grows as a grayish-white, tough coating which shines like tarnished silver, and may be peeled off in long strips.

Blood-serum is slowly liquefied; yet the bacterial growth here offers nothing worthy of remark.

This bacillus grows luxuriantly on boiled potatoes. It spreads out as a creamy-white, tolerably dry coating over the slice, and as a rule spores are formed in great number at a suitable temperature.

The list of substances able to furnish the anthrax bacilli the necessary conditions for undisturbed growth is not yet by any means exhausted. The most widely-differing substances, chiefly of a vegetable nature, it is true, are able to yield sufficient nutriment, and infusions of hay or pea-straw, seeds of all kinds which contain starch, flour, wheat especially, and also turnips, etc., satisfy the appetite of this bacillus, which is not very dainty.

Although after what has been said it may not seem sufficiently clear that the anthrax bacillus is not specially formed for a parasitic existence, yet this view will be corroborated by a very simple consideration. It will be remembered that we regarded the formation of spores as the expression of the culmination of development in a micro-organism; it is at any rate a very important, almost indispensable, part of its development. But the anthrax bacillus forms its spores only outside the bodies of animals, and we may, therefore, conclude that it is originally a genuine saprophytic species, which indeed can occasionally make an excursion into foreign regions, but must return to its proper home to obtain its maximum development.

For us, indeed, the parasitic existence of these bacilli is particularly interesting and important, since it is here that they display their pathogenic qualities. The anthrax bacillus is one of the most infectious species known to us. The very smallest portion of a healthy culture, suitably inoculated into a susceptible animal, suffices with certainty and under all circumstances to produce splenic fever, and as a rule to cause death.

The entrance of the bacteria into the body may take place by any of the ways which are open to the micro-organisms, and there is no definite mode of infection to which the anthrax bacillus is limited.

Thus it may enter through slight injuries to the skin, by inoculation, or by subcutaneous application. In this so-called wound anthrax or inoculation anthrax the bacteria spread chiefly by way of the blood; a simple cell may increase to many millions and so overrun the entire organism. The affection is characterized as a genuine septicæmia, and the post-mortem state of the body confirms this view.

After inoculation of a Guinea-pig with anthrax, the vicinity of the place of inoculation is almost unaltered as a rule. Only in rare cases do we find a considerable extravasation of blood or even gangrene near the point where the infection took place. The virus has not time with this kind of inoculation to produce circumscribed local effects. Marked œdema of the abdominal walls will be seen. The subcutaneous cellular tissue is gelatinous, partially infiltrated with blood, and shakes when touched, but *nowhere* (which should be specially noted) do gas-bubbles form. The surface of the muscular tissue is paler, soft and moist, and looks almost as if boiled. The spleen alone is greatly altered—a fact from which the affection takes its name of splenic fever. The spleen is considerably enlarged, of a dark color, soft, and at the same time brittle. The liver too shows some degree of enlargement, the lungs are pale red, the heart full of blood. Nothing further can be found worth noting.

Next comes the microscopic examination to which the blood and the tissue fluids must be submitted.

The bacilli, as we are aware, take all the different anilin stains well. The double-staining acording to Gram's system requires great caution, and to obtain good results we must carefully regulate the time during which we expose our objects to the solution. The bacilli decolor readily and then take the contrasting stain. Not infrequently this takes place with regard to portions of a bacillus only; one sees one end stained with the first color, the other end with the second color. Sometimes too, under the influence of the iodine, the contents of the cell contract into regular, round grains, so that the bacilli are scarcely recognizable and look more like a chain of micrococci.

The blood or tissue juice almost always contains very rich quantities of bacteria, which are but seldom seen in large aggregations and generally occur in groups of two to five members. The bacilli always lie between the blood-corpuscles or tissue-cells, never in them.

If we examine tissue hardened in alcohol, we find the chief masses of bacilli confined to the vessels.

The principal abode of the bacteria is in the capillary vessels where the passages are widest and the current is slowest, equally distant from the commencement of both arteries and veins, but only a few cells wander into the large vessels, while the capillaries sometimes look as if injected, completely filled out and stuffed full of the foreign intruders. The spleen is equally filled with them in all parts; in the liver they lie particularly in the middle between the capillaries of the veins and those of the vena porta. In the villi of the intestines they take possession of the ends, in the kidneys they fill the glomeruli. Under the pressure of the rapidly-increasing bacilli the capillary vessels sometimes burst, and let blood and bacilli escape at the above-mentioned places—i.e., in the glomeruli and in the villi of the intestines, and also in the mucous membrane of the stomach and in the salivary glands. This is, however, most frequent in the glomeruli. Many of them are burst and the rod-cells pass into the uriniferous tubules, yet they do not penetrate far. Koch says: "I have only found them in the first portion of the convoluted tube, in which they grow to long threads which become felted into each other. In the straight tubes, on the other hand, I have never met with the bacilli."

In a second class of cases the lungs form the door of entrance for the bacteria, and the virus is received in respiration. The particulars of this process we owe chiefly to the investigations of H. Buchner. He caused animals to inhale anthrax spores, and found that the majority of them perished from typical splenic fever. The germs had found a resting-place in the tissue of the lungs, had penetrated the uninjured surface of the alveolar mucous membrane, and then passed over from their first settlement outside the vessels into the blood and juices, where they increased in the manner already observed, and caused a general septicæmic infection of the whole organism. In the lungs themselves only slight alterations are visible in this kind of infection, scarcely sufficient to indicate that the entire affection took its commencement from this point.

But the matter is very different when, instead of spores, bacilli are introduced into the respiratory passages. While the former at first produce no striking effects, the latter (the bacilli) act on the tissue from the very beginning as a decided irritant, and thus a very violent pneumonia is easily produced. The pulmonary vesicles fill with exuded serous or sero-fibrinous fluid, in which large quantities of anthrax bacilli are balled together. The walls of the alveoli are also œdematous, but seldom become a habitation for the invaders. As a rule, no general infection follows the first local infection.

Lastly, the anthrax bacilli are also able to enter with the food and commence their attack from the intestinal canal. The bacilli, indeed, are always killed by the acid gastric juice, and the micro-organisms can reach the intestines only in the form of spores. In the alkaline contents of the intestines they find sufficient nourishment, and also a suitable temperature in which to germinate and increase. They adhere to the epithelial lining of the intestines and become attached to the villi, where they form complete coatings or dense masses. Then the epithelium is pushed aside, the cells penetrate into the deeper recesses, into the parenchyma of the mucous membrane, get into the lymph passages or direct into the blood-vessels, thus gaining the opportunity to exert their terrible action to its full extent.

The animals on which we usually experiment are generally non-susceptible to the intestinal or alimentary infection. It requires very large quantities of spores, for instance, to infect a Guinea-pig in this manner. When the experiment is successful, a hemorrhagically-infiltrated or even ulcerous spot in the intestinal mucous membrane shows the place where the intestinal infection took place. Sheep and cattle, on the other hand, are particularly susceptible to anthrax, and we shall shortly declare that this is the most usual way in which the infection takes place under natural circumstances.

In our transmissions we must, of course, always take note of these differences, and the susceptibilities of the various species of animals to one and the same kind of infection are important parts of the question.

It has been found that dogs, the majority of birds, and the amphibia are almost entirely refractory. If we place even a large quantity of anthrax culture or a portion of infected spleen under the skin of a frog (for example, in the dorsal lymph-sac) it remains unaffected by the attack, and at the place of inoculation will be found a rich store of *phagocytes*, i.e., of white blood-corpuscles, all filled with bacilli evidently on the high road to decay and dissolution. But if we keep our frog at a high temperature in the incubator and employ warmth as a means of favoring the growth of the bacteria, we may nevertheless succeed in bringing them to develop, even on a soil to which they are naturally so averse. The animal dies, and on examination is found to harbor numerous bacilli, mostly arranged in long, curiously-winding, and intricately-tangled threads.

Of other frequently-employed animals, the white rat is usually but little susceptible. The rabbit is very much more accessible; Guinea-pigs, sheep, and cattle are sure to yield to the infection, and at the top of the list stand the white mice, which form a never-failing object of experiment.

The pathogenic effects of bacteria, as stated, are referred to their producing certain peculiar excretions which are the true cause of disease, and are, therefore, the most important item in the whole process. In the case of some micro-organisms we have succeeded, as already noted, in defining these substances in a tangible form and fixing their nature.

The attempts made to reach this point as regards the anthrax bacilli have not as yet been entirely successful. Hoffa, indeed, has reported experiments in which he thinks he has nearly accomplished this; but his results are not sufficiently certain and indubitable to warrant a final decision as to the nature of the anthrax virus. More recent experiments have made it at least extremely probable that, in the case of the anthrax bacilli and other pathogenic bacteria, not only crystallizable substances of a basic character, but also some peculiar derivatives of the albuminoid substances, the so-called *toxalbumins*, play an important part in the pathogenic action.

We have special reasons for the opinion that the excretions are of decisive importance. We have already considered at some length the fact that it is possible by certain measures of an injurious nature to attenuate anthrax bacilli and rob them of their infectious qualities. The best means for this purpose is heat—the breeding of the bacilli at an unusually high temperature.

Toussaint, for instance, was able to render blood from splenic fever harmless by keeping it for ten minutes at 55° C.; Pasteur made use of low temperatures; Koch, Gaffky, and Löffler showed that a high temperature of about 42.6° is the most suitable for depriving the anthrax bacillus of its virulent properties.

One must, it is true, cultivate it for a considerable time at that temperature, and it is not fully harmless till the end of about twenty-four days. One takes a number of Erlenmeyer's flasks with food bouillon, inoculates them with bacteria in possession of their full virulence, and lets them stand about three weeks in the incubator at 42½° C. Cultures which before the end of this period are brought back into natural circumstances show at least a partial attenuation, and it is possible to preserve in a continuous series the intermediate degrees of virulence. For instance, one takes the first flask out of the incubator at ten days, finds by experiment on an animal that the virulence has diminished, and transfers at once to a culture medium which remains at ordinary temperature, in order to breed this variety of attenuated virus.

Behring has found that the attenuated bacilli form different excretions than the virulent ones. With the imperfect means at our command we have only been able to absolutely fix this difference in one point. We have found that the virulent rod-cells develop large quantities of acid, while the attenuated ones excrete substances rather alkaline than acid. The latter, too, have a power of raising the temperature which is wanting in the former. While the heat of an infected animal's body will rise several degrees when infected with the attenuated anthrax bacilli, the inoculation with virulent ones produces no such result, and shortly before death we even notice a very considerable decrease of temperature.

The question of attenuation has not only a theoretical, but also a practical interest, since it stands in immediate connection with the question of artificial immunity.

We are aware that sheep and cattle which it is intended to inoculate protectively on Pasteur's system, are first infected with a weaker material (premier vaccin) and after some days with the stronger material (deuxième vaccin). As a rule, a more or less violent inoculation fever ensues; when this is over, one can inoculate the animals with material of full virulence without its harming them. Against intestinal anthrax they are, however, not perfectly safe, as has been shown by Koch, Gaffky, and Löffler, and this is one of the reasons which make it undesirable to unconditionally recommend the adoption of protective inoculation as a regular practice.

No other species of bacteria has been so carefully studied as the anthrax bacillus with regard to attenuation and immunity, separately as well as in their mutual relations to each other. For both, quite a number of other procedures have been recommended besides the one just mentioned, which, however, is the only one that has any importance for ordinary practical use.

According to Toussaint, virulent anthrax bacilli are rendered non-poisonous by adding 1% of carbolic acid to anthracic blood; Chamberland and Roux obtained the same result by breeding in food solutions containing from $\frac{1}{100}\%$ to $\frac{1}{60}\%$ of bichromate of potassium; Lubarsch and Petruschky found the attenuation took place when the bacilli are compelled to exist in the bodies of non-inoculable animals—of frogs, for instance; Chauveau discovered that an increased pressure of six or eight atmospheres produced the same effect; Arloing was able to diminish virulence by direct exposure to sunlight.

Chamberland and Roux, to obtain artificial immunity, employed sterilized cultures of virulent bacteria instead of attenuated bacteria; Hueppe and Wood performed a successful protective inoculation by means of a perfectly harmless species of bacteria, scarcely allied to the anthrax bacillus; Hankin obtained from anthrax cultures a substance which granted immunity—an albuminoid body of peculiar qualities; Wooldridge, lastly, took aqueous extracts rich in albumin, from the thymus gland and the parenchyma of the testicle of healthy animals, and with this obtained protection against anthrax.

This is the proper place to consider certain discoveries which are more or less connected with the question of immunity. Behring made the rat, which is naturally refractory to anthrax infection, susceptible by diminishing the alkalescence of its body; Emmerich, Pawlowsky, Bouchard, and Freudenreich repressed a commencing anthrax infection and cured it by introducing other micro-organisms, such as the erysipelas coccus, the Micrococcus prodigiosus, and the Bacillus pyocyaneus—facts which, as we have seen, confirm us in the belief that the effects are produced chiefly by the excretions of the bacteria.

By microscopic examination, by cultivation outside the animal body, and by successful transmission from artificial cultures, we know that the anthrax bacillus is the sole cause of anthrax disease. How are the peculiar phenomena of the disease to be explained as resulting from the properties and habits of the bacillus?

It was stated that splenic fever is one of the most wide-spread of all infections. In fact, there is scarcely any country where it is not known, while some are particularly subject to it. In France and Germany, Hungary and Russia, India and Persia it spreads its ravages every year among the most valuable cattle, and the number of its victims amounts to thousands; in Siberia it is such a terrible scourge that it has been taken for an evil peculiar to that country and called the Siberian pest. Only England and North America are comparatively free from it, and it only occurs there in isolated cases.

There are generally certain smaller districts notorious for its prevalence; such in Germany are the Upper Bavarian Alps, and in France, Auvergne, where it is said to have existed thousands of years, and whence, according to Pliny, it had spread, 300 years before his own time, into Italy.

It is at its maximum in the hot summer months from June to September; the coming of winter almost always causes a temporary cessation of it. The special influence of dry or damp or otherwise exceptional weather has not been actually observed.

Taken as a whole, the explanation of these facts is not difficult. We are aware that the bacillus is capable of living saprophytically and of finding suitable conditions of existence outside the bodies of animals. Where it finds these conditions in the greatest perfection, the disease will naturally break out most frequently and be most destructive. That this should generally occur in summer is also natural, since at that season the surface of the soil is, for a longer or shorter time, at about the temperature best suited for the development of the bacteria.

Yet with all this the chief question remains unanswered: How does the bacillus find its way into the body, how does the animal become infected, and how does this plague spread among cattle?

We are not able as yet to give a satisfactory answer to every separate point, for it is, as may be supposed, no easy matter to follow the bacillus in its hidden paths; yet as to the most important matters we are sufficiently informed by precise investigations.

It has been noticed that the anthrax bacilli generally obtain entrance in the natural course of things by the same doors which have been found capable of admitting it for experimental transmissions.

Slight injuries to the integument, scratches, pricks, bites, cuts, and stings admit it in some cases. This mode of infection is perhaps not very frequent, and only in the case of human beings does it occasionally come under notice. People who have to deal with animals suffering from or having died of anthrax are most exposed to infection, and thus the anthrax as a human disease is pretty much confined to certain trades and occupations. Cattle-drivers, herdsmen, butchers, and those who have had to skin and cut up the cattle, sheep, and horses which have died of anthrax, as well as those who have to handle the hair, wool, skin, etc., of such animals are the commonest human victims of the disease.

Yet man is not one of the highly-susceptible classes. As a rule the disease remains local; the local inflammation known as malignant pustule sets in, and but rarely does a general spreading of bacteria throughout the whole body take place.

Particularly with mankind we often observe the second kind of infection, the reception of the virus into the lungs. In England especially a peculiar, violent lung disease has been noticed, which follows pretty much the same course as a severe attack of pneumonia, and which affects workmen employed in sorting rags, teasing wool, etc. It was called the "wool-sorter's disease" and was long a puzzle to the faculty, until more precise investigation at length discovered its true nature.

In Germany, too, many cases have recently come under notice which have been recognized as pulmonic anthrax, and according to the reports of Paltauf and Eiselsberg, there can be no doubt that the German "Hadern Krankheit" is generally quite identical with the English wool-sorter's disease.

The infection takes place by the inhaling of anthrax spores, in the same way, therefore, as in Buchner's experiment with mice and Guinea-pigs. The germs adhere to the hairs (or wool) coming from ainmals which perished of anthrax disease.

The most important way in which animals as a rule and human beings occasionally become infected is the third: infection through the intestinal canal by reception of the virus mixed with food or water.

Now we are aware that the anthrax bacilli cannot pass beyond the stomach of a healthy animal, being killed by the gastric juice, and that only the spores are able to get as far as the intestines. It will be asked: How do the animals get these spores, and how are these facts to be reconciled with the peculiar epidemic nature of the disease?

Long before the anthrax bacillus was known or even dreamed of, people had noticed that the place where an animal had died of anthrax, or where a victim of that disease had been buried, remained a dangerous pasture for sheep and cattle. Very often the terrible scourge broke out again at such a place, and people learned to avoid it without knowing exactly why.

When at length the bacillus was discovered, one endeavored to explain these facts in connection with it, and a very plausible explanation was soon found. In the buried victim, it was said, the bacilli continued to develop and increase, and only required to reach the surface in order to infect any animal that might be there. For this somewhat difficult resurrection from the grave, ways and means were quickly advanced. Either, as Pasteur maintained, the earth-worms aided the micro-organisms, loading themselves deep below with crumbs of earth containing bacilli, and delivering them safe at the surface, or the underground water, that "deus ex machina" for all who search causes for diseases in the lap of mother earth, somehow or other managed to convey the anthrax bacteria to the surface, and the varied conditions of the soil as to warmth and temperature had to appear as witnesses of the process.

For the latter assertion not a shadow of a proof could be adduced, and the earth-worm theory was disproved experimentally by Koch, who showed that the entire supposition to support which these explanations have been conjured up was without foundation.

The preservation and distribution of the virus—i.e., the formation of spores—does not take place underground. The bacilli (the sporeless rod-cells) soon perish at a depth of two or three metres (as may be proved by direct experiment), because even during the warm season the low temperature which prevails at that depth does not allow of any growth, still less of any sporulation, for which, as we know, a pretty high temperature, about 24° C., is requisite.

This temperature, however, is not attained at the depth of half a metre in our climate, and another requisite, the free access to oxygen, is also difficult, to say the least, at such a depth below the surface.

Everything, indeed, leads us to the conviction that the development of anthrax spores—i.e., of the only form of anthrax poison which is specially dangerous to animals in a state of nature—is a process which takes place on the surface of the soil or in its uppermost layers; that it begins and ends there.

In the bodies of animals no spore-formation takes place for want of oxygen. But while still alive, animals suffering from anthrax discharge bloody urine and fæces, both of which contain bacilli; after death bloody liquids also containing bacilli flow from the mouth, nostrils, and anus; and if the body is skinned, opened, and cut up, an immense number of cells are distributed around. These increase, either in the blood and urine which remain in or near the surface or on suitable vegetable food media, and during the hot summer weather they do not fail to produce spores.

And when spores have once been produced, the way is open for the spread of virus for the extension of the disease. Either they are eaten by the cattle while grazing, and so find entrance without the aid of the earth-worm and the underground water; or they get into the hay; or, as Frank has shown, into the clay used as flooring of the cattle-stalls, in which case they lead to sudden cow-house epidemics during the winter; or, lastly, they are carried away by rain into brooks, etc., arrive at distant places where anthrax disease had never occurred before, and there lead to the breaking out of mysterious, unexplainable cases of disease and death.

These facts point plainly to the best and most sensible method of dealing with the bodies of animals that have died of anthrax.

If the diagnosis is already pretty clear, the body should on no account be opened. The best thing to be done would certainly be to burn the body entire. If this is not practicable it should be buried 1½ or 2 metres deep, and that will certainly prevent all formation of spores. Any bloody evacuations must be disinfected without fail, a 5% solution of carbolic acid being the best means. One must also be very cautious with regards to foreign imported hides and hair, with which the virus is often transported, as was recently shown anew by the investigations of Rembold.

With this we will close the subject of anthrax bacillus. All the facts which numerous and precise investigations have elicited of which this bacterium has been the subject have by no means been given, but the chief points have been presented and the impression must have been conveyed that our facts already stand on a firm footing, and that in many respects final, indisputable results have been attained.

II. BACILLUS OF MALIGNANT ŒDEMA.

The great majority of questions more or less connected with the origin of anthrax could not be answered with certainty until recent methods had rendered it possible to judge the properties of micro-organisms with exactitude, and until anthrax disease could be distinguished from other affections which resemble it at first sight, and which are caused by similar species of bacteria.

One of the latter has become more exactly known by the investigations of Koch; we refer to the bacilli of malignant œdema, as he has called them, and which are probably the same which Pasteur discovered among his "septicémie" and described as "vibrions septiques."

Malignant œdema has recently been observed also in human subjects in connection with compound fractures of bones and deep wounds, as also in subcutaneous injections. It produces an extensive emphysema of the skin, putrefaction and œdematous softening of the superficial muscles; in most cases death ensues in a few days. For these cases we must suppose that the injured parts had in some way come into contact with germs of malignant œdema—a supposition all the more probable since these germs are very widely diffused in nature.

At least we can easily produce the disease in susceptible animals with the most varied material of infection. Various decomposing matters, foul water, the dust from between the planks of flooring, the blood of animals that have been suffocated, and particularly the upper layers of garden earth, all serve excellently for this purpose.

If we take a moderate quantity (as much as can be raised on the point of a knife) of the latter material, and place it in a pouch under the skin of a Guinea-pig's or rabbit's abdomen, the animal usually perishes within twenty-four or forty-eight hours, and on examination œdema bacilli are found to be the cause of death.

These are slender, thin rod-cells, considerably narrower than the anthrax bacteria, with rather sharply-pointed or rounded ends. In cultivation, as in the bodies of animals, they tend to unite in long threads which are often curiously bent in the form of a bow.

The œdema bacilli have lively motile power and are among the rod-cells species in which R. Pfeiffer has succeeded, by the aid of Löffler's staining process, in showing the existence of lateral flagella. In the hanging drop, the faculty of locomotion generally ceases after a short time, since this bacillus, as will presently be seen, is killed by the oxygen of the air.

At a somewhat high temperature, above 20° C., sporulation

speedily takes place. The spores are large and central, occasionally rather broader than the bacillus, which consequently has to bulge to some extent.

The œdema bacilli are strictly anaërobic, and thrive only in an atmosphere free of oxygen; they show themselves extremely sensitive to the presence even of small traces of this gas. They grow at usual temperatures and also in the incubator.

They readily take the anilin stains; in cover-glass preparations of tissue-juice the pointed shape of the ends forms a decided contrast to the anthrax bacillus. Gram's method cannot be employed, since the cells lose the first stain under the influence of iodine. The spores are capable of double-staining.

As we have to deal here with an anaërobic species, we can only watch its growth in our artificial media by means of special appliances introduced for the cultivation of these oxygen-hating micro-organisms.

In gelatin the colonies are visible to the naked eye as small gleaming balls with liquid, grayish-white contents. They increase gradually in size without essentially altering their appearance. The microscope shows the interior of such a colony to consist of a close network of long threads, in which the spontaneous motion can be seen, often with a low magnifying power. The edge has a peculiar striped or ray-like appearance, such, for instance, as was described in the case of the hay bacillus.

On the agar plate are seen cloudy, dull-white markings with irregular border. The microscope shows an intricately-branched mass which spreads out like a covering of moss.

In the deep test-tube cultures the restriction of growth to the lower part of the puncture is at first clearly visible. With the growth an extensive decomposition of the gelatin takes place, and it is changed into a grayish-white, cloudy, dull liquid. Almost always an abundant development of gas-bubbles takes place, particularly when the food medium contains an addition of grape-sugar, which allows the micro-organisms an opportunity to display their powers of fermentation. With the accumulation of gas the culture slowly extends upward, till at length the free surface is reached. A peculiar, disagreeable odor is always noticeable.

In agar a bacterial growth takes place with jagged edges and granulated contents, the whole widening out like a club at the bottom and becoming thinner and thinner as it rises. The profuse development of gas which (particularly in the incubator) generally bursts the medium into separate portions and the already-mentioned stench are here particularly noticeable.

It has already been stated that garden earth almost always contains germs of the œdema bacillus, and that it is, therefore, a very good material for infecting susceptible animals, such as Guinea-pigs.

If we examine a Guinea-pig which has died two days after being inoculated with garden earth, the following conditions will be noticed: If the skin be incised and thrust aside the peculiar state of the body becomes visible. The subcutaneous cellular tissue and the superficial muscles for a considerable distance round the point of infection are œdematous and saturated with a dirty-red fluid, and almost everywhere, especially in the axillæ, a stinking, frothy ichor has collected.

But what is seen here is not simply the result of œdema. In garden earth there are always, in addition to the œdema bacilli, numerous germs of other bacteria, some of them anaërobic, and of which the Bacillus spinosus, for example, is already known to us. These develop with the bacilli in the body of the animal, are often recognizable under the microscope by their difference of shape or by their want of spontaneous motion, and complicate the pathological examination very considerably.

That such is the case may be proven by infecting a susceptible animal (rabbit, Guinea-pig, or mouse) with a pure culture of the œdema bacillus.

Here too a strong, bloody œdema of the subcutaneous tissue and of the superficial muscles extends to a considerable distance from and around the point of inoculation, and gives its name to the entire affection. But the liquid which collects is no longer ichorous, but consists of a reddish serum without odor and without very marked development of gas.

The internal organs are but slightly altered. The spleen is generally somewhat enlarged and dark-colored; the lungs are of a peculiar grayish-red.

Cover-glass preparations from the œdema fluid, the heart blood, and tissue juices display a very remarkable fact. While the first show large numbers of rod-cells, the tissue juices of the larger organs show but few, and the blood none at all. This experiment is corroborated and complemented by the examination of sections from the various organs. Whether we take the spleen, the liver, the lungs, or the kidneys, in no case shall we find bacilli in the interior of the tissue; in particular the blood-vessels, the chief seat of alterations with anthrax, will be found perfectly free from them. Only at the edge of the preparations—i.e., on the surface of the organs—large quantities of bacteria will be found in and immediately under the serous covering. Here we may see them isolated or joined in long threads, but scarcely ever does a cell venture deeper down into the tissue.

It is true that these appearances can only be observed distinctly when the animal is examined as soon as possible after death. The longer the time that elapses, the more the picture alters. The bacilli which, as long as life lasted, confined themselves to the surface of the organs, and scarcely ventured to take a few timid steps toward the interior of them, possess the faculty of thriving in the dead body, in which they multiply immensely. "It is evident," says Gaffky, "that, aided by their motile power and by the serous saturation of the muscles of the abdomen and breast, they pass from the subcutaneous œdema, their proper home, into the thoracic and abdominal cavities and thence from the surface into the interior of the organs." And there they are found in great quantities. First they fill the entrances with a close network; they then advance further into the deeper parts (forming long threads in the lungs); they grow into vessels, and at length fill the tissues as completely and in as large masses as do the anthrax bacilli.

In mice this state of things will always and from the first be found, though not in Guinea-pigs and rabbits. With mice the space for action is so small that the differences just described have not room to develop. The bacilli penetrate during life into the interior of the mouse's diminutive organs, fill out the tissues, burst the walls of the vessels, and are thus carried in the blood to the most distant parts.

Moreover, since the serous effusion in the subcutaneous cellular tissue of a mouse is very small, while the spleen is almost always greatly enlarged, dark-colored, and generally softened, it will readily be comprehended that malignant œdema may easily be confounded with anthrax in such cases, and that only careful examination can detect the error.

It was with the œdema bacilli that Roux and Chamberland most conclusively proved the important part which the excretions, the "substances solubles," play in causing immunity. If they kept bouillon cultures for ten minutes at a heat of 115° C., or if they filtered them through porcelain tubes, about 100 c.cm. of the liquid without any bacteria sufficed, when injected in three several doses into the abdominal cavity of Guinea-pigs, to render them proof against inoculation with the bacteria themselves. The success was still greater when, instead of the cultures, the blood-serum obtained from dead victims was filtered. An injection of about one cubic centimetre, repeated seven or eight times on as many consecutive days, produced the desired effect.

III. RAUSCHBRAND BACILLUS.

The disease known by the French as charbon symptomatique, by the Germans as rauschbrand, by us as black-leg, quarter-evil, or symptomatic anthrax, has many points of similarity with anthrax or splenic fever. Like the latter it appears in the summer months, disappears in the colder seasons, and attacks the herds, especially the horned cattle; like anthrax it keeps chiefly to certain clearly-defined districts—for example, the Bavarian Alps, parts of the Grand Duchy of Baden, some portions of Schleswig-Holstein—and within such black-leg districts there are again certain black-leg localities. As the appearance and symptoms of the two diseases are similar, it can surprise no one to hear that for a long time anthrax and black-leg were regarded as one and the same thing or were confounded with each other.

Feser and Bollinger were the first to show distinct differences between them. In black-leg we have as a peculiar feature the rising of irregularly-shaped, strongly-emphysematous, and therefore (when touched) crackling swellings of the skin and muscles, which have their seat chiefly in the quarters. The striking black-leg discoloration of the diseased muscles is also unknown in ordinary anthrax, and lastly, in the serous bloody fluid of the diseased portions a micro-organism of "club-like" appearance is found which is not identical with the anthrax bacillus. This is what Feser and Bollinger report. Since then the subject has occupied much attention, and special efforts have been made to ascertain the qualities of the supposed exciter of this malady. Arloing, Cornevin, and Thomas have discovered a number of important facts regarding the black-leg bacillus, and recently Kitasato has succeeded in breeding it artificially, and by successful transmission of the malady to susceptible animals has given an indubitable proof of its importance.

The black-leg bacillus is a rather large, slender rod-cell with distinctly-rounded ends, which occurs singly as a rule, but occasionally in twos, but never in long threads. It has a lively motile power, which, however, it quickly loses in the hanging drop, because, like the œdema bacillus, it belongs to the strictly anaërobic species, and soon perishes if placed within the reach of oxygen.

Its tendency to produce involution forms is worth noting. Cells that have reached a somewhat advanced age, or which have grown up under circumstances that for some reason or other did not suit them perfectly, nearly always take forms which differ widely from the normal one, and might even raise the suspicion that one had to deal with some quite different species. The cells grow to a gigantic size, become unshapely, and present irregular contours. In particular it will often be noticed that the rods swell moderately in the middle and take an appearance which at the first glance reminds us of the shuttle-like forms of the clostridia. On closer examination, however, it will be found that there is nothing like spore-formation at work. In the hanging drop the whole micro-organism appears equally dull, slightly granulated, without any trace of a special body existing within it as apart from the rest. When we stain the bacilli the only parts which could be taken for spores, those gleaming grains at the poles of the club-shaped rods or the central part of the clostridia, take the coloring matter particularly well, thus giving the clearest evidence of their not possessing the nature of spores.

In fact, the sporulation proceeds in quite a different manner. It is true that the rods widen out at the place where the spore develops, but only to such a slight degree that it can often scarcely be perceived. The spores—large, brightly-gleaming bodies, generally somewhat lengthened and slightly flattened or bent at the sides—lie at the end of the rod-cell generally, rather eccentrically nearer to the one or the other wall. After sporulation has been accomplished the remainder of the bacillus quickly perishes, the spore becomes free, and is then only covered by a tender film of remaining protoplasm.

The conditions of sporulation for the black-leg bacillus are not yet certainly known. According to Kitasato, it does not take place in the living animals, but occurs in the dead body and in our artificial cultures. The spores possess quite a high degree of resisting power; in a dry state they retain their vitality a long time; against the influence of heat and chemical agents they also show themselves very resistant.

The bacillus is, as we have just seen, a strictly anaërobic micro-organism, to which oxygen is as fatal as to the œdema bacillus; it thrives at ordinary temperature, above 18° C., and still better in the incubator.

The staining is performed in the ordinary manner without difficulty. Gram's method robs the cells of their color again. The spores do not take the aqueous anilin solutions, but are capable of double staining.

In gelatin mixed with grape-sugar or other reducing agents, the colonies develop in a few days and show themselves as round masses with irregular borders, which liquefy the gelatin very quickly. Under the microscope one sees a dark opaque centre surrounded by a dense tangle of ray-like threads. The latter sometimes penetrate to a considerable distance all round and give to the whole a thistle-like appearance.

Stab-cultures in deep gelatin show a growth first of all at the bottom of the needle-hole. A stocking-like liquefaction forms with opaque gray contents. It rises gradually higher, but stops short two or three fingers' breadth from the top. Soon a plentiful development of gas begins, and now the culture advances upward till at last only a thin layer of gelatin at the extreme top remains free. While the œdema bacilli create a decided odor, one can here perceive nothing but a peculiar sour smell which is a characteristic of the black-leg bacillus.

In the agar cultures the bacillus thrives rapidly in the incubator; in twenty-four hours the food medium is filled with numberless gas-bubbles, ruptured in many places, while the puncture develops bacilli quite to its upper end.

In bouillon, too, the bacilli can exist. At first they render the liquid opaque, but soon sink in white flakes to the bottom and collect there as a thick sediment; at the edge or the surface numerous gas-bubbles collect.

If portions of the artificial cultures are transmitted to susceptible animals, the latter perish of black-leg. For laboratory experiments the Guinea-pig is the most suitable animal, since it regularly yields to very small portions of the virus. If we inject a drop of a good bouillon culture into the subcutaneous cellular tissue of a Guinea-pig, or if we put a silk thread with black-leg spores into a pouch in the skin of its abdomen, death will take place in twenty-four or thirty-six hours, and the post-mortem state of the body will show clearly the alterations peculiar to black-leg.

The subcutaneous connective tissue, as well as the surface of the muscular system and its deeper layers, are œdematous, saturated with a very abundant bloody serous fluid. Above all, however, the eye is struck by the dark red, frequently almost blackish, discolorations of the muscles in the immediate neighborhood of the point of infection and for some distance around it, while the internal organs offer nothing particular to notice. Examined under the microscope in the hanging drop and in stained preparations numbers of rod-cells are to be seen in the serum, some of which are motile and possess a straight form, while others depart more or less from the normal shape. A club-like swelling of one or both ends is particularly common; sometimes one sees heart-shaped or clostridium-like forms, etc. If but a short time has elapsed since death took place, no spores are visible; after a few hours the first spores make their appearance, and the state of things now alters, inasmuch as the rod-cells are also found in the internal organs, where they were at first absent.

The virulence of the black-leg bacilli is extremely various. In some cases it is subject to natural attenuation. Thus Kitasato found that his bouillon cultures quickly lost their poisonous property, and often at the end of a week could no longer be transmitted to animals, while they had perfectly preserved their vital energy. With frequent change to fresh food media this loss did not take place, and so far at least no such diminution of infectious power has been noticed with gelatin and agar, not even in very old cultures.

Artificial attenuation is easily produced. Here also heat is the best means; breeding at 42° or 43° C. robs the bacteria quickly of their pathogenic force.

It is remarkable that the spores are also susceptible of attenuation. If one exposes the flesh of animals which have died of black-leg and which has been dried at about 30° C. to a dry heat between 80° and 100° C. for several hours, it gradually loses its virulence. It may even be employed as a means of protective inoculation, and is indeed particularly recommended for practical use, since it preserves the diminished degree of virulence unaltered for any length of time. Yet the attenuation, as might be expected from the manner in which it is brought about, is not very enduring with the bacteria. The but partially removed virulence may be quickly recalled by mixing up the bacteria (as directed by Arloing, Cornevin, and Thomas) with a 20% solution of lactic acid and injecting the mixture into susceptible animals. The addition of acid injures the tissue and the vital action of the body, reduces its power of resistance, and allows the bacteria to gain a footing. Wheh they have once done this they quickly recover their former power, and after a few more transmissions regain their lost virulence to its full extent.

In the same position with this striking phenomenon are others noticed chiefly by French investigators, and more particularly by Roger. These observations concern the inoculation of black-leg into animals naturally non-susceptible to it; rabbits, for instance, which at once yield to the malignant œdema, enjoy immunity from the effects of black-leg poison; the same is the case with mice, pigeons, fowls, etc. Yet if one injects them with the mixture of black-leg bacilli already mentioned and solution of lactic acid, or if one also injects at the same time with the black-leg bacilli sterilized or non-sterilized cultures of Micrococcus prodigiosus, Proteus vulgaris, etc., the transmission takes place and the animals die of genuine black-leg.

As may be remembered, we have already pretty fully considered these matters and have attempted to account for them.

Artificial immunity from black-leg may be brought about with extreme ease and in many different ways. First, it can be produced in connection with attenuation. Kitasato rendered Guinea-pigs non-susceptible with his bouillon cultures which had lost their virulence; the bacteria bred at high temperatures produce the same effect, and, as we have said, the inoculation of attenuated spores works very successfully. It is the only method employed in real practice, and, proceeding from Arloing and his collaborators, has been perfected by Kitt. His process is as follows: Muscles of animals which have perished of black-leg are dried at 32° to 35° C. and then divided into two parts. One part is kept for six hours at a heat of 85° to 90° C., the other also for six hours, but at 100° to 104° C. Thus one obtains a premier and a deuxième vaccin; the powdered flesh is mixed with sterilized water or bouillon and injected subcutaneously into the animals at suitable intervals. In other cases—for instance, with cattle and sheep—the inoculation of very small quantities of unaltered rod-cells suffices to produce a local reaction which is also found to grant immunity. Further, the inoculation of the virus into parts of the body—the tip of the tail, for instance—where it finds no suitable point of attack, proves successful, as does also an injection direct into the blood-vessels, which are unconnected with the subcutaneous tissue, the proper bed of development for these bacteria.

In the case of black-leg bacillus we see with special clearness that it is the excretions which play the chief part in producing immunity. Roux and Chamberland found that filtered cultures were successful in causing it, and even granted to the Guinea-pig not only an immunity from black-leg, but also from malignant œdema —a fact, however, which Kitasato denies on the ground of some experiments of his own. The last-named investigator kept bouillon cultures half an hour at 80° C., killing thereby the sporeless bacteria, and produced immunity with this material.

As the protective inoculation against black-leg, if properly conducted, is without danger and, at least as far as has yet been observed, does not occasionally demand a victim (as in the case with anthrax), and as, further, the immunity appears to be enduring and reliable, the inoculation of cattle is unconditionally recommended for actual practice, even by cautious practitioners.

By microscopic examination, by breeding outside the animal

body, and by transmissions from artificial cultures, we know that the black-leg bacillus is the specific and sole cause of the black-leg malady. We must now ask how the symptoms of the disease and the circumstances of its appearance are to be explained in connection with the qualities of the micro-organism which excites it.

Although we are still unable to give an answer that shall be satisfactory in all respects, yet we may say, with a high degree of probability, that under natural conditions the infection of animals, the spread of the disease, is owing to the penetration of the bacteria through small wounds, particularly in the extremities, the thighs, etc., and their further development in the subcutaneous tissue, in which they cause the peculiar changes of which we have spoken. The two other ways by which micro-organisms can obtain entrance to the body, the intestinal canal and the lungs, may also occasionally be taken by these bacilli, yet neither experiment nor observation has yet yielded satisfactory proofs of such being the case.

The black-leg bacillus, in contrast to the anthrax bacillus, is scarcely able to exist outside the bodies of animals, for on account of its anaërobic character it speedily dies in contact with the air. The task of preserving, spreading, and propagating the virus is probably performed by the spores exclusively.

They begin to form in the body soon after death has taken place, and the flesh of the dead victims contain large supplies of them. If the animals are skinned, cut up, etc., the ground is infected at the place in question by means of escaping tissue-juice, and so gives rise to a black-leg locality, and even if the bodies are not opened, the bloody serous fluid peculiar to the disease may nevertheless infect the place where the animals have stood, making it dangerous for healthy animals that may afterward come there. It is, therefore, easy to understand that the malady adheres to certain districts, that it often appears suddenly epidemically, that it sometimes disappears as suddenly (when the cattle no longer visit the place of infection), that it is generally confined to the summer months, etc.

It is to be noted that young and also old animals are but little susceptible; as a rule, those from one to three years old are attacked. Further, the young of animals which have been protectively inoculated usually inherit the immunity of their parents.

IV. TUBERCLE BACILLUS.

When we consider that almost one-seventh of all deaths are due to tuberculosis, and that this disease occurs very frequently among animals, it may readily be understood that for many years efforts have been made to discover its cause and the mode of its diffusion.

This seemed to be a hopeless task as long as science was not united on the preliminary question as to what should be considered tuberculosis, what its limits were, and what were the surest methods for its recognition.

While some endeavored to form a picture of their judgment from the symptoms of the disease by purely clinical criteria, others looked for a picture of the disease in the tissue changes solely.

But even in this narrower sphere there was no agreement Laennec, the great French investigator, saw in caseation the real character of the disease. Virchow, on the contrary, recognized as tuberculous only those changes in which there were present the tubercle nodule, those small, millet-seed growths of gray transparency which were first described by von Bayle in 1810 as being peculiar to consumption.

Villemin, in his observations published in 1865, was the first to open the way out of this controversy. He succeeded by inoculation with tuberculous matter in producing tuberculosis in previouly healthy animals, and thereby demonstrated that tuberculosis was an infectious disease. It was Cohnheim above all to whose keen and experienced eye the significance of these facts was apparent, and who, after his own inoculations into the interior chamber of the eye, repeatedly and emphatically declared that it was a specific infectious disease.

Before his untimely death he saw the correctness of his declarations conclusively shown.

On the 24th of March, 1882, R. Koch, before the Physiological Society in Berlin, made the announcement that he had found the cause of tuberculosis, which was due to a peculiar bacillus of a special shape.

"I have seldom in all my life felt greater pleasure than at the reception of this news," were the words with which Cohnheim greeted the new discovery, and one could see that he spoke those words with the deepest conviction.

The impression which the discovery of Koch made was in fact extremely deep and lasting. The incomparable certainty and positiveness of his investigations were admired by everybody.

In a methodical and conscientious investigation he had paved the way step by step to this knowledge, established his views, and with one stroke disclosed the faultless and thorough character of his labors. So powerful and so free from objection was every argument that no one attempted to combat them, and the concluding sentence of his communication was: "We can with good reason say that the tubercle bacillus is not simply one cause of tuberculosis, but its sole cause, and that without tubercle bacilli you would have no tuberculosis."

Through the microscopical recognition of tubercle bacilli in all properly-examined cases of tuberculosis, and only in them, and through successful cultivations of the germs outside the body and their successful transmission and reproduction of the disease, he proved his assertions, and in this way established a wonderful advance in medical knowledge.

Now, there was no doubt as to what was to be considered tuberculosis and what was not. "In those processes where you find tubercle bacilli, there is true tuberculosis," no matter what the macroscopical or microscopical pathological picture is, or what the clinical evidence may show in single cases. The subsequent advantages from a diagnostic point of view are evident.

Wherever processes of a tubercular nature have occurred the bacilli have been the factors, and therefore they are observed in tubercular tissue and in the excretions of tubercular persons, especially in the sputum of such individuals.

They are very slender, medium-sized rods, somewhat smaller than a human red blood-corpuscle. They have clearly-rounded ends and are seldom perfectly straight, more often bent like a fiddle-bow. Usually they are seen single, less often two together, and here and there are to be seen larger chains of five to six members. They are incapable of voluntary motion.

The question whether the tubercle bacilli form spores has as yet not been positively decided. In the examination by the hanging drop, we do not see in the interior of the rods those sharply-circumscribed glistening bodies of definite shape which are recognized as characteristic of sporulation of bacteria. If we stain the preparations, we observe very frequently in the interior of the bacilli small bright spaces, often in regular arrangement, reminding one of spores. In fact, they are often called such.

Still, one observes by closer examination facts which will not wholly agree with the theory of sporulation. At one time we will not infrequently find in the same rod several of these structures, which in all other instances known up to the present time only produce a single spore.

Further, the outlines are usually not sharp, the boundary between the clear and the stained portion is somewhat obscured, the size of the clear spaces varies, and lastly, their form also differs from those of the usual spore because as a rule they do not appear round or oval, but, on the contrary, contain contracted walls, and do not present bi-convex, but bi-concave surfaces. Since the bacteria found in pure cultures do not possess greater power of resistance than non-sporulating bacteria, but on the contrary are destroyed by a low degree of heat—70° to 80° C.—it will not be erroneous to recognize these appearances as eventually due to vacuolation or something quite similar.

Noteworthy and highly significant in a practical way is the fact that the bacilli themselves without sporulation are capable of resisting destructive influences to an unusual degree. Tubercular sputum—as an example composed of thick albuminous matter which especially protects the bacteria against destruction—will withstand drying for months, temperatures near the boiling point, the action of the gastric juices, and also the influences of the strongest decomposition, without the least curtailment of its infectious activity.

The tubercle bacillus shows in its staining a marked peculiarity referable to its exceptionally compact and impenetrable skin or envelope, which offers an obstinate resistance to the penetration of the staining fluids. This explanation will be permitted because of the obstinacy which the bacillus also shows under other conditions. The actual presence of spores is not necessary to this bacillus in order to perpetuate itself. This need not in any way conflict with the circumstance that we may through further investigations discover actual spore-formation.

The tubercle bacillus is an anaërobic bacterium. It is a strictly-parasitic micro-organism that can only with difficulty be cultivated outside of the body, and is very selective in regard to its culture medium, thriving slowly under all conditions and especially restricted in its development to narrow ranges of temperature. Small deviations of temperature from the required blood heat (37° C.) suffices to wholly retard its increase.

The demonstration of tubercle bacilli in unstained preparations is associated with great difficulty, still it is possible, and undoubtedly at about the same time that Koch published his investigations concerning the causes of tuberculosis, Baumgarten had also seen the same bacillus in unstained preparations and recognized its significance—though of course without being able even remotely to bring conclusive proofs of his discovery, which rendered Koch's communications so highly valuable.

Koch first concluded from the striking relations of this bacillus to staining that it should be regarded as a peculiar kind. With our usual watery or diluted alcoholic anilin stains the bacilli are not stained at all, and it was due to this fact that the germ evaded discovery for so long a time. Koch found that solutions to which had been added an alkali were rendered more powerful, although staining slowly, and that they were not at all or only reluctantly decolored, while on the contrary all other known bacteria were readily stained by this solution and just as readily decolored.

In this way Koch first stained the bacilli with his alkaline methyl-blue, whose method of manipulation has been previously stated, and we will therefore at once consider the complete method first described by Ehrlich.

The essential of the Koch-Ehrlich's staining of tubercle bacilli was the employment of anilin-oil water-color and the diluted acid for decoloring. The basis of proceedings has remained unchanged since that time, even if in some instances deviations in technique have been employed. Thus the anilin-oil mixture has been supplanted by Ziehl's carbol-fuchsin with advantage. The length of time in staining cover-glasses is in general shortened through heating, and the concentration of the decoloring acid is somewhat lessened.

Upon this has been constructed the method of which, because of its practical importance, should be described with special accuracy, and is as follows:

In the examination of cover-glass preparations for tubercle bacilli, they must first be properly prepared before staining. For this purpose pour the entire sputum into a dark glass vessel or plate,* so as to recognize better its constituent parts; for we have not only to examine the excretions of the diseased lung, but also the mucous secretions of the upper air-passages and the saliva, and as a rule it is only the lung secretion that contains the bacilli. Therefore select out of the mixture on the plate one of those yellow conglomerate and tough masses of mucus which undoubtedly have their origin in the lung and were so recognized in pre-bacteriological times as "lentils" in tubercular sputum which called for special attention. Extract one with the platinum needle, put it upon a cover-glass, and then place a second cover-glass upon it, just as in preparing blood and film preparations.

Through pressure upon the upper glass and sliding both glasses to and fro crush the nodule and attempt to spread it between and upon the glasses as evenly as possible. The latter are then carefully drawn apart and permitted to dry in the air, when they are drawn *three times through the flame* and are then ready for staining.

---

* An ordinary soup-plate, painted a dead black, will answer.—J. H. L.

It is needless to say that we proceed in the same way in making any other cover-glass preparation, of fæces, pus, or a pure culture preparation.

Cover-glasses thus prepared should be held only with forceps. Cover the prepared slide with carbol-fuchsin and hold the cover-glasses over a flame until evaporation takes place. Withdraw the cover-glass in an instant, repeat the procedure several times, replacing if necessary the evaporated staining fluid. Now pour off the staining fluid, wash the preparation with distilled water until cleansed.

Now follows the most important step of the process, namely, *decoloration*. The coloring matter is retained with unusual tenacity by the bacilli, but can be removed from the surrounding tissue by means of a 15% to 20% solution of nitric acid. The cover-glasses are moved to and fro in the acid solution a number of seconds until the deep red color has changed to a greenish-blue.

The preparation is next placed in a vessel of diluted (70%) alcohol for the purpose of removing the fuchsin which has been set free by the acid. It leaves the preparation in thick red clouds, the preparation becoming paler and paler until finally it remains as a light film of a rosy hue. After removing from the alcohol the cover-glass is washed with distilled water and then counter-stained with the usual diluted methyl-blue solution. Those parts of the preparation which under the influence of the acid have been decolored now take up the blue; the tubercle bacilli, on the contrary, retain their red color, and consequently are seen with special distinctness. The various other bacteria in the same preparation are colored blue, which distinguishes them at once from the tubercle bacilli.

The expectoration from consumptives is usually very rich in foreign micro-organisms, especially streptococci and bacilli of decomposition, and therefore it is necessary in each case to notice this contrast carefully.

The excess of methyl-blue is in a short time removed with water, and the blue-looking preparation is either at once examined or else dried and then mounted in Canada balsam.

The first procedure is best, because by it we are able to complete our examination more rapidly and overcome difficulties such as insufficient decoloration, and because, furthermore, it retains the shape of the bacilli best, which frequently shrink in the balsam or subsequently lose their color.

Where we do not wish to make permanent preparations, the entire method may be shaped after the following concisely-described rules: After drying the cover-glass and drawing it three times through the flame, add a number of drops of carbol-fuchsin, and heat the same a number of times over a Bunsen burner until evaporation. After a short washing in water decolor in nitric acid and treat further with 70% alcohol, water, methyl-blue and water.

With this method we shall in most cases succeed, and it is unnecessary to describe separately the numerous other methods which differ in some one or other point from the above-described procedure. Only one other for specially staining tubercle bacilli will be mentioned. This was first described by B. Fraenkel and later by Gabbett without essential changes. It involves a new basis of procedure, and for practical purposes, where a rapid and simple examination of the expectoration for bacilli is desired, it may perhaps be recommended as the best procedure.

The manner of decoloring and counter-staining heretofore mentioned is here combined. The second color is united with the diluted acid, and in consequence does away with all the intermediate steps of the former procedure.

Stain with hot carbol-fuchsin in the usual way, then place the cover-glass without further preparation in the second solution. This is a mixture of dilute nitric acid and alcoholic methyl-blue, formed of 50 parts water, 30 of alcohol, 20 of nitric acid and methyl-blue to saturation. The acid removes the fuchsin and leaves it only in the bacilli; the decolored parts, however, now at once take up the new stain (the methyl-blue), and in a short time the preparation appears to the naked eye uniformly blue; it is then washed in water and examined, the rods appearing red upon a blue ground.

The essential point in this, as in the previous procedure, is, of course, decoloration by the acid, which is resisted only by the tubercle bacilli; these are, therefore, distinguished from all other bacteria by a specific staining.

We know but one regular exception accessible to this staining process, viz., the lepra bacilli. They are nearly related to Koch's bacilli, but, unlike the latter, they take the usual watery anilin stains equally as well as other bacteria, and thus present a very remarkable tinctorial difference.

The tubercle bacilli, it is true, are not quite as refractory to simple color solutions as was formerly supposed. By spreading a pure culture upon cover-glasses and treating them (for twenty-four to forty-eight hours) with watery fuchsin or gentian-violet, we will find that a large portion of the bacilli has been stained. Some, it is true, are not stained at all, some but imperfectly; but this fact proves that, as to staining, no very great contrast exists between the tubercle bacilli and other kinds of bacteria. The relative, gradual, or quantitative differences are, on the other hand, great enough, at any rate, to impart to the specific staining of the tubercle bacilli the value of a chemical reaction which enables us to discriminate substances that can be separated only with difficulty.

The tubercle bacilli may likewise be prepared according to Gram's method. The protoplasmic contents of the rod-cells are frequently contracted into small globules under the influence of iodine; they lie one behind the other in a row, like a chain of cocci; inexperienced investigators having, in fact, pronounced them as such.

Gram's method will, however, be used only in exceptional cases, since by its application we lose the important advantage of being able to recognize the tubercle bacilli as such even by their reaction to staining alone.

The question as to how specific staining is brought about and what are the delicate processes which distinguish this variety of bacteria cannot as yet be answered with certainty. But it is at any rate very probable (as stated above) that this is (according to Ehrlich's opinion) to be attributed to the existence of a particular covering surrounding the rods and offering a strong resistance to pigments. It is said to become more permeable under the influence of various additions to the solution, of alkalies, phenol, and anilin; but the acids (decoloring everything else) are unable to penetrate the skin later on; hence the coloring matter once absorbed by the rod-cell is surely and permanently retained in it. We may in this case, therefore, on the whole have to deal with the same conditions we have already become acquainted with in the double staining of the spores.

By the aid of special staining and microscopic investigation, Koch has been enabled to establish the regular occurrence of bacilli in all cases of tuberculosis, and only in this disease, and thus render their specific significance very probable. In order to prove and place it beyond doubt he undertook to cultivate the micro-organisms artificially with the view to transmission.

Many attempts to accomplish this in the ordinary manner had failed, and Koch became convinced that the tubercle bacillus required unusual conditions before it could be successfully cultivated.

We have seen that in this case we have to deal with a strictly-parasitic bacterium, dependent for its development upon the animal body and highly sensitive to changes in its surroundings. Koch thus found that our ordinary food media, the meat-peptone-gelatin and agar, and the bouillon did not suffice, and that only the artificially-coagulated blood-serum could be used which he (Koch) introduced for this very purpose.

Substances containing bacilli were spread on such coagulated, transparent blood-serum and left in an incubator at 37° C. The repeated examination with low magnifying powers resulted, after several days, in the discovery of peculiarly-shaped colonies consisting only of tubercle bacilli, as ascertained by stronger magnification and the color reaction.

It may be thought that there is nothing easier than establishing pure cultures of bacilli according to the method prescribed. But whoever undertakes it and finds out by his own experience the extraordinary difficulties that must be contended against, will admire again the success of Koch, who obtained entirely by his own efforts all the data which we follow to-day.

Pure cultures, according to Koch, are obtained as follows:

Inject into the abdominal cavity of some animals—for instance, the very susceptible Guinea-pigs—some tuberculous poisonous matter, such as consumptives' sputum ground and washed with distilled water. After some time, about three to four weeks, the first of the inoculated animals will die and the post-mortem will show an extensive tuberculosis of the liver, spleen, lungs, etc. Now one of the remaining pigs is killed by strangling and immediately cut open, before putrefaction bacteria or some other foreign micro-organism can settle, in order to obtain available inoculating matter. The skin is turned back with sterilized instruments, a window is cut into the chest-wall, and the root of a lung is drawn out with a platinum wire. (Knives and scissors must be free from germs.) Take from the organ one or more distinct nodules and place them upon glass slides thoroughly sterilized. The tubercles should be firmly crushed between the latter and the bacilli thus exposed as much as possible.

Blood-serum cannot be spread on plates. Allow the fluid serum to coagulate in small glass cups, and distribute the crushed tubercles on the surface of this firm nourishing medium with a strong platinum loop as forcibly as possible; it is best to rub the inoculating matter directly into the blood-serum. The little cups are then carefully covered with glass plates and placed in the incubator.

It is difficult to understand why, in spite of all precaution, the separation of germs in this manner is not as thorough as by the method with the plate. Stress must for this reason be laid upon a careful removal of the matter. We must also bear in mind that the tubercle bacillus possesses extremely little energy of growth, thrives very slowly, and develops exclusively at breeding temperature. Eventually, transmitted foreign bacteria will (even when they were originally in the minority or even when there was a single germ of them) cover the tubercle bacilli with their luxuriant growth and irretrievably ruin the culture.

This danger having been properly obviated, the first commencement of colonies becomes visible in about ten to fourteen days after planting. They gradually increase in size and assume a very characteristic appearance from the end of the third week.

We can see with the naked eye the appearance of grayish-white, lustreless, dry, and small crumbs upon the bottom. We notice under the microscope, in the middle, the remainder of the little piece of tissue from which the colony has started. To this are attached variously entwined or simply undulated curved tracings representing the growth of bacteria. Curled in locks and entangled, they remind us of strangely-fantastic characters and arabesques.

In arranging a "print" preparation of such a colony, we must, of course, stain them in the specific way in order to at once distinguish the tubercle bacilli as such. It will then be found that the tendril-shaped threads dissolve into a very great number of single rods lying all in about the same direction, close to and behind one another and separated by a distinct and ever-recurring interstice. We frequently perceive here the appearance of those roundish unstained and bright spots in the interior of the rods formerly regarded as spores.

As to culture in the test-tube: until recently the exclusive application of obliquely-coagulated blood-serum had to be resorted to. But since Nocard and Roux have ascertained that our common nutrient becomes, by the addition of 3 to 5% of glycerin, an excellent nourishing medium for tubercle bacilli, on which they flourish even better than on the serum, the agar thus prepared is preferred and the use of serum has been almost completely dispensed with. The latter can only be sterilized with difficulty, for which reason failures of culture resulting from an impure quality of the culture medium have formerly been altogether unavoidable. With the glycerin-agar this danger need not be apprehended, nor can it be doubted that it can be used for the preparation of plates and that it is preferable to serum, although such an experiment has, so far as known, not yet been made.

If we inoculate the tubercle bacilli with the platinum needle on the surface of the solid, oblique glycerin-agar, the incipient growth will be perceptible, after about fourteen days, throughout the entire extent of the inoculating line. Remaining for one and one-half to two weeks longer in the incubator, the culture is brought to the height of development. It appears then as a thick, crusty skin of grayish-white color, dry and lustreless, extremely brittle, joined together by numerous little scales, flakes, and nodules. A large quantity of condensation-water always gathers at the dependent parts of test-tubes, and the tubercle bacilli cover this water as a film consisting of single lamellæ without even projecting into the depth of the fluid and without dimming or altering the latter in any manner.

It is a matter of course that by careful transmission upon fresh media artificial cultivation may easily be continued and kept up. We now have a flourishing glycerin-agar culture of tubercle bacilli, which has descended in uninterrupted succession as the one hundred and seventh generation from the first Koch's blood-serum preparation. And with this stately number of ancestors they have preserved almost unchanged all the qualities of their progenitors: they are just as fit for infection as those and assume specific staining exactly in the manner above stated.

Such a success can, it is true, be obtained only by using special precaution. Attention is, therefore, called to a few little manipulations and measures applied with advantage in transmissions.

The inoculating matter must be firmly rubbed and pressed in. A shank of very strong platinum wire may be used, but should be thoroughly sterilized in the flame every time before using.

Having left the tube thus prepared to itself in the incubator, it will soon be seen (unless an unusually well-working thermostat is employed) that an evaporation of the condensation water, a drying up of the culture medium, is taking place, causing a stunted development of the culture. Attempts to prevent this may be made by drawing small and tightly-fitting rubber caps over the cotton plugs; but while doing so we incur another danger. A kind of moist chamber is formed under the rubber cover; the spores of moulds attached to the plug almost always begin to germinate and shoot mycelium threads through the fibres of the cotton; on the surface of the glycerin-agar, where the tubercle bacilli should appear, there will shortly be seen a layer of moulds.

We must, therefore, free the cotton from these undesirable inmates; this is best done by clipping the plug in the test-tube with scissors after inoculation and by burning the surface in the flame until it is carbonized and turns black. Now carefully drop on the top of the plug one or two drops of a 1:1,000 sublimate solution, and finally draw over it the rubber cap which had previously been lying in sublimate.

The tube may now be returned to the incubator. In fourteen days afterward the beginning of development is perceptible; it has reached its height after four to five weeks, and the cultures may now be removed from the incubator and preserved at the usual room temperature. Inoculation on fresh culture media is done about every six weeks, thus securing transplantation of cultures capable of living.

Bacilli prosper likewise in a bouillon containing 3 to 5% of glycerin. Pawlowsky finally pretends to have cultivated them even on potato slices prepared according to Globig's or Roux's method, and protected against exsiccation by a subsequent hermetical sealing of the test-tube. This procedure has as yet, however, not been confirmed.*

Koch succeeded, by the use of his artificial cultures in all cases, in a very large series of experiments, in reproducing in susceptible animals typical tuberculosis with all its clinical and anatomical symptoms, and thus furnishing the valid proof of having found in the bacillus the genuine sole exciter of the disease. His transmission was successful in 217 animals, mostly rabbits, Guinea-pigs, and field-mice. A small quantity of culture was removed by the platinum needle from the surface of the culture medium, and rubbed to a thin fluid by sterilized water or bouillon. Small quantities of the latter introduced into the body proved invariably successful.

Koch had caused the poisonous matter to be absorbed by subcutaneous application, by inoculation into the interior chamber of the eye, by injection into the large cavities of the body or into a vein, and, finally, by inhalation, and he has evoked the outbreak of tuberculosis in every way.

Other investigators after him have proved that by our food, too, containing tuberculous material, the disease can be artificially produced, and it is no longer to be doubted that the bacilli are capable of entering the body in all possible ways.

The changes occurring within the organism in connection with tubercular infection may be generally characterized as follows:

The disease develops in the first place in the immediate neighborhood of the spot where the bacilli found entrance, the affection being in the beginning merely local. Only later—with Guinea-pigs, for instance, after several weeks—a more general infection takes place gradually from spot to spot. Only when the poison from

* The editor has seen some very flourishing pure cultures of tubercle-bacilli grown on potatoes after this method in the Loomis laboratory. They were prepared by Dr. J. M. Byron of this city. He has also employed this method successfully himself. J. H. L.

the beginning has entered the blood-current and can be distributed in it, a more extended miliary appearance of the tuberculosis, a sudden inundation of the body by the micro-organism, is noticed at once.

We shall, therefore, generally have to deal with cases of the nature first considered, and the pathologico-anatomical lesions will be in conformity with it.

The macroscopic picture differs greatly according to the animals infected. We may find an extensive necrosis without actual cheesy degeneration (liver and spleen of Guinea-pigs), or rapid softening and formation of thin, liquid purulent secretion (tubercle of monkeys), or simultaneous consolidation and cheesy deposits (murrain of cattle), or formation of compact tumor masses with imbedded lime concrements (tuberculosis of the hen), etc.

But these general apparent differences appear alike on microscopic examination, and the histological structure is in all cases the same.

The distinguishing criterion for the existence of tubercular changes is, of course, the presence of bacilli. It is, then, obvious that we must properly prepare the tissues in order to prove the existence of these micro-organisms.

The staining of sections is done essentially according to the principles followed with the cover-glasses. First place the sections in a little cup with carbol-fuchsin and let them remain for about one hour. The duration of staining must be correspondingly longer, because we must not, in this case, heat the fluid to accelerate the process, for fear of destroying the sections. Next follows the decoloration in diluted nitric acid (10%), into which the sections are dipped with a needle for about one-half to one minute, according to the thickness of the preparation and intensity of the stain.

The red color will change into a distinctly green or greenish-blue tint. Seventy-per-cent alcohol will wash out the coloring matter dissolved by the acid, and the fuchsin leaves the section slowly in dense red clouds. This process having terminated and decoloration accomplished, the sections will show only a rose-red shade, almost completely fading, especially in the thinner parts of the preparation. The counter-staining is done, as with the cover-glasses, with ordinary methyl-blue. The sections remain here about two to three minutes, are decolored and deprived of water in absolute alcohol, brought into the clarifying oil, spread on the slide, and, finally, imbedded in Canada balsam.

On examining such a preparation with a high magnifying power the bacilli will be recognized as red rods on a blue background, but we will see at the same time a series of histologically striking changes in the tissue, long regarded as characteristic of tuberculosis. This affection produced, in the first place, a new growth, appearing mostly in the form of those small, grayish-white, transparent tubercles from which the disease derives its name and which were called by Cohnheim "infection tumors" according to their origin and kind. The tubercle is composed of an accumulation of round cells similar to lymph-corpuscles; besides, there are more or less numerous, somewhat larger, so-called epithelioid cells and also some giant cells lying in the middle or near the edge. The bacilli are especially abundant in these latter (but are also found outside of them), and there can be no doubt that the entire formation is caused by the action of the bacteria.

We do not yet positively know how this is produced. While it had formerly been supposed that emigrated white blood-corpuscles had exclusively formed the tubercle and had then united with epithelioid and giant cells, Baumgarten has assigned an important function to the cells of connective tissue and of epithelial origin. The bacilli coming in contact with these cells excite an irritation, which results in a proliferation of the cells to the extent of a division of the nuclei. But this irritation is not sufficient to cause the formation of an entire new cell. Hence there remains only the former division, and a giant cell is, according to Baumgarten, not formed by the union of several epithelioid cells, but by a simple germ division. The "epithelioid-cell tubercle" is developed, after which, under the influence of continued irritation, migration of white blood-corpuscles from the vessels takes place, changing the epithelioid into a lymphoid-cell tubercle.

The series of changes is not yet completed. A process takes place in the interior of the giant cells, an accurate description of which we owe to Weigert. Here the bacilli present produce a degeneration or coagulation necrosis, especially in the middle of the tubercle. The result is a uniformly dull, non-nucleated mass which the anilin colors do not stain. This mass consists of degenerated, practically necrotic cells, and very soon the bacilli disappear from it. But frequently we find in the same giant cell, the interior of which has undergone a retrograde change, a plentiful nuclear division resulting in the progressive development of tissue resembling granulation tissue in which are found large numbers of tubercle bacilli.

Thus we have here in both of those processes which are characteristic of the tubercular process, neoplasm (production of tissue on one hand) and retrogressive metamorphosis (destruction of tissue on the other hand). Finally, sooner or later the structure of the tubercle, the infection tumor, undergoes degeneration and perishes, the final result being, as a rule, the formation of a tubercular abscess. The blood-supply is shut off, or very greatly impeded by an obliterating endarteritis, which is caused by the proliferation of the endothelial cells lining the vessels.

The distribution of bacilli in the tissue corresponds to the above facts and conditions. They are seen, in the beginning, lying singly between the cells without the appearance of any other changes. But the first settlement is soon surrounded by the lymphoid elements, which now commence the formation of the real tubercle, and the rods appear at the same time within the cells which have received the foreign intruders, perhaps with the hopeless view of destroying them. If we regard the white blood-corpuscles as precursors and preliminary steps in the formation of epithelioid and giant cells, the presence of bacteria in these latter is not remarkable. But if, with Baumgarten, we consider that the connective-tissue cells essentially give rise to the histological neoplasm, it is difficult to understand how micro-organisms destitute of voluntary motion can get into the interior of the cells.

The tubercle bacilli generally occupy a quite characteristic position in the giant cells. While the centre consists of a non-nucleated region destroyed by coagulation necrosis, there are found at the margin of one side of the cell many nuclei arranged in a wreath, and exactly opposite to them, at the other side of the margin, the rods. This position is by no means regular. Here, too, the peculiarities of the different kinds of animals play a part; with the marmot and the Spermophilus guttatus (a rodent very common in southern Russia, which, according to Metschnikoff's investigations, is exceedingly susceptible to infection with tubercle bacilli) the rods also lie in the very numerous giant cells, irregularly scattered and often exhibiting very remarkable forms of degeneration. They are swollen, club-shaped, and change finally into amber-yellow, flaky masses hardly indicating the origin of tubercle bacilli.

But there are also transition forms met with which are of a peculiar nature. Metschnikoff regards this annihilation of the bacterial protoplasm as an immediate sequel of cellular activity, and sees in the giant cells even phagocytic elements.

The bacilli perish in the course of the tubercular process accompanying the progressive decay and increasing necrosis. Their number is thus frequently small, especially in cases of extensive and radical tubercular changes. The bacteria may, finally, disappear even completely, and leave their traces only in the consequences of their destructive action. The quantity of rods will, therefore, by no means always correspond to the severity of the disease, and the practical hint conveyed by this statement is obvious.

The bacilli appear within the vessels only exceptionally; they are noticed in the blood only when the poisonous matter had entered the circulatory system from the beginning.

We know by the microscopic investigation, by cultivation outside of the animal, and by the successful transmission from artificial cultures that the tubercle bacillus is the specific exciter of tuberculosis. How, then, can the peculiar manner and appearance of the disease be explained by the living qualities of the original microorganism?

It will be interesting, above all, to ascertain the way by which the bacillus finds, under ordinary conditions, entrance to the body. All the doors of entrance at all possible have, in experiments, proved accessible, and it was natuarlly supposed that these ways were significant for natural infection, as with anthrax bacilli.

Experience and clinical observation have vindicated this supposition. Infection occurs from the surface of the skin through wounds by contusions, cuts, or otherwise. Thus the well-known corpse-tubercles of the pathologists are small, well-circumscribed, locally-limited settlements of tubercle bacilli surely established, as Karg and others have found. Thus infections have occurred on the fingers of people who had been injured by glass vessels and other objects soiled by phthisical sputum. Some cases of tuberculosis have become particularly famous which have been observed several times in connection with the ritual circumcision of Jews. The wound made by circumcision is closed and sucked by the operator's mouth for the purpose of hæmostasis. If this is done (as has happened sometimes) by a man afflicted with phthisis, tubercle bacilli may get into the lesion, healing is delayed, the neighboring lymphatic glands swell, and finally a general tuberculosis of the children is developed.

Lupus is a tubercular affection of a peculiar character confined to the skin. It is in its clinical symptoms but little similar to other tubercular changes, and impresses one as a peculiar, unique disease. Expert investigations have proved that it is unconditionally a tubercular affection; and Koch especially succeeded not only in proving the existence of bacilli in the lupus nodules, but also in obtaining from them pure cultures of the bacilli. It cannot as yet be definitely stated what the causes are for these differences in the clinical attitude of ordinary pulmonary tuberculosis and lupus.

We know that animals—for instance, rabbits—may be infected by feeding them with tubercular sputum. The mesenteric lymphatic glands are first attacked, then the intestine, spleen, liver, etc., in succession. Similar changes are frequently observed with phthisical patients who swallow their sputum, thus affecting the intestinal canal. A very important kind of transmission is that by the unboiled milk of cows afflicted with murrain. The investigations of Bollinger, Hirschberger, and others have shown that even if the udders themselves are not tubercular, the secretions of the mammary glands may contain tubercle bacilli, and also that in nearly half of all tubercular animals the milk proves to be infected. The bacteria will, then, enter the human body with the food; by virtue of their covering they resist the acid gastric juice (without the aid of spores) and then get into the intestine, from which they are carried to the lymphatic glands.

The danger of drinking unboiled milk, according to these observations and experiences, appears very great, and it is the duty of every sensible physician to strictly forbid the use of unboiled milk, especially with children.

But all the possibilities of transmission of tuberculous poison are surpassed in importance by the infection by respiration. We know that, as a rule, tuberculosis does not appear as an early general infection of the body, but that the bacteria usually produce merely local changes at the spot where they found entrance and the first opportunity of showing their dangerous nature. The lungs are, however, in man as well as in animals, the organ in which the morbid processes, if not exclusively, at least principally and primarily, are manifested, and this very fact points to the absorption of the poison at that place.

But, it may be asked, how do the bacilli gain access to the lungs? The tubercle bacillus is a strongly-parasitical bacterium, to which the conditions necessary for development are offered nowhere outside the bodies of man and the warm-blooded animals. A transmission of the disease can, therefore, take place only from one individual to another. That it may occur from direct contact has been proved; for instance, cases of tuberculosis produced in connection with circumcision.

But the same or similar conditions never pertain to the lungs. We know, indeed, only one possibility of the entrance of bacteria into the respiratory organs, viz.: the medium on which the microorganisms have developed must dry up, disintegrate into powder or dust. All the former hypotheses, according to which the bacteria free themselves from their surroundings and ascend, or are said to be lifted upward during the evaporation of fluids, or, finally, to be detached from the surface of their place of colonization by strong currents of air, have been shown to be erroneous, the last-mentioned supposition especially having been disproved by Naegeli's investigations.

It may be stated from the start, therefore, that only such bacteria can enter the body by way of the air or by respiration as do not succumb to desiccation.

We have already seen that the transmission of spores (for instance, with the anthrax bacillus) by respiration is not surely established, but that the bacteria nevertheless possess a very considerable resistance to the influence of desiccation. It may be that this power must be credited to special forms or that the rods are in themselves such resisting structures; but it is sufficient that the fact itself is established beyond doubt, thus furnishing the first condition for the infection presumed to take place.

Where is an opportunity offered to man to inhale dried bacilli? This question is easily answered. If it be remembered that the very expectoration of tubercular persons usually furnishes the richest supply of rods, and if it be borne in mind how carelessly and heedlessly this dangerous matter is almost everywhere treated, how it is strewn and scattered about, it will be found a source of infection flowing, unfortunately, so copiously that other sources need hardly be looked for.

Cornet's beautiful and significant investigations have proved that we are not dealing with a possibility, but with a fact founded on actual conditions. Cornet ascertained that the tubercle bacilli are by no means scattered all about us without choice or difference (as was formerly supposed); that they are not ubiquitous; but that they are only met with in definite, narrowly-circumscribed regions, the centre of which is regularly a tuberculous and phthisical person.

Cornet examined the dry, powdery dust usually settling on the floor and in the recesses of our dwellings. A small quantity of it was injected into the peritoneal cavity of Guinea-pigs, which are so highly susceptible to tuberculosis. If the matter came from places where consumptives had been dwelling the animals succumbed, after the lapse of a few weeks, almost without exception to a pronounced tuberculosis. Tuberculosis was mostly confined to the large organs of the abdominal cavity, and as a rule the way in which the spread of the infectious matter had taken place could be established. Local changes had first (as always) occurred in the immediate region of the spot where inoculation had been made. The neighboring lymphatic glands were swollen and had in part already become caseous; the poison had afterward, by slowly creeping on, conquered the territory step by step. In very pronounced cases infection had attacked the lungs through the diaphragm. But the respiratory organs were never the chief seat of pathological change (as happens with the spontaneous tuberculosis of those pigs), and in the anatomical picture supplied a definite conclusion concerning this kind of transmission.

But when the inoculated dust came from places where there had been no consumptives, the animals did not fall a prey to tuberculosis.

Having succeeded in discovering the recesses usually concealing the infectious matter, Cornet further showed the manner in which the tubercle bacilli generally get into our apartments. It may be thought that the direct discharge of the expectoration upon the floors plays the principal rôle. But this bad habit is, after all, not so general, and the aversion of even common people to soiling their own home is so pronounced that we cannot attribute the regular occurrence of bacilli in places serving as abodes for consumptives to this cause. Cornet ascertained that the truly dangerous procedure is the reception and preservation of the sputum in pocket-handkerchiefs. It finds there the best opportunity of quickly drying and being turned into dust after a repeated use of the cloth. Especially the bedclothes on which the pocket-handkerchief lies during the night ready to be taken up during paroxysms of coughing proved, in Cornet's investigations, to be a fruitful place of deposit for the bacilli, and a series of particularly interesting instances showed the real significance of this fact. The rooms in our hospitals in which the consumptives are so often accommodated along with other patients, and, strange to say, especially with persons whose lungs are otherwise affected, proved to be infected. A hotel room in which a phthisical actress had lived but a few weeks contained large quantities of bacilli, and so on.

The sensation caused by these discoveries has been quite extraordinary and justifiable. The view is gaining ground that tuberculosis is an infectious disease, that man belongs to a readily-susceptible species, that the bacilli are the sole cause, and that the dried expectoration especially causes a spread of the affection. Every phthisical person indicates, therefore, an immediate danger for those around him, and we should bear in mind that in intercourse with tuberculous presons we stand nearer to fate than otherwise. The very places where a continual aggregation of people takes place and where a patient may distribute the requisite infectious material for months, thus exposing healthy persons, are the favorite seats for tuberculosis; hence barracks, lunatic asylums, jails, etc., are its most favorite seats.

But Cornet's observations have pointed out a way by which we may prevent (at least to a certain degree), and without great trouble, the unlimited extension of the evil and with a decided chance of success. Little as we should succeed, even by the strictest measures, in banishing anthrax altogether (whose bacilli find opportunities for their development outside the body and in a thousand uncontrollable spots), tuberculosis would, according to theory, be prevented from the moment that all men and animals affected by this disease should be properly quarantined.

We are far from the realization of this prospect, but much can always be done toward accomplishing it. By merely preventing the sputum of phthisical persons from drying up, we shall render harmless by far the most important kind of infectious matter. Admonish patients (as suggested by Cornet) to empty their discharge into a vessel filled with water or into a closed spittoon (as proposed by Dettweiler) even for use outside the house, in the street, etc.; point out to patients the danger they may inflict on their family and all around them; see to it that too intimate intercourse between tubercular and healthy persons be restricted as much as possible. Thus every one will find ample occasion in his sphere to effectively aid in combating the pernicious evil.

Do not object that there is, after all, no absolute protection thereby afforded against the transmission of the disease, and that the tubercular person continues to be a focus of infection. This is certainly correct, and an isolation of the consumptive would undoubtedly be a better and surer means of prevention. But as long as such a radical procedure is impossible owing to feelings of a humane and social nature, do not let us relinquish the half because the whole is impossible.

Tuberculosis is a contagious disease caused by a specific bacillus. We should ever remember this dictum as the Alpha and Omega of our knowledge, while touching briefly two questions that cannot be properly ignored in discussing this matter. A number of investigators believe that the appearance of tubercular infection always depends on a previously-existing disposition of the body afflicted. In considering this question we are vividly reminded of the famous quotation, "wherever definitions are wanting, a word comes forward at the proper time," and we are looking in vain for a concise explanation. We will not, however, dispute the fact that a series of circumstances can certainly favor the transmission of tubercle bacilli, and that a general debilitation of the organism, imperfect respiration, catarrhal affections of the upper air-passages, etc., have their influence. But no observation known to us, no theory free from objection, speaks in favor of any condition as absolutely necessary for the success of infection. This whole question is, besides, of secondary importance to us. Even the strictest adherent of the doctrine of predisposition has to deal not only with the disposed individual, but also with the infectious matter, the bacillus whose action is required; and it seems idle to lose many words concerning the greater or less amount of susceptibility to be brought forward as a cause to produce an effect.

Let us pass to the second question. If tuberculosis is not considered as an acquired, but as an inherited disease (whether transmission is presumed to occur during fecundation or only later, during intra-uterine life), all our discussions concerning the development of infection from the inhalation of bacilli, successful preventive measures, etc., must appear incorrect and fall to the ground. But not a single indubitable case of congenital tuberculosis (established before or during birth) has thus far been observed in man. Johne and Malvoz have, it is true, found tubercle bacilli twice in cattle in the organs of embryos. The hereditarians of the strictest order have for years danced most enthusiastically around the calf described by Johne.

But we can surely object that these observations are decided exceptions to the rule, and, moreover, that we should take care not to transfer conditions found in cattle to man. We are far from saying that such a thing cannot happen; but it has as yet not been established, and all cases of tuberculosis occurring during the first months of life thus far communicated have proved open to the suspicion that they were the result, not of an inherited affection, but of one acquired at a very early period—i.e., a genuine infection.

How can we reconcile this theory of inherited tuberculosis with the fact (familiar to everybody) that the disease so frequently afflicts men in middle life? Baumgarten attempts to account for this by tracing the late appearance of tuberculosis to a long-continued latency of the tubercle bacilli in the body. Their germs are said to be transmitted by way of inheritance and to remain dormant for years before they break forth at a time when the power of resistance of the living tissue has become less, and then commence their destructive activity. Analogous conditions with tardy hereditary syphilis are referred to as supporting this view. But apart from the circumstance that in the latter disease most of the cases are by no means of a uniform character, we must not draw conclusions from one infectious disease to another as long as their conformity or similarity has not been established beyond a doubt. Clinical experience and, above all, direct experiments showing no latency of tubercular germs lately received, prove the view represented by Baumgarten to be entirely untenable. If we may state our view of these things in brief, we declare that tuberculosis is an infectious disease, caused by a specific bacillus and transmitted to man mostly through inhalation of dried sputum of the lungs of phthisical persons.

IV. LEPRA BACILLUS.

Leprosy is a disease possessing several points of resemblance to tuberculosis, though in every case distinguished from it by very important and unmistakable indications. In Germany it is as good as extinct, and is only rarely seen now and again in the hospitals as an exotic rarity, yet it has persisted in certain districts even in Europe, and is still prevalent in Southern Spain and on the coast of Norway.

In consequence of the peculiarities of its development and occurrence it has always attracted the attention of the learned, and has provoked great differences of opinion as to its nature and its causes. In the year 1880 Armauer Hansen, a physician at Bergen, proclaimed as the result of many years' investigation that he had succeeded in many cases of leprosy in recognizing the presence of bacteria. These, he said, were to be found chiefly in the nodules characteristic of the disease, and usually possessed the shape of rod-cells. Hansen's statements were corroborated by Neisser, and the lepra bacilli have since become generally recognized.

They are slender, moderately-large rod-cells, with sharp ends almost identical with the tubercle bacilli in appearance; perhaps a little shorter. The lepra bacilli, like the tubercle bacilli, have no voluntary movement. Whether the oval or round spots which in stained bacilli appear as light uncolored portions in the interior of the cells are to be regarded as spores or not cannot as yet be decided, and there are no facts known which would point to the necessary existence of enduring forms.

As to the staining of the bacilli, we already know that they are the only species of bacteria as yet known to which the special procedure employed for tubercle bacilli is also applicable. Yet the staining of the lepra bacilli is performed with much more ease and rapidity, and Ziehl's solution penetrates without difficulty into the bacterial protoplasm. Gram's method is also recommended, as it brings out the cells beautifully and is specially adapted for accurate investigations.

The lepra bacilli and the tubercle bacilli are at once distinguishable from each other by their behavior when treated with the ordinary aqueous anilin solutions. The former are as sensitive to them as are the majority of all micro-organisms. With fuchsin or methyl-violet in particular it is easy to obtain good preparations.

When we speak of the rod-cells regularly observed in leprosy as "lepra bacilli," we are raising to a certainty that which in point of fact is only a very high degree of probability. It is true that these bacilli are found in all cases of leprosy, and generally, too, in very great numbers, and also that they are found only in cases of leprosy. If we then take into account the similar symptoms and other conditions peculiar to tuberculosis, one is perhaps justified in concluding that here, too, the bacilli are the cause of the affection. But this opinion cannot be proven.

For there is as yet no well-established case in which it has proved feasible to cultivate the bacilli outside the body, and still less to reproduce the disease by the aid of such cultures.

Some recently-published facts would seem to contradict this assertion, and an Italian investigator, Bordoni Uffreduzzi, has reported some experiments which certainly deserve attention. He succeeded in obtaining from the marrow of the bones of a man who had died of leprosy a rod bacterium which at incubator temperature grew slowly on hardened blood-serum with the addition of peptone and glycerin, and which he proclaimed to be the lepra bacillus. It stained in the specific manner—i.e., it was susceptible to the ordinary staining matters—but when treated with anilin-fuchsin did not lose its color again under the influence of acids, thus completely fulfilling in this respect the conditions which we should expect to find in the real lepra bacillus.

The micro-organism grew, as already mentioned, but slowly on the artificial culture medium, and it was not till after several days had elapsed that a commencement of development was visible in the incubator. The colonies appeared as little round plates, of whitish-gray color, with thickened centre and irregular, jagged edges. The stab-culture showed wax-like, slightly-yellow coating, which did not liquefy the serum.

Attempts at transmission to animals, which were made in the most varied manner, all remained unsuccessful. Bordoni explains this by the supposition that the bacillus outside the body very quickly becomes a prey to natural attenuation; that in exchanging its parasitic for a saprophytic mode of life it loses all its virulence.

Several circumstances, indeed, would seem to favor the correctness of this view. Though the bacillus at first developed only at incubator temperature, on a food medium specially and carefully prepared for it, and then with but very little energy of growth, yet soon a diminution of these peculiarities was observable, and after a few generations a luxurious growth took place on ordinary gelatin and at ordinary room temperatures.

All these facts show it to be quite possible that it may have been the genuine lepra bacillus. But, on the other hand, some serious objections cannot be suppressed. The marked difference between a cover-glass preparation of Bordoni's bacillus and one of lepra bacilli obtained directly from a genuine case of leprosy is at once apparent. In the large, thick, swollen rod-cells of Bordoni's bacillus we will scarcely find a trace of resemblance with the slender, elegant forms of the other. Of course the thought naturally arises that the artificial medium has produced only involution forms, and that this may explain the suspicious appearances. But when we weigh against this the fact that all the endeavors of very numerous and experienced investigators, armed with all the means and appliances of modern science, have hitherto failed in arriving at the same results with Bordoni, it can but seem fair to suspend judgment and not to regard the artificial cultivation of the lepra bacillus as a definitely-solved problem.

On the other hand, the reproduction of the disease with all its peculiarities, together with the occurrence of the bacilli, has in many cases been accomplished with undoubted success by inoculation with portions of the diseased tissue.

Arning has experimented on the human subject. He had for several years made leprosy his special study on the Sandwich Islands (one of the chief seats of the disease), and there he found an opportunity to make the experiment on a criminal who had been condemned to death. The man in question was not from a leprous family and was in good health. He was inoculated with portions of freshly-taken lepra tubercles by means of subcutaneous application, and further development was then watched. After some months typical leprous changes were visible near the point of inoculation on the upper arm. These spread gradually, and in the course of five years he died of undoubted general leprosy.

With animals, too, investigators at Königsberg, Melcher and Ortmann, obtained positive results. They transferred lepra tubercles immediately after excision from a human subject into the anterior chamber of rabbits' eyes, and found that the animal died after some months. On dissection, an extensive leprosy of the entire viscera was found; the cæcum in particular, but also the lymph glands, the spleen, and the lungs were full of tubercles varying in size from a pin's head to a millet seed, in which the lepra bacilli were clearly recognized. Any one who could have seen the preparations made from these experiments can scarcely have a doubt but that it was a case of genuine lepra and not a case of mistaken tuberculosis.

None of these experiments decide in what manner the distribution of leprosy takes place under natural circumstances. We only know that man himself is the chief means of conveying the virus; but as to the important question whether an infection from one human being to another generally occurs or can possibly take place, and, if it does, how it is accomplished, opinions differ very widely.

Like tuberculosis, leprosy attacks almost all organs and parts of the body; yet it exhibits a preference for the skin and the peripheral nerves.

We may, therefore, suppose that the symptoms of the disease and the conditions revealed by post-portem examination will be different. Only those tubercles which have already been frequently spoken of are usually found. Examined under the microscope, they appear almost exactly like those of tuberculosis in their composition, and macroscopically they can hardly be distinguished from the latter at the beginning of their development. Giant cells, it is true, appear very rarely in the lepra tubercles, and in their structure the inflammatory cells, the lymphoid elements, are almost exclusively concerned.

It is chiefly in the tubercles that the bacilli are found, but they are also found in the skin and the connective tissue surrounding the nerves, and in the lymph glands, the spleen, and the liver, but are generally absent in the blood.

As to the precise distribution of the rod-cells in the tissue there has been of late years much difference of opinion, and the question, "Where do the lepra bacilli lie?" has often been asked. The opinion long prevailed that the inflammatory cells of which the tubercles are composed were to be regarded as the chief seat of the micro-organisms, and the presence of bacteria in them was regarded as their characteristic and hence they were often called "lepra cells." Unna opposed this view with the assertion that the lymph ducts of the glands were the true seats of the bacilli, and that by means of his desiccation method (the main facts of which have been given)* the truth of his statement could be proved.

If he dehydrated his sections, after the decoloration in nitric acid and distilled water, by heating over a flame instead of by

---

* See page 57.

alcohol, and if he further clarified not with oil of cedar or cloves, but with xylol, he saw, as he maintained, that the agglomerations of bacteria which had always been taken for lepra cells were in fact no cells at all, but only a deception caused by the wrong treatment of the object, and that the supposed cells were only free, ball-shaped assemblages of rod-cells in enlarged cavities of the lymphatic vessels.

We know, however, that just complaints are made against the drying system on account of its destroying the transparency of the tissue, and it has been suggested that Unna's lymph passages are nothing but artificial productions. In fact, the greatest authorities on the lepra question, Neisser, Touton, Arning, etc., all persist in the opinion that the chief mass of the rod-cells lies within the lepra cells, while a certain portion of them is also to be found distributed in the tissue.

V. SYPHILIS BACILLUS.

If important factors are wanting in lepra to positively prove bacilli the cause of the disease, this is still more the case in another disease which somewhat resembles tuberculosis and lepra, namely, syphilis.

As everybody is aware and as experience abundantly shows, syphilis is a very infectious and easily-transmissible disease. But we know hardly anything about the sources of its infectious nature and the causes of its peculiar phenomena. Although we might be inclined to suspect a bacterium as the bearer of its virus, yet our knowledge is not yet sufficient to enable us to prove the correctness of such a supposition.

The first approach to a satisfactory explanation is perhaps to be found in the observations published by Lustgarten a few years since. He announced that he had succeeded, by means of a special staining process, in finding a particular species of bacillus in syphilitic lesions and the secretion of syphilitic sores, the occurrence of which was restricted to these places named, and which was, therefore, peculiar to syphilis.

The thinnest possible sections must be treated as follows according to Lustgarten: First they are to be stained for twelve to twenty-four hours at ordinary room temperature with anilin-gentian-violet, and then the process is continued for about two hours at 40° C. in the incubator; next they are washed for several minutes in absolute alcohol, and placed in a 1½% aqueous solution of permanganate of potash for about ten seconds, and at last well rinsed in distilled water.

The process is now repeated a few times, beginning with the decoloration in potassium, always shortening the time (for instance, only three or four seconds for the solution of permanganate of potash), till the sections are quite colorless. Then comes alcohol, oil of cloves, and xylol-Canada balsam.

"Smeared" cover-glasses are treated in the same manner, only that after the staining in gentian-violet distilled water must be employed instead of pure alcohol, and the different operations must be conducted more rapidly.

Besides this method several others have been recommended which claim to attain the same results more easily. We will mention that of de Giacomi, by which cover-glasses and sections are stained in anilin-water-fuchsin, which is employed hot only a few moments for the former and twenty-four hours cold for the latter, and then decolored with a solution of perchloride of iron, at first much diluted, then quite saturated. Cover-glasses are rinsed in water, sections in alcohol. The after-treatment is as usual.

In such preparations, Lustgarten discovered peculiar rod-cells which resembled the tubercle bacilli in appearance, but were decidedly curved more frequently than the latter, and are further remarkable for slight knob-like swellings inclosed singly or in groups in large cells, which have no visible connection with their surroundings.

It is of course natural to attribute a special importance to these bacilli, which stain in such a peculiar manner and have such a remarkable position in the tissue, but we have no proof of their being the cause of syphilis.

In the first place, we must object to the process by which the bacilli are rendered visible as being unsatisactory. It is not alone extremely complicated, but its results are wanting in certainty. Although Lustgarten assures us that he has regularly found the bacilli in the cases he has examined, many who have wished to confirm his statement have been less successful. In the cover-glass preparations the bacilli are, indeed, often found, but in sections they have been found only by a small number of investigators, even when the instructions given were followed out with the utmost exactitude.

Lustgarten himself found the micro-organisms in question only in very small quantities, and neither their numbers nor their position and distribution in the tissue seem calculated to account for the violent changes which are the peculiar result of syphilis.

Lastly, the value of the method (and, therefore, also the importance of the rod-cells) has been seriously doubted, since bacilli have been found which behave very similarly, though not precisely in the same way as regards staining.

That the tubercle and lepra bacilli would also stain on his system was noticed by Lustgarten himself. Yet Lustgarten's bacilli lose their color rapidly in hydrochloric, nitric, and sulphuric acids, while the tubercle and lepra bacilli do so only after long exposure to their action.

Then the discovery was made simultaneously by Matterstock, Alvarez, and Tavel, that in the preputial and vulvar smegma rod-cells occur which may be stained precisely according to Lustgarten's directions, and which in their appearance are scarcely distinguishable from the supposed syphilis bacilli.

The correctness of this observation has been corroborated on all hands, and has often been adduced against the claim that Lustgarten's bacilli were the cause of syphilis. At first, it is true, a slight but regularly-occurring difference was thought to exist between the smegma bacilli and those of Lustgarten. The former were said to lose their color much more quickly than the latter under the influence of alcohol. Further experiments, however, have not confirmed this statement, and there remains but one argument in favor of the syphilis bacilli—their occurrence in the tissue. Here there is no chance of their being confounded with smegma bacilli, as there was, perhaps, in the case of cover-glass preparations of ulcerous secretions, for it is not conceivable that smegma bacilli could penetrate into the deeper parts of syphilitic scleroses or even into gumma.

A definite solution of the question is clearly not yet possible. Very distinguished investigators—as for example Doutrelepont, who has paid special attention to the subject—are of the opinion that Lustgarten's bacilli do in fact stand in some connection with syphilis. Others are of the opposite opinion, but all are convinced that further progress can only be made by the substitution of a better method, and that the discovery of such an improved or a radically new method must be our immediate object of search. Especially should our efforts be directed to the artificial cultivation of syphilis bacteria outside the body—an undertaking which has as yet baffled all attempts.

VI. BACILLUS OF GLANDERS (MALLEUS).

Tuberculosis, leprosy, and syphilis stand nearly related in regard to the pathological changes which they produce in the tissues, and are in this respect akin to a fourth affection, which, however, occupies no such prominent place in human pathology as do the three others. We refer to glanders or malleus.

This disease, which was known and feared in ancient times as particularly fatal to horses and asses, sometimes infects human beings, and generally ends in death. In the bacteriological study of this affection, therefore, and especially in handling the micro-organisms of which we are about to speak, caution is of the first importance. Almost half a dozen cases could be cited in which the neglect of necessary care has led to the death of the investigator.

Although there were plenty of opportunities to study the disease and watch its progress, its nature and its causes long remained mysterious, and even down to the middle of this century doubts were entertained as to whether it was an infectious disease or not. Then, indeed, the conviction gained ground that glanders spreads only by infection from animal to animal, and people began to seek after the causes of the infection.

In the year 1882, soon after the discovery of the tubercle bacillus, Löffler and Schütz recognized as the bearer of the infecting virus a definite species of bacterium, the glanders bacillus, which they found within the infected tissues. This was cultivated outside the animal organism, and at last transmitted successfully from the artificial cultures, so that its specific importance no longer admitted of a doubt.

The glanders bacilli are small, slender rod-cells with rounded ends, somewhat shorter and decidedly thicker than the tubercle bacilli. They usually occur singly or in pairs, never in long threads. They possess no motile power, though the extremely brisk molecular motion which they frequently display in the hanging drop gives them the deceptive appearance of spontaneous movement.

The occurrence of spores has been positively proved by the experiments of Baumgarten and Rosenthal, who succeeded in making them visible by means of double staining. The genuine, sharply-defined spores are not to be confounded with the frequently-occurring light spots and gaps in the stained rod-cells, which by their irregular shape and arrangement are clearly seen to be something else. Löffler regards them as indications of involution and indicating an incipient decay. It is remarkable that the bacilli without the help of spores retain their vitality almost three months in a dry state.

The glanders bacilli, like the majority of pathogenic bacteria, belong to the semi-anaërobic species. They require for their development a tolerably high temperature, and are, therefore, at least by preference, parasitic. They do not thrive under 25° C. nor over 42° C. Their optimum temperature is between 30° C. and 40° C.

In staining, the glanders bacillus appears to belong to that class of micro-organisms already mentioned which absorb the coloring matter quickly, but readily lose it again when we decolor. Many plans have been tried to overcome this difficulty, so that for staining this species a long list of instructions might be formed.

For cover-glass preparations Löffler recommends his anilin-gentian-violet or anilin-fuchsin, allowing the hot liquid to act for about five minutes. Then the decoloration is effected in 1% solution of acetic acid, to which the yellow color of Rhenish wine has been given by adding tropæolin in aqueous solution. In this the preparations remain one hour and are then rinsed in distilled water.

The same result may be more readily obtained by first treating the cover-glasses with warm carbol-fuchsin or Kühne's carbol-methyl-blue, and then only with distilled or slightly-acidulated water—ten drops of hydrochloric acid to 500 of water.

The double staining of the glanders bacilli has not yet proved possible, and neither the process adapted to the tubercle bacilli nor Gram's method is applicable.

As the glanders bacilli only thrive at high temperatures, they cannot be cultivated on gelatin. On plates of ordinary agar, or agar with an addition of 4% glycerin, and kept at about 37° C., the colonies are found abundantly developed on the second day, appearing as light-yellow or whitish transparent, roundish accumulations. The microscope shows brownish-yellow, dense, somewhat granulated masses, with edges comparatively smooth and sharp.

In the test-tube on oblique agar, or, better still, on glycerin agar, a clearly-defined whitish, translucent, moist shining coat forms in four or five days along the inoculation scratch at incubator heat. On blood-serum, generally in the same length of time, spots appear here and there, which are roundish, perfectly transparent, more or less yellow, and drop-like, and do not liquefy the medium. Afterward they coalesce into one even, tough, slimy covering.

The growth of the glanders bacilli on potatoes is very characteristic and in many respects worthy of note. Shortly after the inoculation on slices prepared after Globig's method and kept at incubator heat, an amber-yellow, curiously-transparent covering appears, which looks almost like a thin layer of honey, but which increases rapidly in thickness and at the same time assumes a darker tint. At the end of a week the culture is reddish-brown or fox-red, and presents such a peculiar appearance as almost to prevent the possibility of its being confounded with any other species.

From the agar, blood-serum, and also from the potato successful transmissions can be made without difficulty, and a small trace of such a culture applied subcutaneously to a susceptible animal suffices to produce genuine glanders with all its peculiar symptoms.

It was natural that we should at first employ horses and asses for these experiments, since their susceptibility to the disease was already known; but afterward Löffler made the discovery that field-mice and Guinea-pigs were almost equally susceptible, while white mice, the common house mice, cattle, and swine were almost completely exempt. Rabbits were found little susceptible, but cats were very easy to infect. A few drops of a culture well mixed with sterilized water, and put into a pouch in the abdominal wall of a Guinea-pig or injected into a field-mouse at the root of the tail, in all cases leads to a fatal termination.

In these artificial infections the fact is clearly apparent that the virus of glanders at first confines itself still more markedly than that of the tuberculosis to local action, and that it only proceeds gradually to spread its fatal influence more extensively. At the spot where inoculation took place the first changes are to be seen, and from there the poison creeps slowly on. But the spread of the affection is only step by step, not by way of the circulation, as the blood is almost always free from bacilli.

Thus in the case of Guinea-pigs the local symptoms of infection appear four or five days after the inoculation, but as many weeks generally elapse before the general symptoms appear and the death of the animal ensues. The same applies to the horse, but the field-mouse forms an exception, inasmuch as the size of the animal is such as scarcely to admit of a distinction between the local and the general symptoms. Mice, therefore, generally die on the third or fourth day after inoculation, sometimes still earlier, and in them the whole course of the disease differs from that observed in Guinea-pigs.

The latter show as the first result of the inoculation a sharply-defined swelling. This gradually becomes caseous, round, or oval-shaped, purulent ulcers with indurated edges develop, and occasionally, after many weeks, a cessation of the processes ensues, healing follows, the ulcers leaving deep scars.

As a rule, however, the local affection is followed by a diffuse swelling of the glands, which leads to suppuration and a purulent discharge. In male animals the testicles thicken into hard nodular masses, which then also fall a prey to suppuration. Lastly, in addition to the other symptoms, comes a diffuse inflammation of the joints, especially of the feet, and death follows from general exhaustion. It is but seldom that the nasal cavity is attacked.

In mice, on the other hand, the local symptoms are not noticeable. Two or three days after the inoculation the breathing is accelerated, the animals sit still in a corner of their cage with their eyelids glued together, and all at once, without previous warning, fall on one side, dead.

Dissection shows the most extensive tissue changes, chiefly in the spleen in Guinea-pigs, in field-mice also in the liver and occasionally in the lungs. The chief point brought out by the dissection is (as also in the case of tuberculosis and other allied diseases) the presence of tubercle-shaped neoplasms, which in glanders have a decided tendency to degenerate—to soften. Macroscopically regarded they strongly resemble genuine tubercles, and appear as grayish-white grains somewhat smaller than a millet-seed and rising slightly above the surface. In the field-mouse the short duration of the disease only allows the nodules to reach a limited size, and in the liver, for example, one often sees quite small, extremely numerous gray dots, hardly recognizable with the naked eye.

Under the microscope these nodules are seen to consist of dense accumulations of round cells, which also surround isolated masses of greater size and epithelioid in character. From the centre an advancing degeneration of the new formation is seen as in the genuine tubercles. The cells degenerate into an evenly-opaque mass without nuclei; complete dissolution of the tissue afterward takes place, which results in purulent matter.

In the nodules chiefly, but also in other places, we will find the bacilli which are to be regarded as the cause of the changes that have occurred. The finding of rod-cells in the sections is attended with peculiar difficulties. Here, still more than in cover-glass preparations, the bleaching which the glanders bacilli are apt to undergo instead of the desired diminution of color is a troublesome peculiarity, and the demonstration of glanders bacilli in tissue has become a sort of test of the efficacy of the different methods of staining.

Löffler formerly recommended a special method of decoloration: after long exposure to alkaline-methyl-blue, the preparations were to be put into a mixture of sulphurous acid and oxalic acid composed as follows:

Distilled water, ...	10 c.cm.
Concentr. sulphurous acid, ...	2 drops.
Five-per-cent oxalic acid, ...	1 drop.

The process was therefore as follows:

1. Löffler's methyl-blue, . . .	about 5 minutes.
2. Oxalic and sulphurous acid, .	" 5 seconds.
3. Absolute alcohol, etc.	

The acids remove the coloring matter from almost all parts of the tissues, the nuclei as well, and leave only the rod-cells deep blue on a pale background. In fact, this method yields really good results, but it is troublesome and difficult to carry out, since the acid solution must be renewed each time.

The more modern processes which serve for the decoloration of sensitive bacteria in general are, therefore, preferable—i.e., Weigert's anilin-oil method or Unna's drying method. If we wish merely to display the rod-cells without particular regard to the tissue, the method adopted by R. Kühne is the best. Stain the sections for six to eight hours in carbol-methyl-blue, then decolor first in diluted acetic acid and afterward in distilled water, next dry them on the slide with the bulb-bellows, lastly give transparency with xylol and mount in Canada balsam.

Particularly in the nodules the bacteria are then seen in great numbers, chiefly in small groups. The latter indicate by their arrangement that they originally lay together in the interior of a cell which afterward degenerated, and frequently the unmistakable remnant of a cell membrane is visible. As we might expect from their rare appearance in the blood, they are, as a rule, not to be seen in the vessels.

Please notice particularly that it is only in quite recent tissue changes that the bacilli can be observed in large numbers and in their characteristic arrangement, and that they are best seen in quite young nodules in the spleen or lungs. When degeneration has commenced and the destruction of the newly-formed parts has made some progress, the bacteria are destroyed along with the rest, and therefore they are only to be found exceptionally in the ulcerous, purulent, ruptured glands and abscesses of the skin, etc.

Their presence in such places, though probably in very small numbers, is nevertheless proved by successful transmissions by means of the pus. Löffler is certainly right in advising that for a diagnosis of glanders in a living animal less reliance should be placed on microscopical examination than on the result of the inoculations to be performed on animals which prove regularly susceptible, such as the field-mouse or the Guinea-pig. It will often be found that attempts to inoculate glanders from pure cultures will fail. The reason is that the glanders bacillus is one of those which are subject to natural attenuation. While freshly-obtained cultures will kill animals with certainty and within the time already stated, one may often find in the fourth or fifth generation that the bacilli begin to grow less poisonous and larger quantities must be inoculated. The general effects take place more slowly or not at all—i.e., only local effects are produced and the original virulence has almost disappeared. It may be imagined that this diminution of virulence is sometimes a source of embarrassment to us, for it is by no means easy always to obtain thoroughly efficient and reliable material at the moment it is wanted.

This behavior of the glanders bacillus, however, supplies a clear proof that it does not find outside the bodies of animals the conditions which completely suit its requirements and enable it to develop its qualities fully. It is a genuine parasite, and when forced to exist under circumstances foreign to its nature and habits, protests against such treatment, as we have seen.

We have not yet succeeded, by the use of high temperature or otherwise, in producing an artificial attenuation of the glanders bacilli. On the other hand, we have seen that a sort of artificial strengthening has been obtained by H. Leo, who transmitted the glanders to white mice (which are naturally non-susceptible) by feeding them with phloridzin and so making them diabetic.

By microscopic investigation, by cultivation, and by the results of transmission we know that glanders is caused by a specific micro-organism. We must here always consider in all cases what relations the symptoms of the disease bear to their exciting cause, how the peculiarities of the former are to be explained in connection with the properties of the latter, and, above all else, how the bacillus gains entrance into the body, and how it causes the development and spread of the disease.

Our experiments have shown that the glanders bacilli are able to enter by way of the subcutaneous cellular tissue after slight injuries to the skin; and it seems that this is a frequent source of infection under ordinary circumstances. In human beings in particular the virus generally enters at some scratch or slight wound which has come in contact with it. Thus the disease almost exclusively attacks persons whose occupation brings them into close contact with horses, such as coachmen, stable-boys, farmers, soldiers, etc. First, pustules and abscesses appear in the immediate neighborhood of the inoculation wound, and it is not until later that swellings of the joints, ulcers on the mucous membranes, and the other symptoms of a severe affection make their appearance.

In horses, in addition to this source of affection, there is no doubt that the disease is also frequently transmitted by the breath. It is much to be desired that suitable experiments should enlighten us as to the conditions under which this kind of infection takes place. For the present we only know its symptoms, which, however, are generally very significant. As a rule, the nasal cavity is the place at which the disease is first observed. On both sides of the nasal septum and on the mucous membrane of the turbinated bones, diffuse, irregular ulcers with thickened edges are formed. They secrete a thin mucus which runs down the nostrils. Large swellings of the neighboring lymphatic glands follow and extend to the lymphatic vessels, which may be felt through the skin, as thick as one's finger. Here and there a swelling ruptures and deep ulcers form on the skin, and at last the great difficulty of breathing gives a plain indication of the place where the glanders has its special seat in horses, viz., the lungs.

VII. ASIATIC CHOLERA BACILLUS.

In the years 1829 and 1837 Europe was visited for the first time by a disease hitherto unknown, which spread from country to country like an irresistible stream, committed the most terrible ravages wherever it came, and seemed destined to become a more terrible scourge to humanity than even the plague had been. The unwelcome guest had come from India, and from this circumstance it was called the "Asiatic" or genuine cholera.

At longer or shorter intervals this murderous pestilence repeated its visits, never stopping permanently at any scene of its ravages, but always retiring again from the places it had desolated, and disappearing for years.

Science sought for the cause and mode of origin of the mysterious disease in vain; opinions and views sprang up in abundance, but none offered a satisfactory explanation. When, therefore, in 1883, after a pause of nearly ten years, the epidemic again threatened to approach the boundaries of Europe, the different governments, appreciating the gravity of the case, endeavored to unravel the dark secret of cholera and its causes. The German imperial government equipped a scientific mission to investigate the origin of the plague and determine its cause. R. Koch was placed in charge of this expedition. After a short time he was able to report, as a result of his investigations in India, that he had found the cause of cholera asiatica in a particular micro-organism, and that he had obtained a pure culture of the germ.

In all cases of cholera he succeeded in observing in the evacuations of the patients and the intestinal contents of the deceased a bacterium perfectly distinct from others by its remarkable form and definite nature. And this fact enabled him to adduce the proof that the new micro-organism appears in no other disease besides cholera, that it must stand in certain relations to it, and that these relations must be regarded as causal.

The micro-organism of cholera asiatica belongs, by its morphological peculiarities, to the class of spiral bacteria, and is hence consistently called, by strict systematicians, Vibrio or Spirillum choleræ asiaticæ. Its true form appears, indeed, very distinctly as a short, rather clumsy rod, about half as long as the tubercle bacillus, but considerably thicker, with round ends and a more or less pronounced curve along its longitudinal axis. The degree of the curvature varies from almost straight cells to nearly semicircular ones. It corresponds closely in appearance to the comma of certain faces of type, for which reason Koch gave this bacterium the name of "Comma bacillus," by which it has attained general celebrity and which we shall retain on account of its historical significance.

It must, therefore, be borne in mind that the expression "bacillus" is, in reality, not correct. Whenever a bacterium prepares to form groups, it will become evident that it is not a "bacillus." If we had a simple curved rod only, it would have to form a more or less complete circle if it continued to grow long enough. But genuine, neatly turned screws, long spirilla, arise which sometimes attain a very considerable length. Even a single individual must possess, besides the curve, a distinct twist, a torsion, and sometimes by the use of good lenses we are able to see the beginning of a screw, namely, a single comma bacillus. Hence the rods are to be looked upon as *only fragments of a real spirillum.*

In examining cholera preparations the developed spirilla are not, in fact, frequently met with. They seem to arise only under special circumstances, usually only when the development (the rapid, continual transverse division and increase) is in some way disturbed and impeded. In hanging drops, for instance, the screws are very numerous. If the bacteria live under unfavorable conditions, e.g., if either the temperature is unsuitable or if the nourishing solution contains small quantities of alcohol, tincture of opium, etc., thus scarcely permitting their growth, such conditions favor the appearance of spirilla, which are evidence that the regular division of the cells has been retarded, thereby causing the formation of groups.

The comma bacillus appears, as a rule, singly or in pairs (as stated above); in the latter case they grow end to end, the curves facing in different directions, resembling the letter S.

The cholera bacteria possess an extremely lively motion, providing they are cultivated at a suitable temperature and receive proper nourishment. They crowd and whirl through the microscopic field "like a swarm of dancing gnats." The spirilla, too, possess the same faculty: they glide along with short and swift undulations. Löffler has been able, by means of his peculiar staining process, to demonstrate the organs of locomotion of the cholera bacilli; they consist of a single flagellum attached to one end of the cell only and are slightly bent in a wavy motion.

It is not yet fully known whether the comma bacillus forms spores or not. Koch and the great majority of investigators have not noticed the appearance of any special germinal forms; nobody has yet succeeded, for instance by staining, in demonstrating spores. Hueppe, on the other hand, claims to have found a process of sporulation by continued observation of the micro-organism in the hanging drop and on the heated slide table. He describes it as follows: The bacteria grew first in spiral threads; then arose (in no particular places) single, small, shining globules, which refracted light more strongly than the other cellular contents, and could easily be distinguished from them. These globules did not develop as special structures, like the spores in an anthrax rod; the entire limb gradually assumed a new shape; two such globules generally issued from one cell. These are, according to Hueppe, immovable and certainly do not increase by fission; but they can germinate and produce new bacteria as soon as they are brought into fresh media. Their lustre decreases, they elongate, and Hueppe claims to have seen the young cell thus arising from the germ ("spore"). He regards the process as a formation of arthro-spores and considers the globules as such.

But we have seen that very eminent investigators deny the occurrence of this kind of sporulation, and only admit an endogenous form. Hueppe's results have not been confirmed by other reliable investigators, as, for instance, Kitasato, who worked on the same subject, so that we may justly consider the occurrence of spores in the comma bacillus as not yet proven.

All other known facts certainly contradict the supposition of sporulation, if we demand of a spore that it represent a germinating and enduring possibility. The cholera bacteria have no form that possesses more than the ordinary amount of resisting power, which would enable it to perpetuate the species with greater certainty.

It has been ascertained, on the contrary, that the comma bacilli are among the most sensitive micro-organisms known. High temperatures (above 50° C.) kill them with certainty in a short time; they cannot withstand the action of chemical agents, especially acids. The acid of the gastric juice destroys them absolutely; they do not thrive on gelatin containing a trace of acid reaction. An extremely important quality (of which we shall speak again) is their destruction by drying in a very short time; for, while they retain the power of development for months in moist surroundings, they perish in a dry condition frequently within a few hours.

It is true, facts have been observed in rare cases, which apparently contradict this assertion. Kitasato and Berckholtz, for instance, have independently established the fact that cholera bacteria attached to silk threads (especially when moisture had been withdrawn from them in the desiccator) remained alive for weeks and even months. But these are decided exceptions due to the manner of drying, the thickness of the culture medium, etc., and the fact first perceived by Koch and his companions in India, that in drying the comma bacilli die more quickly than most other micro-organisms, is generally maintained.

As to the ability of the comma bacilli to resist the presence of other bacteria and of struggling for existence in rivalry with them, Koch had ascertained that they are overpowered and quickly disappear in putrescent liquids. Kitasato and Uffelmann saw them perish within a few days in artificial mixtures of fæces; but Gruber and others observed them in putrescent evacuations of the bowels and separated them therefrom in pure culture. Here, too, special conditions are important, such as the influence of temperature, the kind of bacteria happening to come in contact with the comma bacilli, the concentration of the media, etc. We must, therefore, beware of establishing absolutely valid laws concerning the capacity of endurance of the cholera bacilli, and only declare in a general way that they are exceedingly delicate structures and quickly succumb, as a rule, to external influences.

The comma bacilli principally require in our artificial cultures the unobstructed access of atmospheric air. This requirement is modified only under certain circumstances. If they are, for instance, grown according to Hueppe in raw eggs, they will flourish in spite of the supply of oxygen. In the intestinal canal of man they are likewise shut off from a supply of oxygen. It is possible that they may obtain some oxygen in these locations by splitting up molecules containing this gas by a reduction process.

The comma bacillus develops equally well at a room-temperature as at the temperature of the incubator; in the latter, however, it grows much more rapidly and luxuriantly. It fails above 42° C. and below 15° C.

The cholera bacillus is stained by the various anilin colors; but a saturated watery solution of fuchsin is the most efficient. But it must be observed that the coloring matter is frequently absorbed with some difficulty, so that cover-glasses must be treated with the stain for at least ten minutes, and even be heated in it in order to obtain perfect preparations. The micro-organisms are decolored by Gram's method.

Upon the gelatin plate, after the usual time, small white dots deep in the gelatin can be seen with the naked eye. These gradually advance to the surface and then cause a rather slow liquefaction of the gelatin. Funnel-shaped depressions are formed in the transparent medium, increasing in depth rather than in circumference, in the bottom of which the colony proper lies as a whitish mass scarcely the size of a pin's head. The plate usually has on the second or third day a quite peculiar appearance; it seems perforated with many small holes or air-bubbles. Liquefaction progresses only later; on the fifth or sixth day the third dilution, too, usually has become completely diffluent.

Under the microscope the colonies present an aspect peculiar to the cholera bacteria. The smaller ones deep in the gelatin have an irregular, receding, and occasionally rough or uneven margin; they are never circular, or at least sharply circumscribed, in the beginning of development, like the colonies of most other bacteria. They are of a bright white or pale yellow color, and in their texture exhibit a remarkably uneven granulation. This will become more manifest as they grow larger. The granulation becomes more and more pronounced; the contents assume a peculiar lustre and glitter; the color looks as if composed of little pieces of glass or crystal grains. Incipient liquefaction is seen under the microscope by the formation of a bright halo around the colony. A pale seam proceeds at a moderate distance from its margin corresponding to the outer limit of the funnel-shaped depression, liquefying the gelatin by the growth of the bacteria. At the same time we see in the colony a roseate light, a reddish hue, to be found in no other kind of bacterial growth.

Development proceeds in the test-tube cultures as follows: A growth occurs along the entire line of inoculation, but liquefaction takes place more extensively only at the surface of the gelatin. A funnel is formed similar to but very much larger than on the plate; a deep depression is formed, appearing in the partly-softened gelatin like an air-bubble, owing, presumably, to the very rapid evaporation of the fluid produced. At this time the principal mass of the culture accumulates close beneath the air-bubble, the central portions of the inoculating puncture appearing as an almost empty, shining thread in the gelatin, like a capillary tube blown out at the end; the bacteria which had been developing have descended into the lower third of the puncture, where they settle as yellowish-white masses loosely curled.

At this period of development (on about the fifth or sixth day) the culture presents an extremely significant and remarkable picture. Its equal is met with in no other kind of bacterium, and a similar one only in a few micro-organisms now known, such as the violet-water bacillus, Deneke's cheese bacillus, and Metschnikoff's vibrio. The former is distinct enough by its color; it possesses also a quality in common with the two latter, that of a much more rapid growth and a quicker liquefaction of the gelatin.

As the culture of the cholera bacillus grows older, the gelatin is more and more peptonized. It gets softened after a few weeks in the upper half of the inoculating puncture, and changed into a turbid yellowish solution. The bacteria settle on the bottom, where the solid layer lies in dense heaps. A whitish film, a kind of mouldy skin, has spread over the surface and consists of thin, brittle little pieces. This is a rich mine for very odd involution forms of the comma bacilli. They are here on the verge of dissolution, and anticipate this event by all sorts of deformities and crippled formations, hardly presenting a similarity with the former figure. Large and small balls, thick lumps, mulberry-shaped berries, most minute débris of cells, and things looking like nails with disproportionately large heads, are found in a motley heap.

The gelatin culture is usually completely liquefied and no longer transmissible after about eight weeks.

The comma bacilli are sustained longer on agar-agar; they have been found still alive on it after almost nine months. They develop on the oblique surface of this medium as a moist film of white lustre along the entire inoculating line. Blood-serum is gradually liquefied.

While (as we have stated) the gelatin, agar, etc., on which the cholera bacilli are expected to prosper must be of marked alkaline reaction, the comma bacilli possess a notable capacity of accommodating themselves also to acid media, provided the acid is of vegetable origin.

The surface of boiled potatoes frequently has a feeble but distinctly acid reaction, and yet the cholera bacilli develop on it, only, of course, by the aid of incubation temperature. Their manner of growth on them is very peculiar. In the vicinity of the inoculating spot there extends a grayish-brown, thin, and somewhat transparent layer reminding one of the appearance of the glanders cultures on the same medium, but they are usually brighter and less tough.

The cholera bacilli thrive rapidly and luxuriantly in the common nutrient bouillon. There is formed on the surface, especially at blood heat, a multifariously-folded, wrinkled, closely-connected skin almost characteristic of the culture of comma bacilli. The mass of the fluid is only very slightly dimmed; only by shaking the glasses there arise from the bottom a few heaps of bacteria, which are then equally distributed throughout the fluid.

It is worthy of notice that the cholera bacteria show a particularly strong growth in a strongly-diluted bouillon, for instance, if mixed with 6 to 10 parts of water. According to Weibel's examinations (mentioned in discussing the Spirillum rubrum), we have to deal in this special case with a quality common to almost all vibriones and spirilli, which has been used even for diagnostic purposes.

The cholera bacteria may also thrive and multiply in sterilized milk without perceptibly changing the fluid—a fact to be considered in connection with their transmission to man. In non-sterilized milk they live for only a short period, as Kitasato has shown; acidification soon appearing destroys the cholera bacilli in a relatively short time, but so much time usually elapses that milk in most cases would be used before this occurs, and therefore cholera bacteria having entered the fluid would be received alive.

The thorough examinations by Wolffhügel and Riedel have finally established the important fact that the comma bacilli sustain themselves in sterilized water, no matter whether it is obtained from river, well, or aqueduct. The increase begins some time after the sowing and reaches its highest point on about the seventh day, but the vibriones may be shown in a state capable of development and in considerable number even after months. This state of things is different, indeed, in non-sterilized water, the cholera bacteria being here almost fully dislodged by existing micro-organisms in a few days.

An observation of Koch shows that the artificial conditions of this experiment do not always correspond to the natural state of things. He succeeded in finding comma bacilli in an Indian tank —i.e., in the greatly-polluted water of one of those marshy reservoirs into which the Hindoos empty their natural offal of any kind and from which they also take their water for drinking and other uses without any hesitation. This significant fact proves that the comma bacilli also thrive in nature outside of the human body, and are able to lead, for a shorter or longer period, a saprophytic mode of life, and are not characterized as genuine parasites, like the tubercle bacilli.

That the cholera bacteria, while they are growing, change those media on which they develop, is shown by the fact that the gelatin is regularly liquefied under their influence. Brieger has been able to show in cholera cultures a definite body of basic character, the cadaverin or pentamethylendiamin, and pointed out the possibility of its being related to certain symptoms frequently manifested in the progress of the disease itself. He and Nencki have subsequently found two or three other bases, proving them likewise to be poisonous by experiments in animals and suggesting the power of transmutation of the cholera bacteria.

We must mention here an observation almost simultaneously made by Bujwid and Dunham which shows that the cholera bacteria are the original exciters of quite important and specific chemical processes. By treating comma-bacilli cultures grown in bouillon containing peptone (or in common nourishing gelatin) with a small quantity of sulphuric acid, there will shortly appear in the solution a reddish-violet (frequently purplish-red) decoloration. This is met with in the same or a similar manner only in the vibrio Metschnikoff, and is absent especially in Finkler's, Emmerich's, and other intestinal bacteria usually cited together with the cholera bacilli. Bouillon cultures exhibit the reaction very distinctly after remaining from ten to twelve hours in the incubator, while the gelatin cultures furnish it only after some days, whenever liquefaction has attacked the greater part of the culture medium.

This specific cholera reaction is, according to Salkowski's investigations, nothing else but the common indol-reaction caused by adding indol to nitrous acid. The cholera vibrios likewise form indol from the albuminates in their cultures, for which reason an addition of peptone to the medium is requisite. We might, therefore, feel inclined to find in this circumstance something peculiar to them; but it can easily be proved that many other micro-organisms produce indol as well. In order to produce the red color reaction, it suffices to mix their cultures with impure nitric or muriatic acid containing nitrites. The point distinguishing the cholera bacteria is rather their production of the nitrous combinations which are necessary for the reaction. These arise (as shown or, at least, rendered very probable by Petri) by reduction of the nitrates traces of which are present in salt and peptone, and especially contained in the crude gelatin. We will now readily understand (after these explanations) that available and really decisive results can be looked for only when using perfectly pure sulphuric acid free from nitrites.

The significance of the cholera reaction is very great, especially for practical purposes, as it affords us a quick and sure means of distinguishing the cholera bacteria from other micro-organisms. Having, for instance, received material suggestive of cholera, we shall first elaborate it by procedure on the plate. If during the first twenty-four hours colonies should develop whose nature is in doubt, but may be comma bacilli, it will suffice to transport one of them to bouillon containing peptone and place this in the incubator. After twelve hours a definite conclusion may be reached by the aid of sulphuric acid.

This method is more effective when applied in connection with a procedure introduced by Gruber and Schottelius. These two investigators availed themselves of the fact already mentioned that the cholera bacilli thrive particularly well in strongly-diluted bouillon. The predilection of the cholera vibriones for this medium is so pronounced that they remain victorious on it even in competition with other micro-organisms, and flourish specially luxuriantly on the surface of the fluid, as may be recognized by the well-known wrinkled film. In the examination of any material, especially the intestinal contents of a corpse or the discharges of a patient, for cholera bacteria, we prepare gelatin plates of it, and at the same time inoculate a number of test-tubes filled with a strongly-diluted food bouillon; then place them in the incubator and examine them in twelve to twenty-four hours. If there has been formed anywhere a film, we must examine it carefully. If it appears suspicious, prepare plates from it and transmit a trace of it to a new tube with diluted bouillon; especially endeavor to ascertain by means of specific reaction whether there are any cholera vibrios present. If the result be affirmative, diagnosis will have been established after twelve to twenty-four hours; but if negative, we must watch the development. In those cases where there were but very few comma bacilli from the beginning, the examination must be conducted with the greatest care by the use of the plate.

This exhausts our knowledge concerning the comma bacilli and their mode of life; but the aim has been to present evidence sufficient to establish the fact that in them we have a definite, well-circumscribed, and, above all, an easily-recognized and distinguishable kind of bacterium.

It was demonstrated by Koch in all cases of cholera asiatica, usually in the intestines in great abundance and even in a state approaching pure cultures. This fact has been verified by all conscientious investigators, and no case of genuine cholera has as yet been reported in which the comma bacilli have been absent.

Koch has shown, furthermore, that their occurrence is absolutely restricted to cholera; that they appear in no other affection; that they appear with the outbreak of the disease and disappear with its cessation. This observation, too, has proved correct in every instance, and the significance of the vibrios in causing cholera is no longer in doubt. The very intimate relations existing between the bacteria and the disease could only be those of cause and effect. Only adversaries not open to conviction could dispute this; they clung to the fact that successful transmissions from artificial cultures to animals have not been obtained. But they overlooked the fact that, according to experience, cholera is a disease peculiar to man which never occurs in animals under ordinary circumstances, but rather spares them without exception in the most virulent epidemics.

Koch had pointed out as early as 1876 that there might be some difficulty in the future in differentiating the germs of cholera and typhus abdominalis, since animals were insusceptible to both these affections.

It is, therefore, not to be wondered at that the experiments with animals failed; this in no way proved or disproved the relation of the bacilli to the disease. We must not demand more than is possible of the experiment under the circumstances, and we may, as far as our knowledge goes and in analogy with other facts, consider the specific character of a micro-organism established if it is found in all cases of a disease and in it alone.

Koch has endeavored, nevertheless, to utilize the experiment with animals to prove his assertions; he has, in fact, succeeded in overcoming some of the resistance and in demonstrating that the cholera bacteria can at least develop pathogenic qualities in the animal body, exert a poisonous influence, and produce symptoms analogous to those observed in human cholera.

No results followed the mere feeding of the cultures to animals in various ways. Only by placing the bacilli directly into the blood-vessels of a rabbit could the animal be successfully infected and the rods afterward found in the organs; but this procedure was so unlike the manner of ordinary infection that it could not properly be regarded as conclusive.

Nicati and Rietsch were able to produce a fatal disease like cholera in Guinea-pigs by putting the comma bacilli directly into the intestinal canal, especially the duodenum, after having ligated the biliary duct.

This fact showed that the cholera bacilli must be placed at once into the intestine, where the reaction is alkaline. The acid gastric juice destroys and renders harmless the most sensitive comma bacilli (similarly to the anthrax bacilli). This was the one obstacle that frustrated all attempts at transmission.

But a second factor had to be considered. The prevention of a flow of bile to the intestine undoubtedly favored the success of the experiment. Bile stimulates the muscular movements of the intestines, and this fact made it probable that by controlling peristalsis sufficient time is given the micro-organisms to grow and multiply and produce their fatal effects.

Koch succeeded under these circumstances in effecting successful transmission to Guinea-pigs.

First administer to the animal 5 c.cm. of a 5% solution of carbonate of soda by means of a pharyngeal catheter, in order to neutralize the gastric juice and the contents of the stomach. Place a wooden gag perforated in the middle between the pig's jaws, lest the catheter be bitten. Next inject the quantity of soda solution stated above; then inject a moderately large quantity of opium immediately into the abdominal cavity, to paralyze the intestinal movements. Generally 1 gramme of tincture of opium to about every 200 grammes of the body weight is used with a Pravaz syringe. The solution is allowed to flow in slowly. This method is necessary, as opium is not readily absorbed in the pig's stomach. The animals, however, bear this manipulation with impunity; they become somnolent, lie down on their side, and fall into a deep narcosis, from which they waken in about half an hour, shortly to become as merry and frolicsome as they were before.

Soon after the opium injection and while the pig is still in an apathetic state the probe is again introduced and 10 c.cm. of a cholera bouillon culture is injected. This ends the experiment. The animal soon recovers, but very soon begins to show signs of discomfort, refuses food, is affected by a sort of paralytic weakness of the posterior extremities, and has a superficial and retarded respiration; it usually dies after forty-eight hours.

It has been shown that these symptoms lack a very essential factor prominent in human cholera, viz., symptoms of intestinal disturbance, and that the Guinea-pigs perish without having vomited or without having watery alvine discharges. Here, too, the fact was overlooked that the conditions of animals differ from those of man. Guinea-pigs, for instance, do not vomit, and the absence of diarrhœa is accounted for by the extraordinary size of the cæcum of these animals, capable as it is of containing considerable quantities of liquid intestinal contents.

The post-mortem appearance, however, corresponds exactly to cholera. The small intestine is congested and filled with a watery fluid containing a great many cholera bacteria.

These experiments on animals show the ability of the cholera bacteria to develop pathogenic properties in the animal body and of causing changes identical with those observed in human cholera. Further information could not be expected from the experiment. At most, all that could be accomplished in this direction was the transmission to man, the only organism susceptible under natural conditions.

The correctness of this supposition has, in fact, been accidentally established. During the so-called "Courses on Cholera" at the Imperial Board of Health (delivered for the purpose of making the existence of comma bacilli more widely known), one of the attendants, having in some way neglected the necessary precaution, was infected by the bacteria and fell sick from a violent attack of cholerine. He had very frequent watery and colorless discharges, great weakness, unquenchable thirst; almost complete suppression of urine; there were shooting pains in the soles of the feet, etc., and quantities of genuine comma bacilli were found in the fæces.

Thus at last the results of microscopic investigation, cultivation, and transmission are in satisfactory accord, and we can no longer doubt that the comma bacilli are the true and sole cause of the cholera asiatica of man.

We now turn again to the question, How does the bacterium get into the body, how does it there produce disease, and how can the peculiarities of this disease be explained by the living properties of the micro-organism?

We are here touching upon a matter engaging the attention of the scientific world and leading to the most adverse opinions.

Discussions concerning the nature and mode of origin of cholera have never been wanting since the time when it first made its appearance in Europe. As long as a quarter of a century ago, Pettenkofer (an eminent authority on hygiene and epidemics) had expressed a characteristic opinion regarding the development of cholera formed on the basis of very detailed and extensive epidemiological observations. This view soon found general recognition.

Pettenkofer's theory is essentially as follows: Cholera is not transmissible from man to man. The poison originates in the soil under certain conditions. The presumable cholera germ develops by virtue of particular properties of the soil, which may be characterized as local and temporal disposition, and from these properties arise the causa morbi. These special properties of the soil, its local and temporal disposition, are mainly founded on changing conditions of moisture and heat, on a physical state, and, finally, on "impregnation"—i.e., on its possession of nourishing substances for the lower organisms.

Pettenkofer intelligently says that "the cholera germ (x) produces, on the ground of local and temporal disposition of the soil (y), the cholera poison (z), just as the torula cerevisiæ (x) produces from the sugar solution (y) the poison of inebriating alcohol (z)."

This cholera poison is transmitted to man by air and absorbed exclusively by way of respiration.

It is thus seen that the soil plays the principal part in this view. The poison must every time be matured there, as it were, before it can attack man, who really stands second, and must, moreover, be especially susceptible to the invasion of the virus in consequence of a certain individual predisposition.

Now, when the comma bacillus appeared on the stage, they tried first to insert it in the place of that x in the solid structure of Pettenkofer's doctrine. And when it did not willingly comply with this demand force was used, and it was declared to have forfeited its right as the exciter of cholera.

It was a completely forlorn undertaking. We should either (on the ground of direct observations) refute the fact that the comma bacillus is the cause of cholera and then remove, one by one, the proofs for the view that the comma bacillus is met with in all cases of cholera, that it goes side by side with the development of the symptoms of this disease, and, on the other hand, only appears with the cholera; or we should bow to the weight of these reasons and recognize the significance of the bacillus. This being so, there is but one proceeding. On the ground of the new discovery we must subject to a new examination all the ideas and views formed till now regarding the disease, as a whole or singly. If they find their explication in the living properties in the special peculiarities of the micro-organism, if they agree with or, at least, do not directly contradict them, their title is proved and they remain intact. But if this should not be the case, they must be cheerfully shelved with the calm acknowledgment that every fragment of our knowledge retains its validity only so long as it is not replaced by something better. But we must not proceed in the opposite direction! We must not attempt, for the sake of a preconceived opinion (however well founded it may seem to be), to trim facts that cannot properly be twisted about or explained away, and, according to epidemiological observations, constitute an artificial micro-organism with prescribed qualities.

But Pettenkofer's view positively could not be reconciled with the living qualities of the comma bacillus.

Not a single fact points to the existence of a special cholera poison produced by the bacterium and absorbed by man, independent of it, as the real cause of the disease. *This cholera poison is identical with the micro-organism itself.*

It is neither directly proven nor very probable that the micro-organism is to prosper in the soil and to find in it the most important seat of its activity. The possibility may be admitted; for we have seen that the comma bacillus is able to exist outside of the human body and to lead a saprophytic mode of life. But the soil itself is scarcely a fit place for it. The ample quantities of bacteria of other kinds dwelling in the upper strata of the earth will make life difficult to the delicate cholera vibrio so little capable of resisting such a rivalry.

The cholera poison (hence the comma bacillus) is, besides, said to rise from the soil to be absorbed by respiration. Now, we know that the bacteria are unable to fly upward independently or to pass from a moist basis into the atmosphere by currents of evaporation. There is but one way in which micro-organisms can be torn and carried off from their culture medium, viz., by the desiccation and subsequent scattering like dust. But the comma bacilli are exceedingly sensitive to desiccation and rapidly perish under its influence. We know, as yet, no persisting form of the cholera bacteria, and as long as such a one has not been ascertained, this kind of transmission is almost unimaginable.

The entrance of the cholera poison through the lungs is, in conclusion, not supported by any fact. Aside from the intestines, the bacteria appear neither in the blood nor in other organs. The disease points altogether to the fact that the intestine stands in the centre of the pathological processes and that there the principal changes take place.

It is thus evident that the comma bacillus is not in accord with Pettenkofer's views. Koch has, therefore, most decidedly opposed the latter in almost every essential point by trying to explain the origin of the disease by the qualities of its cause.

Cholera is *transmissible from man to man.* This does not, as a rule, occur directly, but as follows:

The comma bacillus enters the intestine, develops and increases there, and produces the severe symptoms composing the clinical picture of the disease. It is received in the body by way of digestion, with the food, frequently with the drinking-water, and leaves it only with the discharges. It then spreads, gets again into the water, or on moist food, wet linen, etc., and by means of these agents finds occasion to attack other previously healthy individuals who are susceptible owing to a certain predisposition, especially a weakness of the intestinal canal.

Man evidently stands in the foreground of events, and the soil at most occupies an occasional place in the circulation from the human intestine through the ingesta and to the human intestine again.

This view certainly corresponds to the actual conditions. Numerous single observations have proved with sufficient certainty that man is the real source of infection. The very fact that the contagious spread of the disease follows man and travels with him is a firm support for the view just discussed, whose direct proof can hardly be demanded and furnished for every case.

We know that the bacteria are found in the intestine during the disease; their appearance, in fact, coincides with the commencement of the infection. Their increase very soon reaches its maximum at the same time the symptoms of the disease do. The intestine contains a nearly pure culture of the comma bacilli. They begin to die after two to three days and to make room for the real inhabitants of the intestine, the septic bacteria. This terminates the process essentially and a cure may, under some circumstances, take place.

The existence of bacteria outside the intestine may also be proved in the copious discharges of the patients; it has been observed more rarely in the vomited matter, presumably only in such cases where intestinal contents found entrance to the stomach. The micro-organisms keep their vitality for a long time, as they remain capable of development in a moist condition for months even without possessing spores. They do not fall a prey to resulting putrefaction because man, who throws the excrements into water, dilutes them as much as possible, transfers them with his fingers to new media, soils the linen with them, and thus affords the bacteria protection from harm and opens many ways of spreading the disease.

The comma bacilli have been found in water even under natural conditions; besides, experiments have shown that they do not merely live in it, but can even increase. We may succeed in producing pure cultures on damp pieces of linen; and it is not impossible that even the upper strata of the earth perform the service of the go-between at times, and aid in transmitting the poison; but this may occur in exceptional cases only and on the condition that they are thoroughly moistened; for the comma bacillus thrives only in moisture, dryness being a more impassable obstacle to it and constituting a better protection against advance than institutions of disinfection and weeks of quarantine.

It is true that various facts regarding their appearance and the

18

peculiarities of their progress gathered by close observations of cholera epidemics cannot always be accounted for by the view just discussed. But many of these observations have been made at a time where points now regarded as most essential were not at all known and, hence, could not be considered. It is, indeed, certain that there are places strikingly free from contagion in a completely pestilential neighborhood, and the cause of this immunity has not in every case been explained. Numerous epidemiological problems are waiting solution; even local and temporal disposition may, in a correspondingly altered condition, be of importance in the origin of cholera.

But although the comma bacillus and our knowledge of its nature do not yet account for everything, we may recall the sentences with which Virchow established (in the second cholera conference) his standpoint. In remarking that a particular disease of the silk-worms (muscardine) was the oldest of the mykotic affections thoroughly investigated and that the parasitic cause of an epidemic disease had first been established in said muscardine, he pointed to the fact that this disease had been known so long and studied so zealously and that so many means had already been used to fight it, and added: "And yet we cannot even to-day state with entire certainty what are the reasons for its appearing alternately in greater or less extent, nor can one say what one must do to suppress it." And he declares afterward: "During my studies of natural sciences I have always been inclined, whenever an observation has been made in a single concrete case under every guarantee of certainty, not to make the acknowledgment of the correctness of such an observation dependent on its ability of accounting for everything."

"Cholera is a disease endemic in certain districts of India, especially in Lower Bengal, the Ganges Delta proper, and from time to time has been brought to our country. Its cause is a specific bacterium. This passes from man to man by means of moist agencies, especially the drinking-water; is received with the food, and being developed in the intestine gives rise to cholera. Its entrance and increase are probably facilitated by a certain favorable condition of the intestinal canal, an individual predisposition, to be accounted for, perhaps, by a diminution of the acidity of the stomach and inertia of peristalsis."

The symptoms of cholera admit of no doubt that the intestine is the real seat of the pathological processes; for the symptoms referable to the digestive canal predominate. Frequent watery, nearly colorless and odorless discharges of the appearance of whey or rice-water, frequent vomiting, complete loss of appetite, and violent thirst belong to this group. All this can readily be explained as an immediate consequence of the effects of bacteria.

There will also be regularly noticed the symptoms of severe general suffering, increasing weakness of the heart, terminating in a complete stoppage of the circulation, lowering of temperature, superficial respiration, and spasms of the muscles; all these suggest that other regions are likewise affected. Micro-organisms appear, however, only in the intestine, never in the blood or the internal organs, and a direct connection cannot be established. Koch thinks that the bacteria produce an active poison in the intestine while developing, and that this poison is taken up by the lymphatics or blood-current and distributed all over the body. Such toxically active substances have, in fact, become known as direct products of the metabolism of the comma bacilli, and these appear, together with basic bodies, to exist as peculiar albuminous substances of the class of toxalbumins and play an essential part in the disease.

The post-mortem changes are almost wholly confined to the intestine. The duodenum is more or less filled with a watery, colorless fluid, frequently, however, possessing a somewhat firmer quality and resembling gruel. The intestinal mucous membrane is swollen and reddened, most noticeably in the region above Bauhin's valve; but the reddening and swelling is often restricted to the borders of the lymph-follicles and Peyer's patches.

Here the bacteria are found to have entered the substance of the intestinal wall.

We stain sections for twenty-four hours in common fuchsin or in alkaline methyl-blue solution. The bacteria are decolored according to Gram. In successful preparations the micro-organisms are seen in the tubular glands, and it will also be seen that they have penetrated in part between the basement membrane and the epithelium and forced the latter from its foundation. The bacteria have also their characteristic comma figure in the tissue and lie close together in dense heaps.

The examination of other organs is practically negative.

One more point in conclusion might be briefly mentioned—i.e., the question whether by overcoming cholera once we are protected against a renewed attack of the disease. A sort of immunity seems, in fact, to take place, though it is not of any considerable duration. It is an extremely rare occurrence to find any one afflicted twice in the course of the same epidemic, but this diminished susceptibility does not generally last more than about four or five years.

In speaking of the general qualities of the bacteria we have become acquainted with a number of species differing in some respects, but obviously nearly related and belonging to certain sharply-defined and well-characterized natural groups.

The best proof of this fact is afforded in the following three micro-organisms, resembling more or less closely the cholera vibrio, sharing with it a series of peculiarities and evincing, in many points, a striking similarity with it, although the differences existing are easily recognized.

VIII. FINKLER-PRIOR'S VIBRIO.

The first of these varieties of the vibrio was observed in the discharges of patients having cholera morbus. This vibrio was originally considered by its discoverers to be identical with the genuine comma bacillus. This opinion would, of course, have deprived that vibrio of every significance. But it was soon ascertained that the statements of those investigators were not well founded, and the many points of difference between the two micro-organisms can easily be made apparent.

Finkler's and the cholera bacteria greatly resemble each other in the shape of the individual germs. But Finkler's vibrio is, as a rule, somewhat larger, thicker, and coarser than Koch's. It develops more rarely into spirilla which scarcely ever become as long as those of the cholera bacteria. It increases in hanging drops, at ordinary temperature, to dense swarms; it resists the influence of oxygen like the genuine comma bacillus.

On the plate Finkler's bacterium is distinguished from the cholera vibrio by an extraordinary growth coincident with an extensive liquefaction of the gelatin. The usual three dilutions do not even suffice, and only on the fourth or fifth plate will well-isolated colonies develop.

When observing them with a diffuse light and the naked eye, they appear first as small white dots in the depth of the gelatin. They rapidly push forward to the surface, the dissolution of the gelatin begins, and there arise circular depressions, proceeding toward the middle, in the shape of saucers. On the second day they appear at least as large as lentils, their contents consisting of a turbid fluid of gray translucency. The border is sharply defined from the solid part of the medium, but there are no other peculiarities.

Under the microscope these colonies appear as yellowish-brown dense masses, possessing a very fine but entirely uniform granulation. Even with low magnifying power an active movement, an incessant intermingling of these small objects can be seen. The edge is bordered with quite short, delicate filaments.

The appearance of the colonies differs wholly from that of the comma bacilli, and a Finkler plate does not resemble at all a cholera plate. Their difference becomes still greater in test-tube culture.

In a four-day-old cholera puncture a thin thread clear as glass will be seen, the air-vesicle above and the neatly-formed heaps of bacteria below. In a culture of Finkler's bacterium of the same age the gelatin is widely liquefied along the entire puncture, nearly half of the gelatin being already changed into a turbid gray solution. The form of the liquefied district in the solid gelatin produced by the growth of the bacteria looks like a "trouser-leg" or a "stocking," and the entire contents of the glass become liquefied after about a week. A film of a smeary-white appearance is then formed on the surface.

Finkler-Prior's micro-organism spreads rapidly on agar as a damp, slimy film, coating the whole surface in a short time.

The cholera bacteria thrive on potatoes only at breeding temperature, and produce a very characteristic yellowish-brown growth; Finkler's vibrio grows on the slices at ordinary temperature and rapidly produces a grayish-yellow, slimy, shining layer extending to the edge of the potato. It can live in milk, but soon perishes in water.

Koch has shown and Finkler confirmed that the latter's bacteria can, under some circumstances, also display a pathogenic effect. If they are brought (in the manner mentioned in connection with the experimental infection with comma bacilli) into the stomach of Guinea-pigs, part of them will die. But Finkler's vibrios are not as poisonous as the genuine comma bacilli, their proportion being 35:30 and 15:5 respectively. The post-mortem appearance, too, is different, the intestine looking pale gray and its watery contents developing a penetrating odor of putrefaction not met with in the contents of the cholera intestine.

The differences thus briefly mentioned between the two kinds of bacteria are, in fact, so considerable that they cannot be confounded. The opinion originally held by Finkler and Prior regarding the identity of their micro-organism with the genuine comma bacillus can be explained only by the inefficiency of their mode of examination. They afterward became convinced of their error; but they did not want to drop their protégé altogether, and wished, after the loss of its first claims, to uphold it at least as the origin of cholera morbus.

It is true that the vibrio was first observed in the dejections of a man having cholera morbus. But Finkler and Prior did not find it immediately after the discharge, but only when the watery and offensive excrements had been preserved for nearly fourteen days, and hence subjected to further decomposition. Nor can the fact, likewise reported by Finkler, that he had found his bacterium in seven cases of cholera morbus immediately post-mortem serve as a decisive proof of its significance; for numerous other observers have established the contrary and been unable to find Finkler's vibrio in cholera morbus in spite of all care and precision.

While the appearance of the micro-organism in cholera morbus is by no means universal, several results have, on the other hand, plainly demonstrated that it is found in cases in no way related to the disease just mentioned. Kuisl has discovered it in the intestine of a suicide, and Miller, of Berlin, has obtained from the hollow tooth of an otherwise healthy man a kind of bacterium agreeing in every respect with Finkler's vibrio, and which is regarded as identical with it. We are, therefore, justified in looking upon the vibrio described by Finkler and Prior as a more or less frequent and harmless tenant of the human digestive canal.

IX. DENEKE'S VIBRIO.

Decidedly similar to the genuine comma bacillus is a kind of bacterium cultivated by Deneke (Göttingen) from old cheese and mentioned here on account of its appearance, though otherwise it is unimportant.

They are neatly-curved bacteria, frequently growing out into spirilla, actively motile, and hardly distinguishable from the cholera vibrios by microscopic investigation. They grow like the vibrios, at ordinary as well as at breeding temperatures, they are similarly intolerant of oxygen, and are equally stained by anilin colors.

On the plate, however, their development proceeds very differently. Their growth is considerably more rapid than that of Finkler's bacteria. The colonies first appear to the naked eye as small round dots in the depth of the gelatin. They then rise to the surface and begin to liquefy the medium. On the second day they are about the size of a pin's head, and of a distinctly yellowish color; they lie in the gelatin at the bottom of the funnel-shaped depression caused by them. The plate when looked at from the side appears to be studded with small air-vesicles, and greatly resembles, at the first glance, a cholera plate.

The colonies appear under the microscope as irregularly-formed, coarse-grained masses, of a marked yellowish-green color in the middle, but paler toward the margin and of a peculiar lustre. Around the colony is a thick circular belt, appearing (on changing the illumination) sometimes clear, sometimes dark, and formed by the lateral walls of the hole-like depression in which the colony rests.

Deneke's vibrio is, therefore, macroscopically distinguished on the plate from the cholera vibrio, by a more rapid liquefaction of the gelatin, a quicker growth of the colonies, and their yellow coloring. Microscopically, by irregular form and the thick rampart surrounding each.

In the test-tube the same conditions obtain. The liquefaction proceeds evenly along the entire inoculating puncture, but here too the bacteria sink in coils to the bottom from the central parts of the culture. They do this so completely that only a more exact observation will exhibit development in this condition. There usually arises at the surface a yellowish, thin layer, above which frequently hovers a funnel-shaped depression, a kind of "air-vesicle" larger than that in cholera cultures. Then follows the path of the inoculating puncture, which appears with a diffuse light as a broad, shining canal; finally, there are yellow heaps constituting the principal mass of bacterial growth. The contents of the tube are wholly liquefied in about two weeks.

Deneke's vibrio thrives on agar-agar as a thin yellowish coat in the neighborhood of the inoculating line.

It grows on potatoes at breeding temperature, as a delicate, yellowish film in which may sometimes be noticed beautifully-formed spirilla.

It is remarkable that Deneke's vibrios possess pathogenic properties. With the method of infection used for cholera bacteria Guinea-pigs may be killed, death ensuing in three animals out of fifteen treated.

X. VIBRIO METSCHNIKOFF (GAMALEÏA.)

A micro-organism observed by Gamaleïa and called "vibrio Metschnikoff" is more nearly related to the comma bacillus than the two kinds of bacteria just described. It was discovered in the intestinal contents of poultry, especially of hens, and was represented as the original cause of a special affection of these animals, possessing in its visible qualities much resemblance to chicken cholera. It is said to occur more frequently in Russia during the summer months.

The vibrio Metschnikoff is a curved bacterium whose single links are usually shorter and thicker, but nevertheless more strongly bent, than those of the vibrio of cholera asiatica. Like the latter, in liquid media it forms longer threads—twisted spirilla of varying extent, the coils being generally rather steep. The vibrio Metschnikoff has lively voluntary movement, which is produced by a long, fine, undulating flagellum at the end of each cell, which can be stained by Löffler's method.

The occurrence of spores is no more proved with the vibrio Metschnikoff than with the cholera bacterium, and the positive results of Gamaleïa (who claims to have established in the interior of the single links the presence of structures accessible to double staining) have never been confirmed. Facts are wanting to show that this micro-organism is possessed of great resisting power. It stands in this respect on an equal footing with the cholera vibrio, and, like it, succumbs to the influence of acids, of high temperatures, and especially of drying, just as rapidly and completely. Its attitude toward oxygen and growth at ordinary as well as at breeding temperature is similar to that of Koch's comma bacillus.

Gamaleïa's vibrios are easily stained. It is frequently seen that in treating with watery color-solutions only the two ends of the single links are colored, while the centre remains pale and separated from its surroundings as a bright gap, an appearance met with in the bacteria of hen cholera in a more pronounced manner. The vibriones are decolored by Gram's method.

On the gelatin plate the colonies appear after the average time requisite.

Their appearance is intermediate between those of Koch's and Finkler's vibrio and may be ranged with Deneke's bacterium. It must be remarked, however, that the colonies do not always develop in the same manner.

Take two gelatin plates of Koch's comma bacillus and Metschnikoff's vibrio three days old. They present an altogether different picture. In one of them we see the gelatin closely beset with small depressions, sharply bordered and whitish, similar to air-bubbles; in the other we are struck by capacious saucer-like, circular areas of liquefaction filled with grayish-white, turbid contents, and so strongly reminding one of the appearance of the colonies of Finkler's vibrio that at the first glance we may think that it is in fact the latter. In looking more closely we may recognize, even with the naked eye, between and along the large, strongly-liquefied colonies, a number of smaller ones lying partly at some depth in the gelatin and partly near the surface. They have, however, only caused a moderate softening of the gelatin; they appear as round cavities filled with clear liquid, having a decided resemblance to cholera bacilli.

By resorting to the microscope the same difference will be observed. In one case we have yellowish-brown masses, variously conglomerated into thick granular or crumbling lumps seen in lively motion with a weak magnifying power. The edge, as in Finkler's colonies, is studded with a uniform border of the finest radiating filaments. In the other case we see colonies strongly resembling those of Koch's vibrios—a liquefying funnel outlined against the solid gelatin by sharp lines, at the bottom of which funnel the bacterial growth lies as a glossy, yellowish-white heap, composed, as it were, of small pieces of glass.

These obvious differences might incline us to think that we did not have a pure culture, but rather a mixture of two different kinds. But by taking a small quantity of one or the other sort of these colonies and preparing new plates, we will always see the two forms reappear, being varieties of the same micro-organism and differing mainly by the greater or less amount of peptonizing power. Our judgment is, therefore, confirmed that the vibrio Metschnikoff occupies in its culture on our artificial media an intermediate place between Koch's and Finkler's vibrio, sometimes inclining more to the one or to the other, and that it is similar to Deneke's vibrio as to its energy of growth.

R. Pfeiffer (who has thoroughly investigated this subject) justly points to the fact that it is easy to distinguish a pure plate culture of the cholera vibrio from one of the vibrio Metschnikoff, but it is absolutely impossible to discern and recognize a few colonies of the one among many of the other.

This conformity of the two micro-organisms often becomes manifest. The stab-culture of the vibrio Metschnikoff in gelatin exactly resembles that of the cholera vibrio, except that the latter grows much more slowly, so that simultaneous inoculations can easily be kept apart.

The growth on oblique agar again reminds us of the cholera bacteria; likewise the growth on potatoes, which at breeding temperature are covered with a moderately luxuriant, yellowish-brown or chocolate-colored covering. But on developing in bouillon a certain difference will be noticed, inasmuch as the liquid under the influence of the vibrio Metschnikoff at incubator temperature becomes entirely cloudy after a very short time, assumes a grayish-white color, and gradually allows only a folded thin film provided with numerous wrinkles to arise on the surface, while the comma bacillus keeps the culture solutions clear for a much longer time and never changes them into a thick gray broth.

The vibrio Metschnikoff is stained red by the addition of muriatic or, better still, sulphuric acid free from nitrite or peptonated agents, just as the cholera vibrio, perhaps even more decidedly.

There will now be no doubt that the cholera and Metschnikoff vibrios are micro-organisms very nearly related to one another, which can, perhaps, trace their pedigree to the same ancestors, but have developed in the course of time in quite different directions. While the one accustomed itself to man and obtained pathogenic properties for him, but can be transmitted to animals only artificially, if not forcibly, the other has subjected these very animals to its pernicious influence.

Gamaleïa's and Pfeiffer's experiments have shown that the vibrio Metschnikoff is pre-eminently infectious for chickens, Guinea-pigs, and especially pigeons, while mice, for instance, are almost entirely refractory. Transmission is better accomplished by the subcutaneous tissue, and this always results in success in the case of pigeons. We may infect Guinea-pigs through the intestinal canal by means of the method indicated by Koch for the cholera bacteria.

The symptoms of disease are but little pronounced in every case. In Guinea-pigs, however, we regularly notice in conjunction with inoculation that after a rise of the body's temperature for a short time, a very considerable fall of it—down to 33° C. or even below—occurs. Death ensues in from twenty to twenty-four hours after inoculation.

Upon dissection (after the subcutaneous application) a bloody œdema is found extending over a wide area about the place of infection, as well as a superficial nècrosis of the tissue and vast quantities of bacteria in the blood and all organs, and the entire affection confines itself so entirely to the vascular system that Pfeiffer calls it vibrio septicæmia. Only slight changes and very few or no micro-organisms are seen in the intestine; but if they had first been introduced into the stomach, the digestive canal would be the chief seat of the pathological processes, including an intense inflammation and the presence of large numbers of vibrios.

Though the vibrio Metschnikoff is exceedingly infective to pigeons and Guinea-pigs, it is easy to grant immunity to these animals, as Gamaleïa has proved. This may best be obtained by means of sterilized cultures of the virulent bacteria. We have here an especially distinct proof of the fact (duly discussed above) that the immunity obtained is mainly owing to the excretions of the inoculated micro-organisms. It is a striking circumstance peculiar to this case, that the substance supplying protection is not even destroyed by continued heating to 100° C.—i.e., the nourishing fluids may be changed into vaccines by keeping them in the steam sterilizer for about half an hour at 100° C., thus destroying the existing germs.

Cultures treated in this manner show, according to their age, a very different degree of virulence. Those of about twenty days in quantities of 2 to 3 c.cm. kill Guinea-pigs after injection into the subcutaneous cellular tissue or the abdominal cavity, while those of about five days are still borne in doses of 5 c.cm. In the former case (as in the inoculation of living bacteria) a rapid fall of the body temperature, far below the normal limit, is perceptible; death takes place after twenty-four to forty-eight hours; dissection exhibits especially a somewhat slow pathological process and a very considerable fatty degeneration of the liver. In the latter case a fall of temperature after a few hours occurs at first, but it rises very soon to a feverish reaction. This lasts about a day, after which time the animals will soon recover, and have then acquired immunity. This does not, indeed, directly follow inoculation. As a rule one or two weeks will elapse before complete success is attained—i.e., before the virulent material (for instance, the vibrionic blood of a pigeon perishing after infection with virulent bacteria) is taken in doses of 1 to 2 c.cm. without causing the death of the animals. Arise of the body temperature, lasting several days, and local changes (such as the appearance of a tolerably extensive œdema at the inoculation spot) indicate that the artificially-immunized body feels at least the invasion of the bacteria.

The nature and composition of the really efficient substance are, as yet, not exactly known. It is striking that the toxic cultures possess a very high degree of alkalescence and lose their virulence, and hence their vaccinating power (as Pfeiffer states), by neutralization with sulphuric acid, but not by treatment with muriatic acid.

The experiments concerning the production of artificial immunity with the vibrio Metschnikoff are especially interesting because the discoverer of this micro-organism, Gamaleïa, is inclined to transfer the observations and experiences he made, in an almost completely unchanged form, to Koch's vibrio of cholera asiatica; he pretends to have reached similar results especially as regards protective inoculation.

By repeated passages through the pigeon's body the comma bacilli are thus said to assume an extremely high degree of virulence; injection of sterilized cultures is said to render the animals secure even against this most infectious matter; and thus a "preventive inoculation" is said to have been found "against cholera asiatica." The granting of immunity against cholera bacteria is said to also furnish preventive inoculation against the vibrio Metschnikoff, etc.

All these assertions have been proved groundless by Pfeiffer's and Nocht's investigations. The other observations of Gamaleïa regarding the vibrio Metschnikoff, its mode of infection and distribution under natural conditions, as also the far-reaching conclusions and criticisms of the action of Koch's bacteria, are consequently devoid of interest and reliability.

XI. EMMERICH'S BACILLUS.

We will now consider a bacillus which is, indeed, only remotely and superficially related to the group of micro-organisms above discussed, and which strayed by a mere accident into that distinguished company.

It is easy to understand that the powerful revolution in our views regarding the nature of such prominent diseases as tuberculosis and cholera (brought about scarcely two years ago by the discoveries of one man) would meet with contradiction. The firm structure of the proofs by which Koch established the significance of the tubercle bacillus could not be shaken. But the results of his investigations of cholera seemed to offer an opportunity of finding out defects and mistakes in their connection and of thus calling their value in question.

In the first place (though only for a short time), the regular occurrence of the comma bacillus in all cases of genuine cholera was contested, but soon the most prejudiced among those who subjected Koch's statements to repeated examinations admitted unreservedly this point of his claims.

It was furthermore asserted that the comma bacillus was not peculiar to cholera and that it was also found in other conditions, or that it was merely a harmless and common tenant of our digestive tract.

Especially Lewes and Klein, in England, pretended to have ascertained that curved comma-like rods and even spirilla could be found at any time in the saliva of healthy people, and declared them to be identical with Koch's bacteria.

The observation itself was correct; but these micro-organisms have, in fact, except in form and general appearance, nothing in common with the genuine comma bacilli. They resist, above all, every attempt at cultivation on our artificial media, for which reason there is but little danger of their giving rise to mistake and misinterpretation.

We already know that Finkler and Prior's communications were originally and openly intended to oppose Koch's discovery, and that they expressed the opinion that the bacteria found in cases of cholera morbus and Koch's comma bacillus were identical.

The value of Koch's investigations was most actively disputed in Munich. Emmerich and Buchner published a number of statements apparently strong enough to seriously question the significance of the comma bacillus as the origin of cholera asiatica.

Emmerich, in 1884, had gone to Naples (at the instance of the Bavarian government) during an extensive cholera epidemic, to gather experience concerning the causes and nature of the disease.

He found Koch's bacillus in a number of cases; but he succeeded besides in obtaining from the organs of cholera corpses and from the blood of a cholera patient a new kind of bacterium, in which he thought he had discovered the real cause of the pestilence.

The cholera poison is, according to Pettenkofer's view, taken up by the lungs. But as the intestines are invariably the seat of the clinical symptoms, there remains only the supposition that the infectious matter gets from one place to another by way of the blood-serum, and must be found there as well as in all internal organs. Koch's comma bacillus did not comply with this demand, and therefore, in the eyes of the Munich school, it lost every claim of originating cholera.

The Neapolitan bacillus, on the other hand, was free from such a reproach, and adapted itself more willingly to the epidemiological facts. For it was found in the blood and internal organs, and (certainly a very important observation) it could be at once transmitted to animals, especially to Guinea-pigs, and (according to Emmerich's statements) by subcutaneous application as well as by introduction into the abdominal cavity or directly into the lungs, and thus artificially presented a correct picture of the disease and the post-mortem conditions of genuine cholera.

It is, therefore, desirable to bestow on this bacterium considerable attention.

They are small, short rods, with rounded ends, usually occurring singly, rarely in groups and pronounced threads. They are non-motile. An eventual spore-formation can neither be observed in the hanging drop nor be demonstrated by staining; for the bright gap usually seen in the latter case between the two more strongly-colored ends of the rod does not correspond to any form. The bacilli can live for some time in a dry state—for instance, on silk threads. The bacillus belongs to the semi-anaërobic variety and can prosper even when oxygen is excluded.

On the gelatin plate the bacillus assumes a characteristic appearance in a short time. The colonies appear to the naked eye first as small white dots in the depth of the gelatin. But they soon advance to the surface and spread there as thin yellowish-white aggregations, shining like mother-of-pearl, being irregularly lobulated, without ever liquefying the gelatin.

When examined under the microscope, the smaller, deeper colonies present peculiar, whetstone-shaped structures of a dark brown color and usually characterized by a concentric stratification of their contents. The large, superficial colonies, however, appear as thin membranes of a faint yellowish color in the middle, but fading toward the margin, which recede unevenly or are serrated, and always allow a kind of mesh of lines to shine through as a regular, neat leaf-like design.

A strong tendency to superficial growth is perceived in puncture culture. Although a rather quick development takes place along the whole extent of the inoculating puncture to its extreme end, the culture thrives by far most luxuriantly on the free surface of the gelatin. A dry membrane, of a grayish-white lustre with lobulated edges frequently breaking apart into single pieces, forms on this surface. Although a liquefaction of the gelatin does not occur, we can (most distinctly on oblique gelatin, but also in puncture culture, and even on somewhat older plates) observe a milky turbidity of the transparent medium in the vicinity of the bacterial growth, frequently accompanied by a formation of rosette-shaped salt crystals. Both circumstances are connected with a change in the reaction of the gelatin. The Neapolitan bacillus possesses the ability of neutralizing the existing alkalinity of the gelatin, producing an acid reaction. This can be demonstrated by adding litmus tincture to the gelatin before its inoculation. The blue color will soon disappear and a more or less reddish tint will take its place.

Emmerich's bacillus grows on agar-agar as a whitish, moist covering without peculiarities.

It forms on the surface of the potato a yellowish-brown smeary film of quite characteristic appearance.

Emmerich succeeded, as stated above, in demonstrating pathogenic properties in his bacillus and in producing in Guinea-pigs symptoms said to agree wholly with the picture of genuine cholera.

Weisser has subjected Emmerich's statements to a careful and comprehensive examination, and reached results strikingly contradicting Emmerich's assertions. He could not at all regularly cause the death of animals by the Neapolitan bacillus, as Emmerich did. He used, like the latter, various modes of infection, by inoculating the bacteria subcutaneously and also by direct injection into the abdominal cavity. But Weisser avoided injection into the lungs, and justly pointed out that we will thereby obtain anything else, rather than (as Emmerich thinks) "an imitation of the natural mode of infection by inhalation."

Of the animals thus treated, about one-half perished after infection. Death ensued, as a rule, within the first twenty-four hours without having been preceded by especially striking symptoms peculiar to cholera, "above all, without vomiting and without liquid or even pasty evacuations from the bowels and without spasmodic attacks."

The post-mortem examination shows the intestinal canal moderately filled with liquid. The mucous membrane is of a grayish-red color; the walls have the usual appearance and normal thickness. Peyer's patches are slightly swollen and reddened only in rare cases. This picture is, therefore, not to be compared with that presented with the genuine cholera of Guinea-pigs, as we have here an intestine filled with fluctuating liquid, the membrane being of a lively rose-red color and Peyer's patches swollen and peculiarly altered.

The Neapolitan bacteria can be easily proved to exist in the intestinal contents, in all internal organs and in the blood, and are at once perceived in cover-glass preparations and in sections. The latter are stained with fuchsin in the usual manner; the bacilli are decolored by Gram's method. They are found in the vessels exclusively and do not appear in large numbers. They form swarms in the smallest vessels and capillaries, in whose midst they accumulate so densely that they can no longer be recognized singly, while their arrangement becomes more transparent toward the margin.

The results of transmission are, therefore, certainly not apt to strengthen the belief in the significance of the Neapolitan bacilli in the etiology of cholera asiatica. The other conditions for recognizing a kind of bacterium as specific, according to our view, are still less fulfilled.

Emmerich himself has admitted that his bacilli cannot be proved to exist in all cases of cholera.

Emmerich had obtained the Neapolitan bacillus from the organs of cholera corpses by putting small pieces of tissue into gelatin free from germs and examining the gelatin about one to two weeks later. He thereby renounced every certainty of his observations and sinned against the first requisite of a correct bacteriological investigation, that of conducting the process, above all things, by the approved plate procedure, and of effecting as soon as possible that separation of germs without which we cannot succeed in obtaining reliable pure cultures.

Koch had in the second cholera conference pointed out these defect's of Emmerich's procedure. He said that it reminded him "of Hallier's manner of instituting his cholera investigations, who had a bottle with cholera discharges sent to him from Berlin, kept it corked up till the following spring, and then investigated with all possible caution. Emmerich's mistake is not quite so great, but at bottom it is the same."

This declaration of distrust proved afterward to be amply justified.

Weisser has, in fact, reached results dealing the death-blow to Emmerich's assertions.

When we first considered the plate process in examining fæces, we found a kind of bacterium on the plate whose colonies in appearance fully resembled those of Emmerich's bacilli. This kind of bacterium is an almost regular inhabitant of the human intestines, and was but rarely missed on the many hundreds of fæces plates prepared in our laboratory. The same micro-organisms may be obtained also from corpses of animals which have been dead for some length of time, from putrefying liquids, etc. And Weisser has succeeded, by the most detailed experiments, in establishing that this fæces bacillus agrees exactly with Emmerich's "Neapolitan bacillus" as regards the morphological properties as well as their biological functions and their pathogenic influence upon animals.

Emmerich's bacilli are nothing else than ordinary fæces bacteria, and Emmerich's assertion that "these fungi are in an essential etiological relation to cholera asiatica" falls herewith to the ground.

XII. BACILLUS TYPHOSUS.

There are a number of diseases of infectious origin whose real nature is so apparent that it has never been in doubt; as, for instance syphilis. There are other diseases, on the other hand, in which their true nature was so skilfully concealed that it has only been gradually ascertained; as, for instance, tuberculosis and typhoid fever. A correct idea of the morbid changes has been retarded by the circumstance that the definitions of infection and contagion were not kept apart with sufficient distinctness, and that one was rejected if the other could not be established. A conclusive understanding as to tuberculosis and typhus abdominalis was obtained only after long efforts.

Typhus abdominalis and spotted typhus, now regarded as wholly different, were confounded down to the middle of the present century, and it was long before typhus recurrens was recognized as a special affection. Only when this preliminary question was disposed of, and it was possible to distinguish the "simple" typhus from a disease having similar symptoms, was a firmer foundation gained for investigating the causes of the disease. The infectious character became more and more evident, and efforts were made to discover its relations to definite micro-organisms.

Observations regarding the occurrence of bacteria in cases of typhus abdominalis were variously made without their leading to safe conclusions. Eberth stated, in 1880, that during investigations on the spleen and lymphatic glands, he had succeeded in demonstrating a special kind of rod, differing distinctly from the simple putrefaction bacteria in its appearance, its disposition in the tissue, and especially by its defective staining with our common pigments; he also stated that it was not met with in other diseases in the same manner. Koch had reached similar results before Eberth, and was, therefore, able to confirm Eberth's statements.

Gaffky's publications (1884) have supplemented these statements in every particular, and enlarged them so considerably that the significance of definite bacilli in the origin of typhus abdominalis was placed beyond doubt.

They are small, slender rods with rounded ends, lying in the tissue singly or in pairs, but frequently forming large groups in the hanging drop and extending as long threads through several microscopic fields. They possess a lively and especially developed voluntary movement, whose peculiar character is noticeable in the single rods and, still better, in the longer threads gliding swiftly along in serpentine windings and curves. The typhus bacilli have (according to R. Pfeiffer's investigations) as special members for this capacity, lateral flagella that may be demonstrated by the use of Löffler's method.

It is a matter of dispute as to whether the typhus bacilli form spores or not. Gaffky had decided in the affirmative, and brought forward many weighty reasons for his view. He had perceived in the terminal ends of the rods roundish or oval corpuscles of the breadth of the single cells, characterized by a strong lustre and their ability to absorb anilin coloring matter. The bacilli were said to show considerable durability, especially against desiccation, and to retain their vitality for several months in dense layers and even on silk threads, and to fully germinate when placed on fresh media.

These structures, regarded by Gaffky as spores, were said to appear regularly within three or four days, by cultivating the bacilli at breeding temperature on boiled potatoes or growing them on agar-agar.

These facts rendered the opinion very plausible that in this case we had genuine spores. But there were certain doubts as to this opinion. The shining bodies in the rods lacked the finely-defined, regular form we are accustomed to see in the spores; they could not be stained in the usual way differently from the other cell contents; they were extremely sensitive to the influence of high temperatures and were surely destroyed by heating to 60° C. for about ten minutes, thus but slightly complying with the requisites of a real spore.

Recent investigations, especially by Buchner and Schiller, have proved beyond doubt that the presumptive spores have no such significance. Buchner ascertained that the shining oval bodies lying in the ends of the rods (the "pole-grains") and the spots appearing as bright gaps by staining are two altogether different things.

The former consist of condensed protoplasm, are even highly accessible to coloring substances, absorb them before all other parts of the bacillus, and are to be looked upon as structures connected with a degeneration of the cell—i.e., involution forms. They develop abundantly, therefore, only under conditions unfavorable to the growth of the bacilli; for instance, by exclusion of oxygen or on acid media, especially on sour potatoes. When we artificially impart to the latter an alkaline reaction the pole-grains disappear, and the active multiplication and energetic growth of the microorganisms become manifest by the circumstance that only quite short, single rods, but no threads tending to division, can still be observed.

The unstained gaps, on the other hand, are produced by the circumstance that (during the drying of the bacteria on the cover-glass or under the influence of color solutions) the bacterial protoplasm becomes detached and contracts from the membrane at these places—a process occurring mostly in the ends, but frequently also in the centre of the rods.

As Buchner has shown and Schiller has confirmed, the bacilli provided with "spores" are, therefore, less resistant than those "free from spores," and the power of resisting desiccation is a property inherent to these bacteria.

The typhus bacillus belongs to those varieties which can thrive in the absence of oxygen, as well as under its free access. But its development in the latter case is much more perfect and energetic.

The typhus bacilli in cover-glass preparations are stained without difficulty with watery color-solutions, but they belong to those bacteria which readily fade again under the influence of bleaching agents, and the staining of sections requires especial care. We have already mentioned that within the rods during staining frequently bright gaps (small unstained spots) will appear, and the significance of this occurrence has likewise been discussed.

Double staining has been so far unsuccessful with the typhus bacilli; they lose their color again by Gram's method.

Gaffky first succeeded in cultivating these bacilli artificially on our ordinary media outside the body. There are developed on the gelatin plate in the usual time at room temperature small white, dot-shaped and superficial colonies lying deep, wide-spread, faintly gray, of a peculiar lustre and irregularly bordered. Even with the naked eye they remind one of the colonies of the Neapolitan bacillus, and this similarity appears, perhaps, more distinctly under the microscope. The deeper ones are remarkable as slightly-granulated, sharply-defined, yellowish-brown heaps of whetstone shape, while the superficial ones appear as thin, almost completely transparent membranes of a yellow color in the centre, but fading toward the borders and exhibiting the undulating, leaf-like design, the linear net, noticed in the Neapolitan bacillus; their margin is uneven and serrated. The typhus bacillus (like Emmerich's) does not liquefy the gelatin.

In the test-tube culture there is an ample development along the whole inoculating puncture, especially at the surface of the gelatin. A thin and delicate membrane of a bluish-gray, mother-of-pearl-like lustre spreads to the walls of the tube.

This pronounced superficial growth can be excellently observed especially on oblique gelatin; there arises on both sides of the inoculating line an almost transparent, glossy film of bluish-white color. The gelatin is not liquefied, but there frequently occurs the milky cloudiness in the neighborhood of the culture which is also met with in the Neapolitan bacillus and traceable to the same cause. The typhus bacillus belongs (as Petruschky has clearly established) to the acid producers among the micro-organisms.

On agar-agar and likewise on solid blood-serum there is developed a moist, white covering which has nothing characteristic about it.

It is, therefore, not so very easy to distinguish the typhus bacillus, in culture, from similarly-growing bacteria, such as Emmerich's fæces bacillus and many putrefactive bacteria occurring in water or in the soil. The action of the typhus rods *on the potato* offers, indeed, a secure means for a correct differentiation, since the development on this medium is usually very characteristic and not met with elsewhere. The typhus bacillus produces on potatoes a very luxuriant growth, though almost wholly invisible to the naked eye. At ordinary temperatures after three or four days (at breeding temperature after two days) the surface of the disc has assumed a uniformly moist lustre, and no other changes, decoloration, or agglomeration, etc., can be noticed. Remove a small particle with the platinum needle, and investigate on the cover-glass or in the hanging drop; large numbers of the small typhus bacilli in extremely lively motion will be found. The same holds good even if the potatoes are kept for some time, and there is never found that thick, yellowish, smeary layer that is produced by Emmerich's bacillus. This growth of the typhus bacilli is so peculiar that by it we are able to distinguish it from other bacteria, and we should, therefore, *never judge definitely of their appearance before every doubt has been removed by potato culture.*

But the appearance of the culture just described is, unfortunately, not met with in all cases and under all conditions. E. Fraenkel and Simmonds, Ali-Cohen, Buchner, and others have drawn attention to the fact that the nature, and especially the reaction, of the potatoes is very essential for the development of the typhus bacilli; that they prosper in a "typical" manner only on sour potatoes, while generating on the alkaline ones frequently a yellowish-brown or gray smeary, sharply-circumscribed film, thus completely lacking the characteristic qualities.

In view of this uncertainty and the unusual importance of this subject, other means and ways have been sought for discovering an unerring criterion for the typhus bacilli. Chantemesse and Widal, for instance, have stated that only the typhus bacilli can flourish on gelatin containing 0.2% of carbolic acid, while the majority of all other bacteria do not develop. Thoinot pretends to have seen good results from the addition of 0.25 grams of pure phenol to 100 c.cm. of water, in which there were typhus bacilli together with other micro-organisms. He says that only the typhus bacilli had remained viable after several hours' contact with carbolic acid, and hence could be proved to exist in culture. Petruschky wants to utilize the formation of acidity of the typhus bacillus for its diagnosis, etc.

But all these procedures have not proved reliable. Only one method, recently reported by Holz, seems to furnish favorable results, at least in many cases. Holz prepares from the juice of raw potatoes, by adding 10% of gelatin, a solid, transparent, strongly-acid medium on which the typhus bacilli are said to grow particularly and luxuriantly, while most other micro-organisms fail, especially if there was added to the "potato gelatin" about 0.05% of carbolic acid, or if the culture apparatus had been treated with phenol by Thoinot's method. Finally, Kitasato's observation should be mentioned, which is also designed to facilitate the finding of typhus bacilli. They differ from other similar bacteria by not yielding the red indol reaction (as the cholera bacteria do) on the addition of nitrite of sodium (1 c.cm. of a solution containing 0.02 in 100 c.cm. to 10 c.cm. of the culture fluid) and a few drops of concentrated sulphuric acid.

The growth of the typhus bacillus on gelatin, potatoes, etc., shows that it is not an absolutely parasitic micro-organism. Besides the media before mentioned, it also thrives on other substances, mostly of vegetable nature; for instance, on decoctions of marsh-mallow, carrot-juice, etc.

Wolffhügel's observation that milk is an excellent medium for their development and that they live or even increase in water is very important as bearing upon the possible distribution of the bacteria. It harmonizes in many respects with the statements made during late years as to the discovery of typhus bacilli in suspicious drinking-water, and consequently their occurrence outside of the human body.

Does the micro-organism designated by us as "typhus bacillus" actually give rise to the affection expressed by its name?

It is not found in *all* cases of typhus, and thus far we have by no means succeeded in proving its existence without exception in every part of the body afflicted. But this may be owing to certain defects in investigation—for instance, the failure of a specific staining process—or to some peculiarities of distribution and arrangement of the rods in the tissue. If we consider that these bacilli have as yet never been met with in any other affection except typhoid fever, we must feel justified in considering the bacteria the cause of the disease.

A conclusive proof by means of transmission could not be expected, since it is known that typhoid fever, under ordinary conditions, never occurs in animals. This difficulty was supposed for a time to have been overcome (as in cholera).

E. Fraenkel and Simmonds injected into the auricular vein of rabbits dilutions of typhus culture. About one-half of the animals died after twenty-four to forty-eight hours. The spleen, mesenteric glands, and the follicles of the intestines were swollen; bacilli could always be found in the first-named place, but they had not generally passed into the intestine.

These results were confirmed by C. Seitz and perfected by other experimenters. Seitz rendered the contents of the stomach alkaline (exactly as in cholera), paralyzed the intestinal movements by opium, and introduced a dilution of typhus bacilli by means of a pharyngeal catheter. A majority of the Guinea-pigs died; large quantities of bacteria were found in the intestinal organs, but the blood was free. The intestinal mucous membrane was much altered in all cases, the spleen and glands being somewhat swollen.

A. Fraenkel in a similar way injected the bacilli directly into the duodenum of Guinea-pigs, with or without previous ligation of the ductus choledochus. The animals perished in the course of three to seven days; rods were seen in the intestine and spleen.

These experiments will prompt us to regard them as proof of the specific character of the typhus bacillus. The investigations by Beumer and Peiper, Sirotinin, Wolffowicz, and many other men have shown, however, that the matter is not as clear as might appear at first glance. If in transmission sterilized cultures of bacilli were used instead of viable cultures, the result remains absolutely the same; the symptom of the disease as well as the pathological condition fully correspond to the picture described before. The question is, therefore, only that of the effect of a pure intoxication, while actual infection cannot be spoken of, as the bacilli introduced alive quickly die and disappear even in the body of animals.

This fact in itself would not yet decide against the specific value of the typhus bacteria. Very similar conditions have been found with the comma bacilli, and led to the correct view that the peculiar activity of the excretions of the micro-organisms (most important in all cases) is manifested especially when an increase of bacteria is not accomplished in the altogether insusceptible animal body, but only when extraneous forces are able to act.

This view has been corroborated by the fact that in the typhus bacilli such poisonous substances (easily separable from the micro-organisms themselves) have been produced as belonging partly to the basic substances, the toxines, and partly to the albuminoid bodies, the toxalbumins. We might in this case, therefore, feel inclined to explain things as in the case of cholera.

One observation, however, prevents this. It has been found that precisely the same changes produced by the introduction of typhus cultures free from germs may likewise be developed under the influence of the excretions of many other micro-organisms of any origin—for instance, of simple water and soil bacteria—so that all the experiments just mentioned express nothing for the specific meaning of our bacilli.

But it has already been pointed out that the transmission to animals had no favorable prospects from the start, for which reason we are convinced that, in view of the nearly regular, and especially of the exclusive occurrence of bacilli in typhoid fever, they are in fact the cause of the disease.

How does the bacillus get into the body, and how does it produce in it the disease with its accompanying symptoms? Two totally different views regarding the cause and spread of the pestilence exist, almost in the same manner and sense as with cholera asiatica. Some maintain that the poison is not transferred from man to man, but that it must previously acquire a kind of maturity in the soil and thus be enabled to obtain infectious power. It is then to enter the body with the air and to be taken up by the respiratory organs. Special relations between changing conditions of the soil and the appearance of the disease are said to take place, and local as well as temporal predisposition to be highly important; the fluctuations in the position of the underground water are said to follow the changing degree of the disease like the hands of a registering apparatus, and a proper conformity between these apparently so divergent views is said to be easily established.

And when the cause of typhoid fever, previously only presumed, assumed tangible shape in the bacillus, its vital properties would not harmonize (just as with cholera) with epidemiological facts. Nor could it be proved that the bacillus passed through a special period of transition in the soil, nor could it be at all discovered there. The possibility of its finding the conditions for developing in the upper strata of the earth must certainly be admitted, for we know that it can lead a saprophytic life. But it is (for reasons holding good likewise with cholera) improbable that it should rise into the atmosphere and enter our body through the lungs.

Nor are facts wanting that point to very different ways of spreading. The typhus bacilli can live in the water and its existence there has even been directly proved. Milk, too, is a congenial place for development, and there can be no doubt, after Hesse's investigations, that still other foods offer it a welcome ground. The intestine, moreover, is the spot where the poison of the disease causes the first and severest changes. These observations have led to another view regarding the mode of infection, it being believed that it proceeds similarly as with cholera.

The diseased person casts off in his discharges a number of vital bacilli which, according to Uffelmann's experiments, remain in that neighborhood for many months. The bearers of the infectious matter by various means again get into the digestive canal of previously healthy people. The drinking-water also (as Almquist has recently rendered very probable), the milk, soiled linen, unclean fingers, etc., are the welcome bridge on which they step over the chasm from the first individual to the second and infect the latter, if he is otherwise susceptible, i.e., "individually predisposed." We shall by no means dispute the possibility of this mode of transmission being promoted or obstructed by temporal and local influences.

The bacilli having once been received, they penetrate into the intestine itself by passing the stomach. They settle in the intestinal wall, begin there their pernicious activity, and gradually find access, by means of the lymph current, to the lymphatic glands, first the mesenteric and then the remote ones. They then get back into the blood and are distributed over the organs, preferring spleen and liver. They never get from the placenta of pregnant women to the fœtus—a fact established by Eberth and several other investigators.

Efforts have been made to trace the typhus bacilli in the blood of living persons in order to discover directly the course of their distribution. These have been successful in a few cases, but the procedure has not led to positive results and did not therefore recommend itself for diagnostic purposes.

The morbid changes are developed most distinctly in the intestine, and these are so decidedly important for typhoid fever that they have given the name to the disease (enteric fever). In the ileum and upper cæcum a decided swelling of the solitary follicles and Peyer's patches is found; on the surface of the latter necrotic scabs are formed which slough and leave typhoid ulcers behind. The mesenteric glands and the spleen are also regularly enlarged; the other organs, the liver and kidneys, however, remain macroscopically nearly unchanged. We can find the bacilli at all these different places only by the aid of the microscope, and frequently they can only be determined with difficulty.

We already know that the typhus bacilli are very sensitive to decoloration and readily lose the coloring substance—a fact naturally becoming more prominent in treating sections than in cover-glass preparations. It is best to leave the sections for twenty-four hours in Löffler's solution, to wash and decolor them in simple water, to dehydrate them in anilin oil, dry them on the slide, and clarify them in xylol. The preparation should then be examined with low magnifying power, because the bacilli appear in the tissue (especially of the internal organs) in a very peculiar arrangement. They are not singly or diffusely spread over wide areas, but always occur in dense out not numerous heaps. The latter might easily be overlooked on observing them with immersion, for which reason it is better to search for them with a low power; they will then appear detached from their surroundings as stronger-colored deep-blue and opaque spots. If we examine them with the immersion lens, they appear as irregularly-circumscribed foci of radiating or net-shaped composition, so tightly joined together in the middle that they can be recognized as single rods only toward the border.

In order to facilitate the demonstration of bacilli, E. Fraenkel has recommended wrapping the organs in cloths soaked in sublimate solution and preserving them for some time (up to three days) after death at a high room temperature, because under these conditions, if not an increase, at least a far more vigorous development of the bacillus foci takes place, for instance, in spleen and liver.

Quite recent cases of the disease are best for investigation, previous to the formation of ulcers and the decay of the tissue in which the bacilli perish likewise. There may then be noticed numerous foci of bacteria in the pulpy, swollen patches and glands; the rods will be recognized later only in the deeper, non-necrotic part of the real mucous membrane, in the mucosa and intermuscularis below the ulcers. About a dozen sections of spleen and liver may have to be examined before we arrive at any result and discover the bacillus heaps in their special arrangement. The bacteria have (besides the tissue of the internal organs) been observed in the albuminous and even in the non-albuminous urine of very sick persons, and they have even been seen (according to Neumann) to appear there in especially large quantities. In the blood, too, their presence has been ascertained, and they were likewise found by different examiners in the dejections of patients.

Typhoid fever is one of those affections in which the co-operation of several diverse micro-organisms can frequently be established. The first original bacterium calls forth a number of morbid symptoms, changes in the tissue, etc., which constitute a proper soil for the subsequent settlement of some secondary causes of infection. The morbid picture stands then under the influence of the common activity of both, and we have to deal with a genuine mixed infection. The new micro-organism may finally push the original one so far in the background that it presides exclusively over the scene.

They are usually streptococci which go hand in hand with the other bacteria and move under their guidance into the diseased organism. In typhoid, too, are regularly found such chain-forming micrococci in the tissue of the spleen or liver or intestinal wall, etc.

Sometimes this association becomes manifest in the symptoms of the disease, inasmuch as typhoid in its course becomes complicated with an erysipelas or similar process. On other occasions the true state of things is not so evident, and only direct microscopical examination or cultivation will succeed in establishing the existence of such a mixed infection which is hardly ever absent, especially in more severe cases of the disease.

It is still a matter of doubt whether the demonstration of typhus bacilli can be of any value in dubious cases for a quick recognition of the disease. The proof of the existence of genuine typhus bacilli depends every time on potato (or a similar) culture, and when we have produced such a culture, so much time will usually have elapsed that the clinician or even the pathologist has anticipated the bacteriologist in his judgment.

For this reason we cannot approve of the procedure in which tissue is taken by puncture from the spleen of the patient, to be examined as to the presence of bacteria. The presence of typhus bacilli has, indeed, been ascertained in this way, but the result is not proportionate to the severity of the means by which it was obtained. This strange mode of investigation should not be permitted for humane and other considerations.

XIII. SPIRILLA OF RELAPSING FEVER.

Typhoid fever was not differentiated from other diseases having similar symptoms till about the middle of this century. Henderson, of Edinburgh, in 1843 publicly declared that a particular affection, characterized by the lack of abdominal changes and by the fact that it reappears in sudden relapses after apparent recovery, should be differentiated from the complex of symptoms hitherto known as "typhus." Henderson's view proved to be correct, and the disease was named relapsing fever (typhus or febris recurrens).

Relapsing fever appeared in Germany, in 1868, for the first time in an epidemic form, and Obermeier, of Berlin, furnished in 1873 the final proof of its peculiar nature by establishing, in all cases of relapsing fever, the occurrence of a special form of micro-organism met with in no other disease.

They are long, undulating threads, with numerous convolutions, resembling the cholera spirilla, though thinner and more delicate than the latter. It is a genuine screw-bacterium and named "spirillum Obermeieri." They are highly motile and rapidly glide and turn about. They readily absorb the common staining solutions, and are colored quickly and intensely in cover-glass preparations by the watery anilin colors, for instance, fuchsin.

The spirilla being found in all cases of relapsing fever and in these only, it may be taken for granted that they are the actual cause of the disease. This becomes still more probable by the fact that previously healthy persons can be infected by transmitting blood containing spirilla and a typical relapsing fever produced in them (as ascertained by Münch and Moczutkowsky). Monkeys have also been successfully inoculated by Koch and Carter, the animals being after some time violently affected by fever, and their blood exhibiting at the height of the fever large quantities of characteristic micro-organisms.

Efforts to cultivate these bacteria outside of the body, and thus to obtain a view of their peculiarities and manifestations of life, have thus far failed.

We must, therefore, still decline to answer the question as to the manner in which the spirilla invade our organism and give rise to disturbances. It appears certain (from the experiences in the course of various epidemics) that relapsing fever is a contagious disease, directly transmissible from man to man.

It is a striking fact that the spirilla are met with only during the paroxysms of fever (occurring so peculiarly in this disease) and disappear in the intervals. This process takes place (according to Metschnikoff's investigations) in the following way: at the height of the attack the spirilla are found exclusively in the blood, but not in the internal organs; during the period of the ante-paroxysmal rise of temperature, they gradually leave the circulation and gather in the spleen, within which they die and are absorbed by leucocytes. They are seen lying there densely rolled up, in heaps or singly.

Some individuals that did not perish and were preserved between the tissue elements become a nucleus for a new generation, advancing into the blood after some time and giving rise to the next attack.

The spirilla are absent in the secretions of the body, such as sweat, saliva, urine, etc. In a suspended drop of blood one sees the narrow, delicate threads shooting through the field in rapid rotations between the blood-corpuscles and pushing them in all directions; they appear singly, but frequently form groups by becoming wrapped into dense, closely-encircled heaps. Their number does not seem proportionate to the severity of the attack of the disease; several fields must sometimes be searched before we see a single one, while at other times they fill the blood in masses.

They stray into the different organs by way of the blood-current. Koch has succeeded in finding them in the vessels of the brain, liver, and kidneys of a monkey he had killed at the height of the fever attack.

XIV. PLASMODIUM MALARIÆ.

Statements and observations tending to prove that intermittent fever (malaria) owes its origin to an infectious cause have recently been on the increase.

Such a cause was long since suspected, although malaria is really the most prominent example of a purely miasmatic affection, never appearing (according to all experiences up to the present time) as a really infectious disease and in no case directly transmissible from man to man. Epidemiological facts point very emphatically to the probability that malaria is, above all, due to very peculiar conditions of the soil. It adhere to definites localities with a predilection and obstinacy based, of necessity, on extremely intimate relations which are of an almost exclusive importance for the natural conditions of infection. For experimenters have repeatedly succeeded in inoculating previously healthy individuals with the blood of sick persons, and in thus demonstrating that the presumed infectious matter actually exists in the blood.

A large number of investigators (first Laveran, in 1880, and after him Marchiafava, Celli, Golgi, Guarnieri, etc.) have discovered a peculiar micro-organism in the blood in nearly all cases of malaria and peculiar to that disease alone.

This represents an inferior living organism not belonging to the class of bacteria, but to the animal kingdom, to be included among the protozoa or mycetozoa, and for this reason named by its discoverers the Plasmodium malariæ. It appears in the blood within the red blood-corpuscles.

In the unstained preparation, the hanging blood-drop, and on the warmed slide table one sees the small, roundish or irregularly-formed structures permeate the body of the cells in which they are domiciled, in a rapid, amœboid movement. The parasite grows quickly until it almost completely fills the blood-corpuscle. It has, besides, at the same time absorbed the greater part of the hæmoglobin and changed it into melanin. While the red blood-corpuscle fades and becomes more indistinct, there is in the plasmodium an accumulation of numerous roundish granules or rods consisting of the accumulated black pigment.

The more exact peculiarities of form of this particular micro-organism may be ascertained by staining. By taking from the finger-tips of a malarial patient a drop of blood (by deep puncture with a needle), spreading it on a cover glass and staining the latter simply with watery methyl-blue, or by adding (according to Celli and Guarnieri) to the fresh, moist preparation methyl-blue dissolved in blood-serum or ascites fluid, a great number of delicate peculiarities may (according to the statements of the Italian investigators) be noticed in the plasmodia, of which we shall briefly note only the most essential. The elements found in the interior of the red blood-discs are said to be composed of an exterior, more strongly refractive part, more accessible to staining (ektoplasm), and an interior, paler part inclosed by the former like a ring (entoplasm). There are also seen in the latter special structures lying somewhat eccentrically and representing nuclei or nuclear corpuscles.

Whenever the plasmodium leaves the first vegetative point (in which it has absorbed the pigment) and enters the second reproductive place—i.e., whenever it commences sporulation—the body of the parasite changes more or less regularly into a number of new sections. The protoplasm is frequently disintegrated by many intermediate walls disposed in rows and running toward the centre so that star-shaped forms arise.

Sometimes elongated or spindle-shaped bodies will loom up which, after division, get into the blood-plasma, and are, therefore, also met with outside of the blood-corpuscles.

The same attitude is observed in another condition of the plasmodia, that of the sickle-shaped structures, which may appear sometimes as cylindrical, sometimes as crescentic, oval, or finally, as Laveran has observed, as completely round, flagellate elements between and along the blood-cells.

For the time being, it is still an open question in what connection all these diverse things stand one to another, and whether or how they pass into one another. We only mention Golgi's view that the single forms possess direct relations to definite sections—in fact, to the entire process of the disease. Golgi thinks that the parasite's process of division exactly coincides with the beginning of the fever or immediately precedes it. The newly-formed micro-organisms then enter other red blood-corpuscles and thus carry on the process by calling forth further fever attacks. It is said that we can predict from the presence of perfectly-developed structures and division forms the approaching beginning of a paroxysm, and that by observing the various stages of development of the parasite the eventual outbreak of an attack can be defined within a day or two beforehand, and that we can even ascertain whether conditions exist for each or any number of attacks according as there appear only one or several successive generations of plasmodia.

Golgi is of the opinion that the diverse forms of malaria (as tertian and quartan fever) are not caused by one and the same plasmodium, but by two different, morphologically distinct sub-species. But this is by no means generally admitted or confirmed. It must be confessed that the striking multiformity exhibited by the plasmodia rather contradicts this view. Danilewsky, an expert investigator in this domain, formerly entertained similar views.

Be this as it may, the investigations made at the most different locations and by very reliable men (thus in France and the French colonies by Laveran, in Italy by the above-named observers, in Russia by Sacharoff, Metschnikoff, and Chenzinsky, in America by Councilman and Osler, in Germany by Plehn, in Austria by Paltauf) have shown that there is found in the blood of persons afflicted with malaria a peculiar micro-organism, appearing in the interior of the red blood-corpuscles, which by its morphological and other appearances must be ranged among the sporozoa or gregarines. Its regular and exclusive occurrence in malaria, as also the circumstance that it disappears under the influence of specific treatment—i.e., after the use of quinine—hardly admit of a doubt that we have here the exciter of the affection.

The significance of the plasmodia (especially from a diagnostic point of view) appears to be quite extraordinary, and renders it desirable that our knowledge concerning this important micro-organism be soon perfected. Only when we shall succeed in artificially cultivating the malaria parasites and using them for experiments of transmission shall we be able to obtain a clear insight into the pathology and epidemiology of this disease.

XV. PNEUMOCOCCUS (FRIEDLÄNDER).

Just at a time when a parasitic cause was presumed to exist for many morbid conditions of the human body (science being still quite deficient in tangible, actual proofs of this opinion), intelligent observers asserted that pneumonia, too, the genuine inflammation of the lungs, belonged to the class of such affections. "Catching cold" had always been regarded as a decisive cause of pneumonia. The keen eye and clear judgment of such investigators as Jürgensen deserve great praise for discerning its infectious cause from its appearance and symptoms.

This standpoint appeared to be confirmed by the investigations published in 1883 by Friedländer and Frobenius regarding a special micro-organism observed by them in many cases of pneumonia. Preparations, especially of alveolar fluid, and sections from the changed lung-tissue had first shown them this kind of bacterium; it was afterward again found in the rusty expectoration of the patients. It was, moreover, readily cultivated artificially outside of the body. Successful experiments having been made on animals with these cultures, the discoverers did not hesitate to see in this micro-organism the originator of pneumonia and named it accordingly.

They called it "pneumococcus." But it is really a short bacillus whose rod shape is sometimes very distinct. Especially in the hanging bouillon drops (where growth can take place unhindered and unrestrained on all sides) and in the tissue, too, long bacilli are formed; even the smallest and youngest members show, under high magnifying power, that one of their diameters visibly exceeds the other. The cells are found usually singly or in pairs, extensive threads being sometimes formed.

This is the usual appearance of Friedländer's pneumonia bacteria, and it would be difficult to distinguish them microscopically in a stained preparation from the cells of the Micrococcus prodigiosus. The pneumococci possess, however, still another peculiarity of formation, perceived, it is true, only under certain conditions. In the animal body, their membrane acquires large dimensions; it swells up to an extensive capsule and surrounds the rod as a bright, transparent halo. But one member, as a rule, is found in each capsule, but sometimes several cells just after division are enveloped by a common cover which is then of considerable size. This thickening of the cell membrane, at first scarcely perceptible, was formerly considered to be a peculiarity of pneumonia bacteria alone, and they were therefore named "capsule cocci," but incorrectly so, there being several kinds of bacteria having capsules, like these pneumococci, and resembling them, besides, in the general absence of the gelatinous sheath when growing outside of the animal body; for instance, in cultures.

The capsule is by no means a special peculiarity of Friedländer's bacteria, and hence, *per se*, cannot serve to differentiate them, the less so because it is not uniformly present and cases occur in which it cannot be proved to exist, even by very close examination.

The pneumococci have no voluntary motion; the formation of spores has not been observed; they belong to the semi-anaërobic species and flourish in the absence of oxygen as well as under free access of air. They readily absorb the common anilin colors, but double stainings have thus far failed, as they decolor during Gram's method.

The capsule remains usually unstained and appears in preparations as a pale, glistening envelope in which the bacillus lies imbedded. But we may render the membrane accessible to coloring substances. For this purpose, Friedländer recommends treating cover-glasses and sections for twenty-four hours with an acetic acid gentian-violet solution (conc. alcoholic sol. gentian-violet, 50 parts; aqua dest., 100; acid. acet., 10) and decoloring them afterward in 0.1% acetic acid; followed by alcohol, cedar oil, etc. Even this method is not always successful.

The pneumococci grow on the gelatin plate (even at a relatively low temperature—16° to 20° C.) quickly into extensive colonies. They rapidly advance to the surface of the gelatin without liquefying it, and develop into thick and white accumulations of a china lustre, with a strong central elevation arched like a button, with smooth edges. The microscope exhibits brownish-yellow, clearly-circumscribed, usually irregular colonies of slightly-granular texture.

Growth in the test-tube proceeds in the beginning uniformly along the entire inoculation puncture. But soon there develops on the surface an especially prominent mass glistening like snow, somewhat arched, and imparting to the culture, in combination with the portion extending downward into the gelatin, a certain similarity to a thick-headed nail. In the course of further development, gas-bubbles in more or less great numbers frequently arise in the gelatin surrounding the inoculation puncture. A slight brown coloring of the medium regularly takes place in older cultures.

The pneumonia bacteria thrive on agar-agar as a moist, thick, whitish film. They produce on potatoes a yellowish-white, smeary, and very thick film in which a formation of gas-bubbles is sometimes observed.

Friedländer was able to effect successful transmission to animals from these artificial cultures. Rabbits proved completely insusceptible; Guinea-pigs were also only slightly sensitive; but the desired result was obtained in a large number (thirty-two) of mice, all of which died. Friedländer injected a dilution of his pneumococci through the thorax into the lungs. Section showed the latter to be altered by strong inflammation, reddened, sometimes hepatized and void of air; the tissue contained, besides, large quantities of bacteria available for further cultures.

The direct injection into the lungs permits conclusions as to the greater or less infectious qualities of the micro-organisms used for experiments only with very great reserve. Hence, we must look for other proofs tending to convince us that we have in the pneumococci an actual cause of pneumonia.

Let us be guided by Koch's demands that a specific kind of bacterium is to be found in all cases of the disease, and in it alone, and microscopical investigation having proved this, successful culture must confirm and transmission ultimately settle it.

It is a generally-admitted fact that the pneumococci cannot be observed in all cases of croupous lung inflammation, and the number of positive results is considerably less than that of the negative ones. Nor is the exclusive occurrence of Friedländer's cocci in pneumonia asserted anywhere, many communications having taught us that the same, or very similar, bacteria appear in the saliva and nasal secretions of healthy persons and in the pulmonary expectorations of persons affected with other diseases.

The claims of the pneumococci are, therefore, hardly to be sustained. Their value as a cause of pneumonia is the more doubtful because the way in which Friedländer discovered them was not exactly free from objection. He did not use the plate method, but brought pieces of the lung-tissue, pulmonary fluid, etc., immediately on solid gelatin, thus committing the same mistake, the fatal consequences of which were evident with Emmerich's Neapolitan bacilli.

Neither microscopic investigation nor culture nor transmission having furnished sufficient proofs for the assertion that the pneumococci play a decided rôle in the origin of pneumonia, we cannot recognize them as the causal factors of this disease. But it should not be overlooked that Friedländer's bacteria have been unobjectionably obtained in inflammation of the lungs by numerous reliable investigations. It may, therefore, be assumed that they are at any rate related to the said affection, and it may not be amiss to regard them (like the streptococci in typhoid) as subsequent settlers on a soil prepared and properly fitted by the activity of some other micro-organisms.

XVI. PNEUMOCOCCUS (A. FRAENKEL).

We are sustained in the opinion just expressed by the fact that we already know a species of bacteria radically different from Friedländer's, and, as it seems, taking an exclusive or decisive part in causing pneumonia.

A. Fraenkel observed in the sputum of persons afflicted with lung disease, especially in the rusty sputum of pneumonia, a peculiar micro-organism which proved to be pathogenic for several species of animals, and was called by its discoverer "the microbe of sputum septicæmia." Fraenkel asserted later, on the basis of minute investigations, that "it was to be regarded as the usual exciter of pneumonia," especially after he had succeeded in finding it in the hepatized pneumonia tissue of the diseased lungs and obtaining pure cultures therefrom.

This micro-organism has, according to Fraenkel's description, the appearance of an "oval diplococcus, the links of which possess an unmistakable similarity to the form of a lancet." Under high magnifying power and by examining preparations of the blood or tissue-fluid, it is found that here, too, one diameter of the cells exceeds the other. The forms are never as distinctly rod-shaped as those of Friedländer's bacteria, but the development of the single members is not as uniform and regular as is usually observed in true micrococci.

The bacterium must, consequently, not be regarded as a coccus, but as a bacillus, though it is usually still called "pneumococcus" or "diplococcus," owing to the fact that Fraenkel's bacillus is usually found in pairs, the pointed ends of the rod remaining united by an interstitial layer. Handsomely-coiled chains of five or six elements often develop, but longer combinations are rare.

Fraenkel's bacillus differs, therefore, from Friedländer's, inasmuch as the cells of the former are, on the whole, shorter, and longer members are wholly wanting; but both develop a capsule in the animal body and never outside of it. In both, one cell and sometimes several are enveloped by the same capsule, which appears as a glassy, lustrous halo or border. Practically both bacteria resemble one another greatly, and have certainly often been confounded.

Fraenkel's bacillus has no voluntary motion; it is semi-anaërobic, being able to thrive without oxygen. It is strikingly sensitive to the influence of temperature. It cannot develop at room temperature below 24° C., its optimum being near 37° C.; higher temperatures, say above 42° C., on the other hand, completely impede its growth.

It is to be remarked, further, that, according to Fraenkel's investigations, this bacillus requires a weak but distinctly-alkaline reaction of the culture medium. Even small quantities of acid render its growth impossible, and the degree of alkalescence of the medium is so important that the culture always fails whenever the requisite alkalescence (easily ascertainable) is increased or diminished.

It easily absorbs the common anilin colors, while the capsule remains intact. It is also accessible to double staining and can be beautifully prepared by Gram's method—a very remarkable and practically important difference from Friedländer's bacillus.

Artificial culture outside the body is difficult. Gelatin plates can only be prepared by observing special measures of precaution. By taking a 15% gelatin and not allowing the temperature to rise much above 24°. C., the gelatin will remain firm and colonies will be developed. They appear under the microscope as small, roundish, sharply-defined, slightly-granulated, whitish heaps, growing but slowly, not passing beyond a moderate size, and never liquefying the gelatin.

On agar plates there are developed on the second day at breeding heat delicate, shining, almost transparent, exceedingly small drops, scarcely perceptible by the naked eye.

Puncture culture in gelatin (15% at 24° C.) after a short time assumes a very characteristic appearance. Along the entire line of inoculation large quantities of small white granules arise, distinctly separated and resembling the picture of the streptococci of erysipelas.

On oblique agar and blood-serum, a veil-like, transparent film is developed, apparently composed of "single dew-drops." The bacteria flourish in bouillon without perceptibly clouding it; only a slight mist in older tubes betrays the presence of the micro-organisms.

The continuity of cultures on our artificial media is exceedingly restricted. On agar-agar, for instance, the diplococcus usually dies after four or five days and is no longer capable of development on transmission to fresh media. It cannot hold out much longer on gelatin; we may obtain somewhat older cultures only in bouillon.

Susceptible animals (mice, Guinea-pigs, and rabbits being among them, according to Fränkel's investigations) infected by this micro-organism usually perish after twenty-four to forty-eight hours. It does not matter how the material is obtained; it may be taken directly from the lung tissue, whenever microscopic examination has established the presence of bacteria, better still the agar plates prepared with tissue-fluid may be used, or cultures in gelatin or agar-agar. Young bouillon-cultures are best adapted, of which 0.1 to 0.2 c.cm. is used.

The first symptoms of the disease are perceived soon after injection under the rabbit's dorsal or abdominal skin. The animal does not eat, sits sadly in a corner of its cage, its temperature is clearly increased, and death ensues almost without exception after from twenty-four to forty-eight hours. Post-mortems furnished under all circumstances the same characteristic picture. Very slight if any reaction at the inoculation spot; much swelling of the spleen (frequently increased to twice its ordinary size) which is at the same time hard and reddish-brown; large quantities of bacilli, together with their capsule, in the blood and in all the organs. The bacteria lie only in the interior of the blood-vessels, the whole affection being thus characterized as genuine septicæmia. Morbid changes, such as small round-cell infiltration, incipient necrosis, etc., are nowhere to be found.

The lungs, in particular, show no visible consequences of infection, and they can surely not be considered as a privileged spot for the settlement of the micro-organism.

But when they are introduced directly, by injecting the infectious matter through the thoracic wall, the pleura will, as a rule, be violently inflamed. The lungs, too, are often affected and exhibit more or less considerable consolidations.

The injection of ever so small a quantity of the blood of an animal having thus perished into another of the same species will surely cause it to succumo to the infection. The bacillus, therefore, belongs to the most virulent or infectious micro-organisms known.

A. Fraenkel's observations (as herewith briefly reported) have been confirmed and amplified by a great number of different investigators. Suffice it to mention Monti's discovery that a genuine pneumonia with all its characteristic symptoms can be produced in rabbits by injecting bacteria into the trachea, the shortest way to the lungs.

The results of experiments on animals do not, however, always harmonize, the efficacy of the material frequently showing great variations. Fraenkel himself and, after him, Weichselbaum have pointed to the fact that the bacteria, taken directly from the lung tissue, possess *a priori* a varying degree of virulence and that they, above all when cultivated, rapidly fall a prey to attenuation, both natural and artificial.

Although we may perform transmission from generation to generation frequently and carefully, in order to prevent the premature death of the micro-organisms on one medium, their efficiency will become extinct after a certain period, while the culture, as such, is still capable of propagation. There is but one means of maintaining their activity, and that is by a prompt renewal of the bacteria in the animal body. Infection of a susceptible animal (rabbit) is best made every tenth day; new cultures should then be produced from its blood, and this procedure should be repeated after the period stated has elapsed.

Attenuation may likewise be brought about by the aid of high temperature. By placing Fraenkel's diplococci into bouillon and keeping this at 42° C. for twenty-four hours, the micro-organisms will have become completely harmless. The same is the case with a temperature of 41° C. in five days. Attenuation succeeds only partially when the tubes are removed from the incubator before the proper time has elapsed. By inoculating a number of rabbits with such attenuated material, it will be found that many of them fall quite sick, but that few perish and these only after several days.

Even after subcutaneous application, the lungs are visibly changed and an inflammation of the pleura ensues, with or without consolidation (as previously observed after direct injection into the lungs).

Is Fraenkel's bacillus really the genuine exciter of pneumonia? Considering that this "pneumococcus," owing to its peculiar properties, its growth at higher temperatures, its rapid decay in artificial cultures, and speedy loss of virulence outside of the body, appears necessarily as a strictly parasitic micro-organism, we may well assume that it is destined to play a pathogenic part. The transmissions also point to a special significance, especially if we bear in mind that pneumonia, under natural conditions, passes to animals rather infrequently and, consequently, affords to experiments but little chance of success.

But all these considerations are not of sufficient weight to compel us to believe that we are dealing with the original micro-organism of pneumonia, and we must again recur to the question whether this bacterium is regularly and only found in inflammation of the lungs. It may be stated (especially after Weichselbaum's extensive investigations) that this bacterium can be proved to exist in a very great majority (over 90%) of all cases. The circumstance of its being missed here and there is not difficult to account for. Fraenkel's diplococcus cannot be surely recognized as such by microscopic examination alone. It presents certain difficulties to cultivation (which alone can afford a reliable conclusion) which are not always sufficiently heeded by all observers. Its exclusive growth at high temperatures and its considerable sensitiveness require careful manipulation in cultivation, above all the indispensable preparation of agar-plates. This requirement is still frequently overlooked.

Weichselbaum and Monti have, moreover, emphatically pointed out another circumstance: the diplococci are found in the diseased lung-tissue in much greater quantities, the fresher and younger the tissue is, while in the further course of the affection they gradually diminish and may ultimately disappear altogether. The exact time of observation is, therefore, of the greatest importance and many failures may be attributed to this circumstance.

All this is decidedly in favor of the specific character of the bacteria. There is, however, one fact which prevents our forming a conclusive opinion.

Fraenkel's bacteria are by no means exclusively found in pneumonia; they are, on the contrary, very extensively diffused. They are found in almost all cases of cerebrospinal meningitis (as proved by Foà and Bordoni-Uffreduzzi) and the origin of this affection is reasonably attributed to them. A. Fraenkel has found them present in pleuritis, Weichselbaum in peritonitis, Banti in pericarditis, other investigators have encountered them in endocarditis and otitis media and numerous other affections; they occur especially in the saliva and nasal secretions of healthy persons (as established by Netter) and they may be regarded almost as regular tenants of these localities.

A worse offense against Koch's law of the properties of a specific micro-organism cannot be imagined. Must not our whole structure so artificially built up from observations and reflections collapse in the face of this one fact? Can we seriously believe for a moment that such an ordinary, common bacterium should be capable of producing a disease as typical and as clearly circumscribed as pneumonia?

Still to prove that this is not the case is not so easy as may be supposed. Fraenkel's diplococcus is, indeed, not the exciter of pneumonia alone, its domain is more extensive; it does not restrict itself to this one function.

Fraenkel's bacterium is the principal exciter of inflammable processes of an infectious nature in the human body. Wherever it reaches a serous or mucous membrane and meets with the requirements for its settlement, it commences operations; it causes meningitis on the pia mater, peritonitis on the peritoneum, and otitis in the auditory passage. Whenever it gains entrance into the lungs, pneumonia is developed, the peculiar properties and characteristic process of which depend upon the peculiarities of the organ invaded and upon the extent of the morbid process. Another bacterium may eventually play a similar rôle and give rise to pneumonia; but, as a rule, it is certainly Fraenkel's diplococcus that displays here its energy, for which reason it may properly be regarded as the real micro-organism of genuine croupous lung-inflammation.

But how can this view be harmonized with the circumstance that this micro-organism is also a frequent guest in the healthy body, that in the majority of all persons it is domesticated in the mouth, whence it might easily and at all times undertake an excursion into other regions and thus soon produce meningitis, otitis or something else? Does the "sword of Damocles" actually hang at every moment so close to our head, and must it not appear almost miraculous that anybody at all is spared by this terrible foe? We can account for this certainly very striking fact only by the circumstance that it requires, as a rule, certain preparatory, as yet unknown factors, to enable this bacterium to make its attacks. The healthy coverings and tissues of the body resist the micro-organisms; only when the native powers of resistance are weakened or neutralized will the foreign invaders take a firm footing and begin their pernicious activity. The better the medium agrees with them, the more rapidly they will develop and the greater will their virulence become. All the minute gradations of their infectious power (observed even under the conditions of their natural appearance and surely strong enough to determine the severity of a single case or the character of an epidemic) will find their explanations in this hypothesis.

XVII. DIPHTHERIA BACILLUS (LÖFFLER).

Another kind of bacterium resembles in many respects Fraenkel's diplococcus.

We conceive genuine pneumonia to be of an essentially uniform character, its origin usually being due to one and the same cause; we thus distinguish it from those so-called "secondary pneumonias" not unusually developing in the course of other diseases (such as typhoid, small-pox, etc.) and showing, pathologically, the symptoms of genuine lung-inflammation, but being caused by other micro-organisms (presumably by streptococci).

Diphtheria, another affection of infectious origin, is of a similar nature. We know quite a number of pathological processes occurring with the formation of croupous or diphtheritic changes of the mucous membranes and by post-mortem appearances not distinguishable from the processes accompanying diphtheria proper. The former may, however, arise from various causes, while the latter, as a whole and in every part of its course, exhibits so many peculiarities as an entity, that intelligent investigators have always considered it as a disease by itself and attributed its origin to one and the same cause. As its infectious nature is manifested so decidedly and is transmitted most fearfully from man to man by direct contagion, it can readily be understood that the specific micro-organism has long been looked for.

All reliable observers having thus far agreed that the blood and the internal organs of persons dying of diphtheria are usually completely free from micro-organisms, the conclusion was reached that there must be an essentially local process, whose action upon the entire organism was promoted by the absorption of dissolved and injurious substances produced by bacteria, and that the real bearers of the poison would only be found in the local changes. But these very changes by no means afford a proper field for bacteriological purposes. While the mouth and the mucous membranes of the adjacent regions teem with divers bacteria under ordinary circumstances, their number increases considerably whenever these parts become diseased. The ulcerous processes formed during diphtheria from the destruction and expulsion of the superficial layers, afford to these micro-organisms an excellent medium for undisturbed settlement and limitless reproduction. They gradually advance into the tissue, immigrate into the inflammatory and coagulated accumulations characterizing diphtheria, and, finally cause a motley array of most diversified forms. Errors and misinterpretations were a natural consequence. Only by extreme caution and keen judgment of his own results, did Löffler succeed in finding the way out of this confusion.

In the course of extensive investigation he discovered in the changed mucous membranes of diphtheria patients a kind of bacterium distinguishable from other bacteria hitherto known; he was able to cultivate them artificially and to employ them for successful transmissions. These results determined him (of course, only conditionally and with every reserve) to ascribe to this microorganism particularly close, perhaps causative relations to the development of diphtheria.

Löffler's statements have been confirmed and amplified by further investigations. A great number of observers, among them Babes, Kolisko and Paltauf, Zarniko, Escherich and d'Espine, have established the regular occurrence of Löffler's bacilli in all cases of diphtheria and proved that they alone belong exclusively to this affection. Finally, Roux and Yersin, by means of the bacilli, have been able to produce in animals quite the same symptoms that are peculiar to human diphtheria and they obtained paralysis of a typical kind, so that we regard Löffler's bacilli beyond doubt as the exciter of human diphtheria.

The bacteria occur within the diphtheritic pseudo-membranes and in the oldest parts of these. They are surrounded by a copious accumulation of cells and do not, as a rule, advance as far as the chief mass of the membranous accumulation which is characterized rather as a layer of exudation containing few if any cells and bacteria.

They are little rods of moderate size, usually slightly bent, about as long as the tubercle bacilli, but twice as broad, hence of a rather coarse appearance and usually with rounded ends. But the form of the micro-organisms is exceedingly variable and the differences are striking in appearance. Really normal bacteria of this variety are seldom met with. The bacteria are sometimes seen enveloped in a more or less capacious, glassy membrane; sometimes the contents separate into several pieces divided by a broad, transverse wall; one end of the rods is very frequently thickened like a club; sometimes this change appears on both sides so that there arise dumb-bell-shaped structures usually pointed out as involution forms. The bacilli are immobile and have no spores, but perish rather rapidly when dried, and succumb at 45° to 50° C.

They absorb the common anilin stains only imperfectly, but it is not difficult to prepare them with Löffler's alkaline methyl-blue. The rod is evenly stained along its entire length. The terminal pieces often prove more accessible to stain and stand out as dark spots over the pale centre. Smaller roundish structures, so-called "pole-granules," appear in the extremities and into these the color enters most rapidly.

The diphtheria bacilli are semi-anaërobic; they only thrive at somewhat high temperatures (between 20° to 42° C.).

On plates of 15% gelatin at 24° C., small, roundish, white colonies of a moderate size are developed which never liquefy the medium. They appear under the microscope as yellowish-brown, dense discs of granulated structure with irregular edges.

On plates of agar or glycerin-agar, at breeding heat, millet-seed sized, flat colonies with a very flat border and of a grayish-white lustre form in twenty-four to forty-eight hours; not rarely they exhibit, macroscopically, a ring-shaped stratification. Under the microscope they show pronounced similarity with the colonies of the Bacillus megaterium on the gelatin plate—a gray mass of peculiar coarse granulation, looking like "shagreen."

In gelatin culture, small, white, round globules are formed along the inoculation puncture. They grow but little at first. A phenomenon is noticed on this medium which is still more distinctly seen in cultures on oblique agar or glycerin-agar. With sterilized scissors cut off a piece of diphtheritic membrane about the size of a pin's head, and pass it with the platinum-needle successively through six or eight tubes containing the culture medium; the quantity of the germs distributed on the surface diminishes from tube to tube; the fifth or sixth will exhibit distinctly separated, single colonies, some of which prove, on microscopic examination, to consist of diphtheria bacilli.

By removing a trace and starting a pure culture on agar, it will be perceived that the development, in the beginning, keeps within narrow limits. The medium evidently does not agree with the bacteria; the growth remains restricted to the immediate neighborhood of the inoculating line, and the film (of a whitish lustre) extends but very slowly. But on transferring the colony from the first generation to the second, we will at once notice that the bacteria quickly accustom themselves to the originally unsuitable culture medium; the colonies proliferate more and more, and finally spread over the entire surface.

A thick, whitish, opaque coating arises on blood-serum; the same is the case with a serum suggested by Löffler for breeding diphtheria bacilli. This serum consists of 3 parts of cattle-serum and 1 part of bouillon containing 1% of peptone, 0.5% of sodium chloride, and 1% of dextrin.

In bouillon, at breeding temperature, the bacilli form white, very small, firmly cohering, peculiarly gritty grains which quickly sink to the bottom or adhere to the walls of the tube, while the fluid itself remains clear, so that the appearance of such cultures is usually quite characteristic. The diphtheria bacilli develop likewise very readily in sterilized milk.

In experimental transmission, we must never overlook the fact that animals under natural conditions are never attacked by human diphtheria, and hence but little adapted for our purpose. All the affections running a similar course (such as the diphtheria of calves, pigeons, and chickens) are originally altogether different diseases and arise from other causes.

Inoculations with diphtheria bacilli have, however, been rather successful in spite of these unfavorable preliminary conditions and Löffler obtained some very important results. He found that mice and rats were very refractory, while rabbits and Guinea-pigs, and most of the birds, such as finches, sparrows, and especially pigeons and chickens, are always accessible to the influence of the bacilli.

After injecting Guinea-pigs subcutaneously, purely local changes first develop at the point of infection. Grayish-white pseudomembranous masses are formed. More general disturbances soon appear; an extensive hemorrhagic œdema in the subcutaneous cellular tissue usually arises all around the point of inoculation. This may be followed by effusions into the pleural cavities and lobular consolidation of the lungs. The bacilli also produce pseudomembranes whenever introduced into the opened trachea of rabbits, chickens, and pigeons; also on the superficially injured conjunctiva of rabbits and the lacerated introitus vaginæ of Guinea-pigs. Local symptoms in all cases are followed by bloody œdema, hemorrhages into the tissue of the lymph glands, and effusions into the pleural cavities.

The animals are sick only a short time, guinea-pigs perishing after twenty-four to forty-eight hours. The process in rabbits is usually somewhat slower; days and even weeks elapse before death ensues. The general symptoms are thus allowed time and opportunity to develop. Inoculation is succeeded (in rabbits most frequently, but also in Guinea-pigs and pigeons) by paralyses reminding us of those observed in human diphtheria. Soon after injection into the trachea the animals begin to have a rattling in the throat; the breathing becomes difficult and more rapid; food is refused and the temperature rises considerably; when death occurs the walls of the trachea are lined with membrane down to the larger bronchi. If the end is delayed, or the animals recover, symptoms of incipient paralysis will be perceived on about the sixth or seventh day. The extremities (first the posterior, then the anterior) are dragged along and gradually become paralytic. Disturbances of co-ordination will finally supervene and will quickly lead to the end, even in animals apparently cured.

The rods are regularly found only in the immediate neighborhood of the point of infection, while the blood and internal organs are always entirely free from micro-organisms, as is the case in human diphtheria.

The results of transmission do not, however, correspond to this description under all circumstances. A difference in the pathogenic effect has been observed under natural conditions (just as with Fraenkel's pneumonia bacteria) and it has been ascertained, moreover, that the diphtheria bacilli very rapidly succumb to natural attenuation. Its adaptation to an artificial culture-medium (mentioned before) goes hand in hand with a loss of virulence. The more the bacteria extend on the agar-surface, the more their strength will diminish; although exceptions to this rule are occasionally met with and a still undiminished virulence may be perceived in old cultures.

The purely local occurrence of diphtheria bacilli and the severe general symptoms caused by them (as pointed out above) can only be explained by the supposition that the bacteria generate a soluble substance which spreads from their locality through the body and thus affects even remote parts. This view has, in fact, been proved correct by recent investigations.

Roux and Yersin found that the germ-free filtrate of somewhat older bouillon cultures of the bacilli (of a strongly-alkaline reaction) possesses a very considerable degree of virulence. On injecting it into the subcutaneous cellular tissue or directly into the blood current of rabbits, Guinea-pigs, and pigeons, it produces essentially the same changes observed after inoculation with living micro-organisms. Pseudo-membranes on the mucous surfaces do not form in any case, but in connection with transmission paralyses are developed which resemble those described above and are just as fatal. Post-mortem examination after subcutaneous application showed a hemorrhagic œdema of the abdominal walls and effusions into the pleural cavity; after injection into the jugular vein, generally acute nephritis and a very pronounced fatty degeneration of the liver are manifest. The greater the quantity of poisonous fluid employed, the sooner the end will come; but even very small doses usually prove effective, except that the result may be delayed for many days and even weeks.

What is the nature and property of the toxic substance generated by the diphtheria bacilli? Löffler, and afterward Roux and Yersin, have attempted to answer this question, and believe that we have here a body related to diastase or enzyme, since it quickly decomposes and is destroyed at temperatures but little above 55° C., is insoluble in alcohol, etc.

Further investigations have shown that we have to deal with an especially typical representative of the toxalbumins. On evaporating some of the filtrate to one-third of its volume in a vacuum at 40° C., and dropping it into absolute alcohol to which a few drops of acetic acid have been added, a grayish-white, flaky deposit is formed which sinks to the bottom in the course of a few hours. This sediment is very soluble in water; it is precipitated again by a new addition of alcohol. By repeating this procedure, together with filtration and dialysis, it may be cleared of all admixtures so that, on drying it in a vacuum at 30° C., it will appear as a snow-white amorphous and very light mass showing the most important reactions of the albuminous bodies. It is at once decomposed by high temperatures and possesses highly-virulent properties. Even small doses give rise to the symptoms noticed after injection of the filtrate; here too, the result is frequently retarded for weeks.

We have, then, presumably a direct derivative of the normal tissue-albumin, which, decomposed by the bacteria in some particular manner, acquires a toxic power manifesting itself far and near. We now understand why the rods in the pseudo-membranes are only found on the surface, while the deeper parts (by the action of the excretions just mentioned) are transformed into a decayed, coagulated layer of exudate; we understand, likewise, why local changes gradually affect the entire organism, the nervous system, etc.

But all this only disposes of a very small portion of the subject matter. We have not yet ascertained (and it might be a task worthy of continued investigation) how the bacteria enter the body of man, in what manner they are transmitted from one individual to another, and whether they require a special preparation of the mucous membranes before they can take root on them. Löffler's observation may throw some light on the subject. He found his bacilli, though only once, in the saliva of a healthy child—a circumstance reminding us of Fraenkel's pneumonia bacteria.

It is also a remarkable fact that the diphtheria bacilli hardly ever appear alone in quite recent cases, but are usually associated with streptococci which follow and support them in their attack on the tissue. This mixed infection is surely significant for the process of the disease. Many investigators even incline to the opinion that the specially severe cases owe their malignant character essentially to these concomitant micro-organisms, the pathological efficiency of which is well known.

XVIII. BACILLUS OF RHINOSCLEROMA.

Let us in this connection briefly consider another affection, even though it is not closely related to the diseases above described. Frequently in Austria-Hungary and Italy, and occasionally in Germany, an affection is met with which is characterized by the development of tubercular thickenings on the external skin, especially by the formation of extensive swellings in the nares and, therefore, called rhinoscleroma. Frisch first observed in the newly-formed tissue, micro-organisms distinguishable from others by their shape and distribution, which fact has been universally confirmed by subsequent investigations.

They are quite short, moderately broad rods with rounded ends, resembling Friedländer's pneumococci, especially as they are likewise inclosed in a capsule.

They readily absorb anilin stains and are accessible to Gram's method, unlike Friedländer's bacteria.

They are frequently found in large cells peculiar to this affection (more minutely described by Mikulicz), which are hyaline and without nuclei; but not infrequently they lie free in the tissue and within the lymphatic vessels.

Paltauf, Eiselsberg, Dittrich, and others, have been able to cultivate these micro-organisms in many cases. On the gelatin plate and in the test-tube, the bacillus exhibits a pronounced similarity with Friedländer's pneumococcus, except that the head of the nail-shaped culture appears more transparent and milky than the thick, white aggregation of a china lustre of the pneumococci.

The bacteria thrive on agar, blood-serum, potatoes, and in bouillon and form luxuriant colonies usually consisting of capsulated rods in the solid media, while the gelatinous envelope fails to develop in the nourishing fluid.

Experiments with animals have shown that the bacilli possess about the same pathogenic properties as the pneumococci. No symptoms, however, resembling the picture of rhinoscleroma have developed in connection with transmission, and it will be well to refrain from any final decision regarding the specific significance of this micro-organism and to await the result of further investigations.

XIX. MICRO-ORGANISMS OF WOUND INFECTION.

Eminent physicians have for a long time recognized that the majority of processes retarding the healing of wounds must be attributed to external influences and are caused by the access of foreign and injurious substances. A better insight into the significance of the peculiar activity of micro-organisms removed any doubt of their important character, and it was generally admitted that the circumstances above mentioned owed their origin to some infection.

Lister succeeded in reaching these conclusions before they were proven, and in thus removing the worst enemies of wound-repair. Wherever these teachings are ignored, the bacteria raise their head ominously and begin their fatal activity.

The really severe wound-poisonings (such as hospital gangrene, etc.) have nowadays nearly disappeared. Of the less dangerous consequences, only erysipelas is still of frequent occurrence. Its "traumatic form," caused by lesions, etc., was formerly distinguished from an "idiopathic" erysipelas regarded as a peculiar, well-defined disease. But this distinction can no longer be maintained, as both forms have but one cause.

Many observers had established the presence of micrococci in the erysipelatous tissue, especially in the margins of the affected skin. Fehleisen succeeded in growing these bacteria artificially outside of the body, and by transmitting them to previously healthy people, he reproduced a typical erysipelas, thus proving conclusively the specific significance of the micro-organisms.

The streptococci of erysipelas are small, completely round, globular cells possessing a pronounced tendency to grow into long chains. They always appear, in the culture as well as in the tissue, in extended, bead-like chains, embracing generally six to ten, but frequently even hundreds, of members. These chains often interlace in a dense coil or form elegantly arranged bundles. The single cells are of uniform size, though once in a while some member, on the point of dividing, is of a somewhat greater size.

The erysipelas cocci are immobile and have no spores. They thrive at common room-temperature, more rapidly of course at higher degrees (30° to 37° C.). They are not particularly sensitive to want of oxygen, but thrive best on the surface of the artificial media with free access of air. They are easily stained by the various anilin dyes and, like most micrococci, prove readily accessible to Gram's double staining.

Their growth on the gelatin plate is rather slow and circumscribed. They may usually be noticed with the naked eye, on the third or fourth day, as small white dots in the depth of the gelatin; they never grow beyond the size of a pin's head, never liquefy the gelatin, and do not even advance to the surface.

The colonies under the microscope appear as round, yellowish-brown heaps, with sharp and smooth edges and of a strangely-granular structure of layers sometimes distinctly ring-shaped.

Their development is most rapid on agar-plates kept at breeding temperature. As early as the second day there arise extremely delicate, transparent aggregations of gray color and in the form of drops, generally not exceeding a moderate size.

The growth of erysipelas cocci in the gelatin culture is quite characteristic. The entire extent of the inoculation puncture is slowly studded with very small white and globular granules generally remaining isolated and presenting an appearance nearly like Fraenkel's pneumonia bacteria. The inoculation on oblique gelatin or agar appears the same; in its neighborhood numerous tiny round drops appear that do not coalesce and reach only a small size.

The same is the case on blood-serum. A distinct development cannot be observed on potatoes.

For the purpose of obtaining large quantities of erysipelas cocci and keeping on hand sufficient material for cultures, cultivation in bouillon is preferable; in this there quickly develops, at breeding temperature, a very luxuriant growth of beautiful chains of micro-organisms sinking to the bottom in whitish, crumbling flakes, while the nourishing fluid remains clear. Only a vigorous shaking of the tube will disturb the bacterial masses and cause a transitory cloudiness of the bouillon.

Erysipelas can be reproduced from such artificial cultures in susceptible animals. Fehleisen, and many others after him, have performed successful transmissions to man for a particular purpose, surgical practice having often shown a striking improvement in malignant tumors which could not be operated upon (especially sarcoma and carcinoma), and whenever the tumors came within reach of erysipelas arising from some other cause, they improved. This experience was utilized by attempting an artificial production of the curative erysipelas; the results have been rather beneficial.

Among animals, mice are usually completely refractory to subcutaneous applications, while rabbits prove susceptible. After inoculation at the ear, there arises a progressive, erysipelatous, inflammatory swelling, rapidly spreading from the point of infection, but usually not going beyond the ear and subsiding in a short time. Suppuration and a severe general sickness with rise of temperature, etc., and even death occur only in young animals. A direct introduction of cocci into the blood current will produce no disturbance.

The experiments with erysipelas cocci present the same result that was observed with Fraenkel's pneumonia bacteria and the diphtheria bacilli: the virulence of the micro-organisms is unstable from the beginning, even if taken up directly from the diseased tissue (their natural habitat), and succumbs more or less rapidly to natural weakening in our artificial cultures. The varying degree of original virulence is certainly significant as to the character and progress of isolated cases, and the general subsequent decrease of the infectious force accounts for the differences in the results of experiments with animals.

We know that the cocci just described are the original exciters of erysipelas, and we must again turn to the question as to the manner in which the micro-organisms enter the body and their influence upon it.

It is certain that, in the great majority of cases, infection commences from lesions which are sometimes scarcely visible, and from wounds of the integument having in some way come in contact with streptococci. The conditions of this mode of transmission have as yet been but little investigated. We presume, but cannot assert on the strength of reliable facts, that in this case we have an accidental infection by micro organisms, largely diffused in the outside world, rather than a direct infection of a healthy individual by a sick one.

The cocci having once entered, the local changes characterizing erysipelas are produced—i.e., progressive redness and swelling of the skin. But severe general symptoms (gastric disturbances, fever, nervousness, etc.) will sometimes appear from the beginning and during the whole process of the disease, which circumstance points to the action of a special bacterial poison, distributed over the body through the blood or fluids, which we have not yet succeeded in separating.

A microscopic investigation of the diseased tissue reveals the cocci generally in considerable quantities at the edge of the inflamed region, which latter usually remains free from them.

The micro-organisms can be stained by Gram's method and exhibit their peculiar distribution in the tissue distinctly. They are confined almost exclusively to the lymphatic glands and vessels which they sometimes completely fill, while the adjacent parts remain free. Occasionally individual cocci appear between or within the cells. They are rarely noticed in the blood and organs.

PYOGENIC BACTERIA.

As diseases from wound-infection were prevented, by the aid of antiseptic treatment, with increasing success, it became more and more probable that the cure of wounds was in reality an exceedingly simple process, even when unaided by a "reaction," formerly regarded as indispensable. Suppuration was considered to be the principal "process of reaction." The question whether suppuration could originate at all without the action of micro-organisms (even if not directly in connection with wounds, since the latter were so successfully kept from complications) was soon answered by the dictum: no suppuration without micro-organisms. The correctness of this assertion was subsequently disputed. A great many theoretical discussions and experimental elaborations followed, some of them being distinguished by the exhibition of admirable care and skill. We have now reached a positive conclusion regarding some important points.

The investigations of Scheurlen, Steinhaus, Kaufmann, and especially Grawitz and de Bary, leave us no longer in doubt as to the fact that many germ-free chemical substances (such as nitrate of silver, oil of turpentine, liquor ammonia caustici, digitaline, cadaverine, etc.) can produce an acute suppuration in the subcutaneous tissue. The sterilized cultures, too, of various micro-organisms act in the same way. The cadaverine (pentamethyldiamine) belongs to the series of well-known bacterial excretions. It is just as certain, on the other hand, that, under natural conditions, suppuration in man is always regarded as a special reaction of the tissue to the presence and activity of micro-organisms, and that, as a rule, certain bacteria act in this manner as specific exciters of infection, and are found in all cases of suppuration, no matter whether we have an extensive and severe phlegmon or a slight paronychia, metastatic pyæmic abscesses or a simple furuncle.

XX. STAPHYLOCOCCUS PYOGENES AUREUS.

Ogston, Rosenbach, Passet, Lübbert, etc., have given us accurate information regarding the properties of these pus bacteria. A species of micrococcus, called by Rosenbach Staphylococcus pyogenes aureus, appears to be the most common. They are roundish, minute cells and smaller than the erysipelas cocci. They too are apt to form groups, though never in the shape of chains. They are arranged rather in dense irregular heaps, the appearance of which, especially in the tissue, remind one sometimes of a close cluster of grapes, hence their name (σταφυλή, grape).

Sporulation has not yet been observed in the staphylococcus (nor in any micrococcus), but it exhibits a very remarkable power of resistance. Drying on the cover-glass for ten days does not destroy its capacity of development; chemical agents kill it only when rather concentrated, and boiling water needs several minutes to destroy it. In gelatin cultures it keeps fresh and capable of reproduction for nearly a year.

The Staphylococcus aureus thrives at ordinary room-temperature, although better and more luxuriantly at a higher point (30° to 37° C.). Its need of oxygen is not very urgent and it even thrives under a restricted access of air. It readily takes up the common anilin stains and is exceedingly well adapted for treatment by Gram's method.

On gelatin plates, on the second day, small, white dots appear deep in the culture medium; they advance rather rapidly to the surface and begin to liquefy the surrounding gelatin. An orange-yellow color, especially apparent in the centre of the colony, is produced at the same time. The liquefaction of the gelatin is usually rather extensive, and the single colonies but rarely exceed a certain size. Under the microscope they appear as roundish discs with sharp, smooth edges of a dark-brown or yellow color, markedly granulated.

The formation of the pigment is still more distinct on agar plates. The superficial colonies, being in constant contact with oxygen, soon assume a beautiful golden-yellow coloring and can thus be recognized at the first glance.

In the test-tube the growth proceeds along the entire inoculation puncture. The gelatin is generally entirely liquefied, most rapidly in the superficial layers. The cocci slowly sink to the bottom and gather there as a distinctly-yellow, crumbling sediment, the upper portion of the gelatin appearing but slightly cloudy. A strange, acidulous odor, like that of paste, may soon be noticed in the culture. The Staphylococcus aureus develops most characteristically on oblique agar-agar. There appears along the inoculation line and restricted to its neighborhood, an orange-yellow, moist, glistening film, looking as if the surface had been coated with oil-paint.

The pigment is particularly beautiful when the cocci are cultivated outside the incubator. In the latter, the growth is so luxuriant and rapid that the production of coloring matter is retarded and the edges of the culture frequently remain nearly white.

The Staph. aureus thrives excellently on potatoes, and at a high temperature a thick, juicy, yellow coat with the peculiar odor is formed.

Bouillon becomes evenly and thickly cloudy; in sterile milk the casein is precipitated and slowly peptonized.

The fact that the Staphylococcus aureus is not a regular and harmless concomitant of purulent inflammatory processes, but their cause, has been demonstrated by successful transmissions. The results corresponded to the natural conditions inasmuch as they produced the most varying forms of suppuration, and thus explained more fully the occurrence of this coccus under so many various conditions.

Inoculations on man were performed by Garré, Bockhart, Schimmelbusch, Bumm, and others. Garré experimented on himself. He once applied a pure culture of this coccus to small wounds on the edge of a finger-nail and noticed a progressive suppuration around it; at another time, he rubbed larger quantities of the coccus on the sound skin of his forearm and caused by it the appearance of a very large carbuncle which required weeks to heal and left behind distinct scars. The aureus was again obtained from the contents of the abscess.

The other investigators reached similar results, while the results of experiments with animals were not so uniform.

The mode of infection has a determining influence; the action of the bacteria appearing under some circumstances in a very different light. Simple inoculation does not succeed with mice, Guinea-pigs, and rabbits. Subcutaneous application may cause the formation of abscesses which subsequently heal or may lead to general sickness and even death. Injections into the abdominal cavity may give rise to severe phlegmonous suppurations. The direct introduction of the cocci into the blood-vessels is still more efficient, and this kind of transmission is the most remarkable in its consequences. The cocci can be found in the blood as well as in all organs, though in small numbers and only demonstrable by culture. They also cause purulent inflammations of the joints, and especially small metastatic abscesses in the heart muscle and the kidneys. In the latter, bean-sized, whitish foci or extensive pyramidal-shaped infarctions are met with, their origin being due to the displacement of large portions of the cortical substance. The cocci stop up the capillaries and even the smaller arteries, and thus give rise to considerable disturbances. They are also found in the cells.

The experiments of Orth, Wyssokowitsch and Ribbert are very striking. The first two ascertained that, after injuring an animal's cardiac valves by catheterization from the right carotid (according to O. Rosenbach's method) and introducing the Staph. aureus into the blood-current, a typical endocarditis ulcerosa appears at the injured places. Ribbert discovered that the same result may be obtained without preparing the valves, by simply taking the material for transmission from potato cultures of the aureus. The thicker particles of this inoculating substance are swept along by the blood-current and brought into the cardiac muscle and deposited especially on the valves, where they cause inflammatory changes.

This staphylococcus having been demonstrated by culture in cases of spontaneous endocarditis ulcerosa and even in verrucosa, it may be regarded as a cause of this disease, although it should not be forgotten that we have become acquainted with another exciter of inflammation in Fraenkel's pneumonia bacterium which also not uncommonly takes part in the development of endocarditis. It may be well then not to regard the latter, with Weichselbaum, as a process due to one cause, but to consider it an event occasioned by one or another micro-organism possessing the power and ability of infecting and inflaming the delicate covering of the cardiac valves.

On introducing the Staphylococcus aureus into the blood-vessels of an animal, a subcutaneous fracture or contusion of a long bone having previously been inflicted, there will frequently occur in these "predisposed" places, symptoms of osteomyelitis sufficiently severe to cause death. This fact is significant, since Becker, as early as 1883, and therefore previously to Rosenbach and Passet's, statements, had obtained from osteomyelitic pus a micro-organism, called by him the "micrococcus of acute infectious osteomyelitis," but which was undoubtedly identical with the Staph. pyog. aureus which was discovered later.

All these experiments and conclusions have been subjected to criticism. Grawitz, above all, on the strength of his own experiments, in which he succeeded in injecting large quantities of living staphylococci into the abdominal cavity of animals without any subsequent symptoms of affection, positively asserts that the pyogenic bacteria are "not specific exciters of infection" like the anthrax bacilli, which thrive in the susceptible organism and infect it, but that certain preparatory factors are required to enable the staphylococci to enter the body and produce suppuration. The special condition of an open wound, and mechanical as well as chemical influences, must be considered; the excretions of the pyogenic bacteria themselves were influential in the positive experiments with animals, and were really the cause of suppuration whenever the cocci were reproduced on a proper soil under natural conditions.

Much may be said against this view. A special predisposition of the tissue, its peculiar tendency to retain bacteria, has also been pointed out as a necessary condition for a perfect infection with other micro-organisms, such as Fränkel's diplococci, the diphtheria bacilli, etc. In other micro-organisms, too, such as the cholera vibrios, the typhus bacilli, etc., the symptoms of the disease as well as the pathological changes were traced to the excretions of the bacteria, but nobody denied their character as specific exciters of infection.

Leaving aside, however, these theoretical considerations, we do not consider Grawitz's "facts" securely established. We should be careful not to apply to man results obtained in animals, and especially if we have to deal with a species *a priori* but little liable to purulent processes.

We must finally consider a circumstance wholly overlooked by Grawitz. The staphylococci possess, like other micro-organisms, a very varying degree of virulence and, besides, quickly succumb to natural attenuation in our cultures.

This circumstance is certainly of great importance in considering the striking diverse influences of the staphylococcus, and we can more readily understand why the same cause sometimes produces a furuncle, at other times endocarditis, and then again an osteomyelitis. Other factors are also significant. The spot where the micro-organisms entered, or the quantity absorbed or, finally, the varying susceptibility of the individuals stricken, will of course frequently decide the result.

Be this as it may, we believe these bacteria to be the specific exciters of suppuration, and suppuration to be a specific reaction of the tissue to the presence and activity of these bacteria. Here, as well as in other cases, the pathological changes and their general sequelæ are due to bacterial excretions—toxines and toxalbumins (among them especially some albuminous substances difficult to dissolve).

As to the way in which the staphylococci invade the organism under natural conditions, a suitable portal is often furnished by small lesions, scratches, etc. Experiments of Garré and others even proved that these cocci do not require any such open passage, since they were seen to penetrate the uninjured skin. No particular tendency to harbor the infectious matter need be assumed in view of the extraordinary diffusion of the staphylococci; Ullmann, for instance, having shown that they are almost regularly met with in the healthy body, the saliva, pharynx, on the skin, also in the water, air, in the dust of rooms, etc.

XXI. STAPHYLOCOCCUS PYOGENES ALBUS.

The Staphylococcus pyog. aureus is the species of bacteria most frequently occurring in pus of various origin, it having been observed in about 80% of all cases examined. But it is often found together with other micro-organisms. The latter have become better known through the researches of Rosenbach and Passet; their properties prove them to be likewise causative of inflammatory processes terminating in suppuration.

One of them, the Staphylococcus pyogenes albus, altogether resembles the aureus above described, being distinguished from it merely by the absence of the yellow coloring matter.

It is less common than the aureus and appears to be (as the experiments of transmission would prove), somewhat less virulent, it giving rise less readily to severe consequences.

XXII. STAPHYLOCOCCUS PYOGENES CITREUS.

Passet has demonstrated in two cases a third species, the Staph. pyog. citreus, which is distinguished by its beautiful lemon-yellow pigment and which liquefies gelatin more slowly than aureus and albus, with which it fully agrees in all other respects.

XXIII. STREPTOCOCCUS PYOGENES.

The Streptococcus pyogenes is a species of bacteria which plays an important part in producing suppuration. It is met with frequently alone, more rarely with staphylococci, in abscesses, etc. An exact and detailed description of its properties is superfluous, as it would be necessary to repeat all that has been said regarding Fehleisen's streptococcus of erysipelas. In fact, both micro-organisms cannot be satisfactorily distinguished. Neither their appearance nor the mode and rapidity of growth on culture media, etc., supplies any distinctive criterion, and experiments on animals lead to surprisingly similar results. Most investigators, as Baumgarten, E. Fraenkel, and others, are, for this reason, of the opinion that erysipelas cocci and streptococci are identical and should be regarded as such.

This view is, however, somewhat objectionable. How can it be imagined that the same micro-organism causes at one time a typical, well-defined disease, and another time, the establishment of purely purulent changes?

This apparent contradiction may, however, be explained. When discussing Fraenkel's pneumonia bacterium, it was found to be a micro-organism appearing as the cause of croupous pneumonia and of otitis media, and considered able to accomplish diverse results on the supposition that we had to deal with a widely diffused exciter of inflammation whose action terminated differently, according to locality and mode of entrance.

The streptococcus is very similarly situated. On the one hand, it is found very frequently in the saliva, in nasal secretion, in vaginal mucus, and in the uretha of healthy individuals; on the other hand, it regularly appears whenever the normal conditions of the tissue is disturbed by some morbid process. It was seen to appear in typhoid and diphtheria as a "secondary" bacterium concomitant with certain changes; it gives rise to a mixed infection in pneumonia, tuberculosis, pleuritis, scarlet fever, etc.; it may, in many cases, be the cause of more severe conditions than the legitimate micro-organism.

Fnally, the streptococcus, by itself, may produce sharply-characterized inflammatory processes. When it reaches the valves of the heart, it causes a typical *endocarditis* (hence a third exciter of infection); if transplanted to the endometrium of lying-in women, it causes *puerperal fever*, and it does so exclusively, according to all investigations hitherto made; on entering the lymphatic vessels of the cutis, it gives rise to *erysipelas;* if admitted to the subcutaneous tissue or the serous cavities, it produces *purulent changes* distinguished by a pronounced inclination to spread slowly and to continue its existence without "breaking down" and often manifesting an especially malignant character.

Quite a number of other micro-organisms have been obtained from pus; they are, however, but rarely observed, are only of subordinate significance, and prove harmless by the result of experiments on animals. It suffices to mention their names: the Micrococcus pyogenes tenuis, the Bacillus pyog. fœtidus, and the Staphyloccocus cereus albus and flavus.

XXIV. BACILLUS PYOCYANEUS.

Another pyogenic bacterium, the bacillus of green or blue pus, Bacillus pyocyaneus, deserves a more detailed notice. The pus from a wound and the bandages sometimes become suddenly discolored green or blue, in the majority of cases without disturbing the healing process. The cause of this striking phenomenon has been found by Gessard to be a particular bacillus often found in non-purulent serous wound secretion and even in the sweat of the skin.

It is a small, slender rod, of the same shape and appearance as the bacillus of blue milk, but rather narrower. It shows distinctly-rounded ends, frequently unites in groups of four to six members, but forms long threads only exceptionally. It is exceedingly mobile; sporulation has not been observed. It thrives at ordinary and at breeding temperatures, and belongs to the semi-anaërobic species.

On the plate the colonies appear to the naked eye as small white dots at the bottom of the gelatin; they rapidly advance to the surface and extend there as rather flat, moderately large, and irregularly-circumscribed aggregations. The medium soon assumes a green fluorescent color in the neighborhood of the colony. The gelatin begins to gradually soften and the plate is usually entirely liquefied on about the fifth day.

Under the microscope the smaller and deeper colonies present roundish, coarsely-granulated heaps with serrated borders of a yellowish-green and shining hue. The superficial ones form delicate laminæ with a smooth depression of finely-granulated texture, distinctly greenish in the centre and paler toward the edges. They then sink into the gelatin, become surrounded with a liquefied region, and are transformed into a dense and indistinct mass.

In the test-tube, growth takes place almost exclusively in the upper portions of the inoculation puncture. On the surface of the gelatin a flat, saucer-like hollow develops, with a distinctly glistening pigment of green fluorescence around it. Liquefaction progresses gradually and advances to the walls of the tube. At the same time the chief mass of the bacterial growth sinks to the bottom in thick, slimy threads, the layers at the top clear up, and a delicate, yellowish-green film appears on the surface. The entire culture glistens with a green glimmer visible at some distance.

A moist, rather thick, yellowish growth is developed on agar-agar. It imparts a greenish hue to the medium.

On potatoes there is formed a yellowish-green and smeary growth which imparts to its neighborhood a peculiar pigment, like the bacilli of blue milk.

This material is probably generated by the bacteria as a colorless product, and becomes a real pigment only on contact with the oxygen of the air, for which reason it is, for instance, only observed at the free edges of the bandages. It is, according to the investigations of Ledderhose, who calls it pyocyanin, an aromatic compound, related to anthracene, crystallizable, and without pathogenic properties.

The bacilli themselves, however, and their excretions are undoubtedly injurious to animals. On injecting into the subcutaneous cellular tissue of Guinea-pigs or rabbits about 1 c.cm. of a fresh bouillon culture, there will develop from the point of injection a rapidly progressing œdema and a purulent inflammation of the adjacent parts causing death in a short time. The bacilli can be ascertained in the affected parts, in the blood, and in all internal organs. After an injection into the abdominal cavity a pronounced purulent peritonitis arises, and the characteristic rods are again found, at the places just mentioned, mostly in dense heaps.

If we take smaller quantities of the infective fluid, purulent foci will be formed with narrower limits, and no fatal termination will ensue. Animals having overcome this invasion will now bear doses otherwise positively fatal. An injection of sterilized cultures will also accomplish such a protective inoculation, which may be regarded as being a result of becoming accustomed to increasing doses of a poisonous substance, although, perhaps, those processes, too, may play a part which otherwise lead to artificial immunity.

The pathogenic character of the Bac. pyocyaneus was, until recently, unknown and has been ascertained only by the investigations of Ledderhose and several French investigators, such as Charrin and Bouchard. The last two have also established the significant fact (already mentioned) that an incipient anthrax infection may be stopped and cured by aid of the pyocyaneus.

XXV. BACILLUS PYOCYANEUS β (ERNST).

Ernst has described a special variety of the bac. pyocyaneus (called by him Bac. pyoc. β) producing, as he states, a blue pigment, the blue pus, while the other variety exhibits the green fluorescent pigment. Both are said to appear often in common under natural conditions, and to develop a mixed hue intermediate between the colors just mentioned.

XXVI. GONOCOCCUS.

Gonorrhœa is a disease principally distinguished by the copious secretion of pus. Everybody knows, and daily observation confirms, that this pus differs very clearly from the products of other processes of inflammation by peculiar contagious properties. There are, in fact, but few affections so strongly marked as being of infectious origin, for which reason efforts have long been made to discover the micro-organism causing gonorrhœa.

Neisser, in 1879, called attention to the circumstance that peculiar cocci are regularly found in this pus, differing from similar bacteria by their appearance and shape. Their occurrence proved to be exclusively restricted to gonorrhœa, and Neisser did not hesitate to declare them to be the cause of a specific inflammation of the urethra, and called them gonococci.

They are large micrococci, almost always appearing as diplococci. Their areas of contact are usually strongly flattened so that each pair looks like a breakfast "roll." There is frequently seen, as a sign of commencing division in single members, a shallow furrowing, destined to separate the body of the coccus into halves which are not always quite equal. Groups have not been observed unless we designate as such the dense heaps in which the gonococci usually aggregate.

The cocci prove accessible to the common anilin colors and furnish very plain pictures with methyl-blue. Gram's method is not applicable, the bacteria again decoloring when coming in contact with iodide of potassium. An excellent method of preparing them consists in treating the cover-glasses for a few minutes with a concentrated alcoholic solution of eosin (by heating the staining fluid), in absorbing the surplus of eosin by blotting paper, and at once allowing concentrated alcoholic methyl-blue to act (for fifteen seconds at most), and then washing it with water. The cocci will then be seen stained blue on a red ground; the cellular elements of the blood or pus have eagerly absorbed the eosin, while the nuclei and micro-organisms appear in blue, the latter manifesting a noticeable relation to the white blood-corpuscles.

The bacteria have swarmed into the pus cells and occupy their entire protoplasm excepting the nucleus only—an occurrence peculiar to the gonococci and hardly ever found in the other genuine pyogenic bacteria.

The significance of this invasion of the micro-organisms into the tissue-elements is still subject to doubt. Some perceive in it a proof of an independent activity of the parasites, but others, on the contrary, regard this phenomenon as a visible expression of the attempt of the body to ward off the foreign intruder with its most efficient weapon.

The artificial cultivation of the gonococci outside of the human organism has been successful in but few cases, in spite of all care and the numerous and persevering efforts of very skilled experts. The gonococci do not grow on our common culture media, such as gelatin, agar, blood-serum, potatoes, and all statements to the contrary are erroneous.

The gonorrhœal pus contains, in addition to the specific diplococci, a number of other bacteria regularly supplanting one another in the culture experiments, and looking so very similar to the micro-organisms sought that they are but too apt to be confounded with them.

So far as our present experience goes, the gonococci thrive only on human blood-serum. They form there, at breeding heat, an extremely delicate, almost colorless coating of small extent, with sharp edges, and hardly perceptible even on close inspection. This covering reaches the height of development in about three days and appears then as if composed of numerous and very tiny drops. Transmission to a fresh medium must be undertaken at that time, if the culture is to be preserved, as it deteriorates on the artificial medium with surprising rapidity and becomes incapable of development. The gonococcus, therefore, belongs to the most incarnate parasites inhabiting the human body, and the conditions of its existence outside of the latter are at any rate very restricted.

Is the micro-organism first described by Neisser and designated as gonococcus actually the original cause of gonorrhœa? This question seemed to have been settled in the affirmative years ago, until recent observers disputed its claim. It was stated that the form and other attributes of the cocci in stained preparations were not calculated to prove them with certainty as such, and that the microscopic examination afforded, therefore, no sufficient ground for maintaining the regular presence of these bacteria in the specific inflammation of the urethra. The healthy urethra also contained micro-organisms in no way distinguishable from the so-called gonococci, so that their exclusive occurrence was doubtful and could, at any rate, not be utilized to settle their etiological significance.

But all these objections must be disclaimed as unjustified. Any individual property of the gonococci, their roll-like form, their position within the cells, their decoloration by Gram's method, or their failure on our ordinary media, is, of course, unable by itself to characterize the bacteria with certainty. But considering all the criteria together, we have sufficient means of distinguishing the gonococci from other bacteria. The result of observations hitherto made, pointed to the probability that the bacteria found by Neisser are the original exciters of gonorrhœa.

This supposition has become a certainty by positive experiments of transmission made by Bockhart and Bumm. The difficulties of artificially cultivating the gonococci and keeping them capable of living outside of the body has been stated. Gonorrhœa being, besides, a disease exclusively afflicting man and promising successful transmission only in transplanting its presumed exciters to man, it will be easy to understand why the number of such experiments has, as yet, been so small. Some of them, however, undertaken by Bumm, are apt to remove the last doubt of the specific nature of the gonococci. The twentieth generation of a culture on human blood-serum inoculated on the sound urethra of an incurable paralytic person produced a typical gonorrhœa.

It is unnecessary to discuss in detail the absorption of the infectious matter—the transmission of the cocci. Be it remarked that only certain mucous membranes are accessible to the settlement of these bacteria. Gonorrhœa, in man, has its seat in the urethra; so it has in woman, but in her it is also located in the cervix uteri and Bartholin's glands, while the vagina, at least in adults, remains regularly free, and is more frequently attacked only in childhood, during which infectious vaginitis is a widely-diffused disease. The conjunctiva must finally be mentioned as a privileged point of attack on the part of the gonococci, which settle there chiefly in the most superficial layers of the mucous membrane, whence they pass into the secreted pus.

XXVII. TETANUS BACILLUS (KITASATO).

We conclude this part by discussing the causative micro-organism of a particular disease of wound-infection, distinguished by very remarkable symptoms, viz., the bacillus of traumatic tetanus. The real character of this strange affection has long been in doubt, and was attributed to diverse agents, such as extraordinary conditions of the weather, colds, and other more or less incomprehensible influences.

Carle and Rattone, and afterward especially Rosenbach, showed that tetanus was transmissible from man to animals, and hence of infectious nature. Other investigators noticed the fact that, by inoculating small quantities of garden-mould, especially on white mice, often a morbid picture was produced which strikingly resembled the one observed in experimental tetanus. Nicolaier was able to prove in such cases the regular presence of a peculiar "bristle-shaped" rod, with round spore-heads at the ends. The same structures were found again in tetanus naturally developed, and it was but natural to suppose that we had to deal here with originally similar objects. The conclusive proof of the correctness of this supposition, as well as further information regarding the presumed exciter of tetanus, were frustrated by the circumstance that artificial pure cultures were produced with considerable difficulty, and hence a more accurate investigation of the vital properties of this kind of bacterium was impossible.

The micro-organism was doubtless strictly anaërobic and appeared always in company with other bacilli likewise anaërobic, from which it could not be differentiated in spite of all caution, so that a kind of symbiosis was seriously thought of.

By means of skilful manipulation of our culture procedure, Kitasato has recently succeeded in proving that the micro-organism hitherto claimed as tetanus bacillus can be isolated from its companions and grown by itself. Kitasato placed a small piece of tissue from a man dying of tetanus, and from the immediate neighborhood of the suppurated wound, upon the usual media and observed that a luxuriant development of divers bacteria took place in the incubator, but that the species forming end-spores proceeded very rapidly to sporulation, while the other species approached it only some time afterward. Before they did so, Kitasato heated his mixed cultures to 80° C. All bacilli not having sporulated were destroyed; these latter however, withstood the manipulation successfully so that he was enabled to obtain further pure cultures without difficulty and to remove, finally, by transmission to animals, every doubt of the existence of a genuine tetanus bacillus.

The germs of the tetanus bacillus seem to be very widely diffused in nature. It is known that they are found in garden-mould; they have also been met with in ruined walls, in putrefying fluids, as well as in manure. With regard to the last fact it must be stated that French investigators, above all Verneuil, think that tetanus develops only in persons having in some way come in contact with horses.

The tetanus bacillus is a large, slender rod, with rounded ends, frequently forming long threads, showing distinctly the separating points of the single members. Sporulation (as already said) takes place in the ends of the rods, this part of the cell swelling up like a drum-stick and developing the form of music notes or pins. Sporulation takes place in thirty hours at breeding temperature and in about a week at room temperature. The tetanus bacillus is motile; it grows at common and breeding temperatures (better in the latter) and belongs to the strictly anaërobic species, since it not only does not thrive in contact with the oxygen of the air, but even perishes quickly so that the rods in the hanging drop, for instance, very rapidly lose their capacity of voluntary motion.

The rods are readily accessible to staining. Gram's method is also available. The spores may be made visible in the usual manner.

On the gelatin plate, in an atmosphere of pure hydrogen, small, radiating colonies slowly arise, which gradually liquefy the gelatin. Under the microscope they appear as dense, firmly-compacted masses with a delicate border bearing numerous very delicate processes and ciliate filaments—a picture resembling that of the hay-bacillus.

The puncture culture in grape-sugar gelatin presents a strange sight. The upper parts of the medium remain sterile, but deeper the inoculation puncture is surrounded by a rapidly-increasing bacterial proliferation, sending out thousands of small, pointed shoots into the surrounding medium, so that the culture, in this stage of development, toward the end of the first week, looks like a wide-branched fir-tree and resembles a young brood of the root bacillus. The liquefaction of the gelatin follows later; it causes the delicate detail of the growth to disappear and gradually progresses until finally the whole medium is involved, which then becomes a turbid, whitish-gray, viscid, and slimy mass.

The growth in deep agar at breeding temperature is considerably more rapid and luxuriant. In twenty-four to forty-eight hours a culture, nearly reaching the free surface, develops in which an abundant formation of gas is noticed, which has a peculiar unpleasant, if not putrid, odor, characteristic of the tetanus bacilli.

In grape-sugar bouillon, the development is particularly energetic, causing at breeding temperature so considerable a production of gas that the flasks, if tightly closed, are sometimes burst and shattered.

On transmitting a small amount of such a culture to a susceptible animal, for instance a mouse, the first symptoms of disease may be noticed after twenty to twenty-four hours. They always appear at first in the parts adjacent to the point of inoculation, hence mostly at one or the other posterior extremities, frequently at the tail, and these parts are affected by a more or less pronounced tetanus. The affection grows apace and usually soon terminates in death. Guinea-pigs and rabbits are somewhat less sensitive than mice; they require large quantities of infectious matter and several cubic centimetres of a bouillon-culture to effect the purpose, the outbreak ensuing only after a longer period (about two to three days). The process is, otherwise, entirely the same.

On dissection the point of inoculation itself and its immediate neighborhood prove but slightly infiltrated and free from obvious changes. In the internal organs, no pathological condition can be demonstrated even by the closest investigation. The bacilli are absent in them under all circumstances, while they may still be found, sometimes at least, at the point of infection. They are usually, however, missed even at that spot, and their number is never proportional to the severity and extent of the sequels due to their absorption.

This striking phenomenon can be explained only by the circumstance that the bacteria at first multiply at the point of inoculation and generate an extremely virulent poison which spreads over the whole body. It then gives rise to a series of changes which become distinct only after the rods have perished and disappeared. Brieger has, in fact, been able to prepare, from artificial cultures of the tetanus bacilli as well as from one extremity of a man dead of tetanus, a number of poisonous substances of a basic nature, called by him tetanin, tetanotoxin, etc. Besides, the bacteria regularly form toxalbumins of an easily soluble kind, which act even in small quantities and give rise to symptoms characteristic of tetanus.

It has as yet not been established with certainty how infection occurs under natural conditions and what particular circumstances prevail. But the great diffusion of the infectious substance occasions its reception so frequently that we scarcely need to look far to find the causes. The contamination of a skin-wound, or a small lesion, with earth, crumbs of sand, stone-splinters, soiled fingers, etc., is, in fact, of great importance in all the cases more accurately examined, and it is to be supposed that our knowledge in this direction will be perfected in the near future.

XXVIII. CHICKEN-CHOLERA BACILLUS.

We have thus far discussed only diseases which affect man exclusively, like the cholera, or affect him principally, as tuberculosis, or under certain circumstances, as anthrax. We will now consider a series of affections likewise caused by micro-organisms but confined to the animal kingdom.

There appears among the poultry kept in yards, especially among the fowl and geese, a very destructive, murderous plague, the symptoms of which remotely resemble those observed in genuine cholera of man, for which reason it is called chicken cholera (choléra des poules). Perroncito, and after him Pasteur, demonstrated the presence of bacteria in the blood, organs, and excreta of affected animals. Pasteur cultivated them artificially outside of the body and, being able (in 1880) to reproduce the disease from cultures, he furnished the incontestable proof of the causative significance of the micro-organisms.

They are small, rather short but broad rods with rounded ends, immobile, frequently met with in unions of two, more rarely in long threads of greater numbers. Pasteur described them as micrococci and it requires really good lenses and staining to ascertain their true form.

A very peculiar property of the bacilli will be noticed in staining: the single cells take up the colors readily only at the ends, while the middle part remains unstained and appears as a bright gap between the two darker poles. Only close observation will show the existence of this intermediate connecting link and prevent the error of regarding the stained ends as independent structures —as micrococci. This phenomenon is most clearly perceived when the preparations are stained with methyl-blue. The bacteria cannot be prepared by Gram's method, since they are decolored in contact with iodine.

Sporulation has, as yet, not been observed with certainty in this germ; but it is a remarkable fact that they possess a rather high power of resistance to external influences. They can, for instance, pass the stomach without being destroyed by its acid.

These bacilli thrive at ordinary, as well as at breeding temperature and belong to the semi-anaërobic kinds.

On the plate the colonies appear about the third day, as small white dots at the bottom of the gelatin. They advance to the surface but slowly, and never reach a large size. The gelatin is not liquefied. Microscopic examination exhibits them as irregularly-rounded discs with sharp, smooth edges of yellowish or yellowish-brown color, particularly distinguished by a concentric stratification and a slightly-granular texture.

In the test-tube there is gradually developed, along the entire inoculation puncture, a white and delicate streak which proves later on to be composed of single small granules. They grow slightly on the surface. But surface cultures on oblique gelatin extend rather widely; in the neighborhood of the point of inoculation a dry, grayish-white, elevated coating arises, which adheres firmly and tenaciously to the medium.

On oblique agar, and on solidified blood-serum, a whitish, shining, moderately thick coating is formed.

The bacteria do not thrive on potatoes at ordinary temperature; a scanty, yellowish-brown, transparent growth is developed in a few days.

Successful transmissions from such cultures to susceptible animals may be made in various ways. By inoculation or subcutaneous application, not only fowls and geese, but also pigeons and sparrows, mice and rabbits can be infected; but Guinea-pigs are rather insusceptible and only succumb to large quantities of the poisonous substance introduced directly into the abdominal cavity or blood-vessels. Mixed with the food of chickens, pigeons, mice, and rabbits, the disease may be produced in the most pronounced manner, and the symptoms referable to the alimentary canal (prominent under natural conditions) will appear with great distinctness.

Post mortem the bacteria (no matter how they were introduced) are found in the blood and all organs, and the affection due to the subcutaneous application is plainly characterized as genuine septicæmia. The subcutaneous cellular tissue in the neighborhood of the point of inoculation is, besides, usually in a state of hemorrhagic inflammation and infiltration; if transmission has been effected by feeding, the intestinal mucous membrane is the chief seat of the changes.

One of the new procedures should be applied to demonstrate the rods within the tissue, since the bacilli of chicken cholera belong to the micro-organisms which readily fade under decoloration. In good preparations large quantities of bacilli will then be seen lying in the smaller blood-vessels, especially the capillaries. The rods never invade the cells.

Pasteur, as will be remembered, made his first observations regarding the process of attenuation on the bacilli of chicken cholera. He perceived that cultures exposed for some length of time (for months) to the influence of the oxygen of the air (i.e., preserved with simple wadding without any other means) lost their virulence more or less and were no longer injurious to animals.

New generations could even be obtained at will, all of which preserved the same attitude. Whenever Pasteur inoculated such material into the breast-muscle of chickens, for instance, a mere local inflammation ensued, which generally became rapidly circumscribed and terminated in the expulsion of the altered tissue by suppuration, without any other disturbances.

Pasteur's interpretation of this phenomenon as the consequence of the unobstructed access of oxygen has been disputed and refuted by many authors. Cultures of these bacilli on oblique gelatin (hence with a purely superficial growth) and propagated in the same way from generation to generation, always retain their virulence, for which reason some authors ascribed the attenuation to the action of the breeding heat.

Pasteur's experiments of attenuation were followed by his significant experiments with artificial protective inoculation. By means of this inoculation, at first with a greatly-weakened infectious substance ("le premier vaccin") and subsequently with a much stronger ("le deuxième vaccin"), even highly-susceptible animals, such as fowls and pigeons, may be secured against infection. A practical utilization of this fact has been attempted, but veterinarians generally are not in favor of such a procedure.

The results of microscopic examination, artificial cultivation, and transmission prove that the bacilli of chicken cholera are the sole cause of the plague.

How do the micro-organisms enter the animals and cause the peculiar disease?

Experiments and careful observations of the natural conditions have furnished a satisfactory answer to this question. It is pretty certain that, in most cases, infection ensues from animal to animal and is brought about by bacilli contained in the excrements of diseased individuals, being reabsorbed with the food by previously healthy birds.

Successful inoculations also show that the poison may be transmitted, too, from slight lesions of the integument etc. The vital properties of the bacillus rendering it probable that it may thrive, or at least continue to exist, outside of the animal body, the occasion for infection is easily afforded.

The micro-organisms having once effected an entrance, they multiply rapidly, and thus cause the group of symptoms appearing in the course of the affection. The chickens frequently sink at once into a state of great debility and apathy; they remain motionless in one spot, as if paralyzed, double themselves up with bristling feathers into a rigid ball, close their eyes, and fall into a death-like sleep from which they awaken no more. At the climax of the disease, which terminates fatally in about twenty-four to forty-eight hours, the animals discharge very copious, liquid or slimy, whitish-gray excrements containing large quantities of bacilli.

The greater part of these disturbances is doubtless to be attributed to poisonous products of metabolism. Allusions in that direction have already been made by Pasteur. He filtered bouillon cultures of these bacilli through clay or gypsum cells, and large quantities of the fluid, free from rods, produced coma or somnolence in animals without any other injuries.

The post-mortem examination exhibits, among the more decided changes, a rather considerable swelling of the spleen (sometimes also of the liver), hemorrhagic, circumscribed infiltrations in the lungs, and especially a very intense inflammation of the small intestine, particularly in its upper portion. The mucous membrane is greatly swollen and reddened, frequently interspersed with small hemorrhages, or ulcers in cases progressing somewhat more slowly. The rods are found, microscopically, in the blood and in all the organs of the infected animals.

XXIX. BACTERIA OF SEPTICÆMIA HEMORRHAGICA.

The bacilli of chicken cholera are the first and most important members of a large group of micro-organisms, so slightly differentiated in part, that they may properly be considered together as belonging to the same species. They have the same appearance in stained preparations (the more distinct ends and the pale middle portion) and the same features of growth on our artificial culture media, as regards the shape of the colonies, the development of puncture culture, time of growth, etc.

But a definite number of these bacteria have a special peculiarity not present in the chicken-cholera bacilli. This circumstance is important enough to induce us almost to banish these micro-organisms from this class. The bacillus of the ferret plague, discovered by Eberth and Schimmelbusch; the bacterium described by Billings as the cause of the swine plague and by Salmon as causing hog cholera; and also the bacillus of the Danish hog plague, the hog pestilence, are all of them motile and supplied with distinct flagella. Fully identical with or very nearly related to the chicken-cholera bacteria are the bacillus of the hog plague discovered by Löffler and more closely studied by Schütz; the bacillus of duck cholera investigated by Cornil; the bacterium of the game plague minutely examined by Kitt and Hueppe; finally the bacillus of septicæmia of rabbits discovered by Gaffky.

All of them differ slightly in their effects produced in experiments on animals. The bacilli of rabbit septicæmia, for instance, are as virulent for mice, chickens, pigeons, and rabbits as the bacilli of chicken cholera.

The bacteria of the game plague kill pigeons, etc., but not chickens or Guinea-pigs. The bacteria of duck cholera affect only ducks, not chickens and pigeons.

The bacteria of the hog plague, finally, do not affect chickens and pigeons, but are exceedingly pathogenic for Guinea-pigs and swine. The former, which succumb but very rarely to the bacilli of chicken cholera and rabbit septicæmia, die in consequence of a simple inoculation after one to three days and show, above all, a very pronounced, bloody-serous œdema of the subcutaneous cellular tissue and the superficial muscles. Swine regularly perish in from twenty-four to forty-eight hours after infection. On dissection, there is found an extreme distention and œdematous infiltration of the subcutaneous cellular tissue for a considerable distance around the point of inoculation, a swelling of the lymphatic glands, especially of the spleen, and a moderate inflammation of the intestinal mucous membrane. Bacilli are present in the blood and all the organs.

Schütz thinks that these bacteria are the cause of a peculiar disease of swine formerly often confounded with erysipelas. Extensive additional investigations proved to him that the absorption of the poisonous substance of the bacilli is, under natural conditions, mainly effected through the lungs.

These differences, briefly mentioned, are certainly worthy of notice and practically important. But we must refrain from using differences in virulence or efficiency as grounds for the separation of otherwise harmonizing species. It will be well to regard these micro-organisms as identical, without, however, overlooking the special conditions and properties.

Hueppe proposed the name of "bacteria of septicæmia hemorrhagica" as indicating the pathological character of the affection.

XXX. BACILLUS OF HOG ERYSIPELAS.

The genuine erysipelas of hogs ("rouget" or "mal rouge des porcs") is an epidemic disease which carries off in Germany more than half of the animals affected and does great harm, being restricted, almost exclusively, to the superior English breeds. Only younger individuals (up to three years at most) are attacked and usually perish after from twenty-four to forty-eight hours.

Löffler found in the blood, in all organs, in the muscles and skin of diseased or dead swine, a peculiar micro-organism which he was able to cultivate artificially outside of the body and the pathogenic properties of which he ascertained by experiments on mice and rabbits. His observations were fully confirmed by Lydtin and Schottelius and particularly by Schütz. They also produced typical erysipelas from cultures by a successful transmission to hogs. It is, therefore, no longer doubted that this special kind of bacterium is the cause of the hog erysipelas.

They are very small, slender rods, looking like delicate bristles or tiny needle-shaped crystals. Though usually single or in pairs, they sometimes form long threads, and may interlace into a pretty network. They have the power of voluntary motion; it is as yet unknown whether they form spores. They grow at either room or breeding temperature, belong to the semi-anaërobic kinds, thrive rather better in the absence of oxygen, stain with the usual anilin colors, and can be excellently prepared by Gram's method.

On the gelatin plate there appears, on the second or third day, at the bottom of the medium a peculiar cloudiness of grayish-blue or silver-gray color, distinctly perceptible only on a dark background. Wherever the colonies have grown to a certain extent, extremely delicate, greatly ramified, and mistily transparent masses are recognized with the naked eye; on the whole, they resemble somewhat the appearance of a "bone corpuscle" with its tiny processes and shoots. The colonies grow on, coalesce, and impart to the entire plate a dim, grayish glimmer. They do not, as a rule, advance to the surface of the medium. The gelatin is not liquefied. Microscopic examination adds nothing, and is of little avail on account of the extraordinary fineness and transparency of the colonies.

Test-tube culture shows, in the neighborhood of the inoculation puncture, dense masses of the appearance noticed in the colony on the plate. Development usually commences but a short distance below the free surface of the gelatin and becomes strongest in the deeper layers. It increases but slowly and gradually, until, finally, the entire gelatin appears traversed by dim, gray clouds. A very slight softening of the gelatin is noticed in the course of several weeks which is followed (in consequence of evaporation and simultaneous drying) by the formation of a funnel.

On agar and blood-serum (especially at breeding temperatures) a very delicate, hardly perceptible coating is formed along the inoculating line. No development takes place on potatoes.

Experiments of transmission of such cultures proved that hogs, pigeons, rabbits, domestic and white mice were accessible to infection produced in the body-cavities by inoculation, subcutaneous application, and injection, while Guinea-pigs and (strange enough) also chickens were refractory. The poison was not absorbed from the alimentary canal even by hogs, if administered by feeding.

Under natural conditions, this way must be open to the invasion of the bacilli; for, according to the observation of veterinarians, infection occurs almost regularly by the excrements of a diseased animal getting into feed and being eaten by healthy ones.

The symptoms of erysipelas seem to be essentially the same in all cases, without being affected by a possibly varying mode of origin. The outbreak of the disease is generally very sudden. The hogs become faint, show a paralytic weakness of the posterior extremities, refuse food, and their bodily temperature is considerably elevated. There appear at the same time, on the skin of the abdomen and chest, irregular red spots which, immediately after artificial infection, are confined to the neighborhood of the point of inoculation, but soon spread and meet in large dark-red areas, not particularly swollen nor painful. Debility increases and death ensues on the first or second day.

A very violent inflammatory œdema and a reddening of the point of infection is observed in rabbits after inoculation at the ear. The changes extend rapidly, frequently attack the head and trunk, and occasionally cause the death of the animals. Domestic mice die on the second or third day; before death they present symptoms of a severe disease and usually squat in a corner of their cage with closed eyelids glued together by a purulent discharge.

The post-mortem examination will almost always show the same characteristic picture, whether the mode of origin of the hog erysipelas be artificial or natural. The spleen is greatly swollen, hard, and of a thick, brownish-red color; the liver is moderately enlarged; the lungs are peculiarly marbled and spotted. The mucous membrane of the stomach and intestine is reddened and the seat of small hemorrhages; the villi are especially altered, the follicles and mesenteric glands swollen (the latter becoming usually brownish-red); the subcutaneous tissue is rather strongly reddened, bloody, and œdematous.

The bacilli are met with in all organs, especially in the liver and spleen, more sparsely in the blood; they stain beautifully, even in sections, by Gram's method. They lie in masses in the vessels and occupy preferably the walls of the smaller arteries and capillaries, but they are also found outside of the blood-vessels distributed in the tissue, and usually inclosed in cells. They inhabit (usually in small groups and dense little heaps, rarely singly) the interior of the lymphoid cells, whose body is more or less rapidly destroyed by the foreign intruders.

Hog erysipelas is one of those diseases in which Pasteur has succeeded in producing artificial immunity by inoculating with attenuated poison. He has also two "vaccins," one of them being altogether harmless ("premier"), and a stronger one ("deuxième") which is said to be applied twelve days after the first and to surely protect the animals against the plague. Schütz has shown that Pasteur's inoculating matter contained, in fact, the bacilli of hog erysipelas. He tested the efficiency of the French "vaccin" and found that it renders the hogs immune; he has finally been able, under the influence of high temperatures, to obtain attenuated descendants from vigorous bacilli, these descendants possessing the same properties as Pasteur's inoculating matter.

Veterinary surgeons have long known that hogs are protected against a second attack of the disease after they had recovered from the first attack of erysipelas. But most of them express themselves with reserve regarding the practical value of protective inoculation. The "deuxième vaccin" is by many charged with affecting the animals with chronic hog erysipelas and giving rise to an unintended and exceedingly dangerous, permanent spread of the bacteria. Additional investigations will determine the correctness of this assertion.

XXXI. MOUSE-SEPTICÆMIA BACILLUS.

The bacilli of hog erysipelas greatly resemble, in their growth upon solid media, and in their action on various animal species, the bacilli of mouse septicæmia, first observed by Koch and more fully described in 1878.

Koch found, by inoculating domestic or white mice with putrefying fluids (especially blood), that a certain number of the animals perished, and there could be detected in the blood and in all the organs quantities of exceedingly fine rods successfully transmissible to healthy mice.

An exact description of the bacilli of mouse septicæmia would necessitate almost a repetition of what has been said in reference to the erysipelas bacilli and we will merely call attention to the differences undoubtly existing between the two micro-organims.

The bacilli of mouse septicæmia are generally somewhat narrower and thinner than those of erysipelas; they seem to be incapable of voluntary motion; roundish and shining corpuscles often appear in the interior of the rods, which are regarded as spores.

They belong to those bacteria which thrive as well, perhaps better, in the absence of oxygen as with a free access of air and can, therefore, be numbered among the semi-anaërobic species.

They are readily accessible to staining with anilin colors, especially by Gram's method.

The development of the colony on the gelatin plate is nearly like that of the erysipelas bacilli, but their growth is not so dense, for they reach a greater extent very much sooner and cover a greater region of the medium, thus imparting to the colony an especially delicate transparent appearance.

This difference is, perhaps, still more perceptible in the test-tube. With the erysipelas bacilli, we find a dense culture restricted to the vicinity of the inoculation puncture, while here the bluish-gray, dim clouds traverse nearly the whole gelatin from the beginning. This peculiarity is unmistakable, above all in young cultures (up to one week); later on it begins to lose its clearness and is finally lost.

The bacilli of mouse septicæmia prove, in experiments on animals, infectious for domestic and white mice, pigeons, sparrows, and rabbits. Chickens, Guinea-pigs, and field-mice are perfectly insusceptible, as Koch has especially emphasized.

When inoculated into the ear of rabbits, they produce an erysipelatous inflammation of the subcutaneous tissue, which usually heals and protects the animals against repeated infections. Mice fall sick, just in the same manner as after inoculating with erysipelas; the pasty closure of the eyelids is again noticed.

The post-mortem appearances, the swelling of the spleen, etc., correspond in every respect to the picture we have seen in hog erysipelas. There is, likewise, the same distribution of the bacilli in the tissue. But the rods of mouse septicæmia seem to occur more copiously in the heart-blood of the animals than those of erysipelas, but more sparsely in the lungs. They are very frequently inclosed in cells, either singly or in pairs.

XXXII. MICROCOCCOCUS TETRAGENUS.

Koch observed a peculiar kind of bacterium, first in the contents of a tuberculous lung-cavity, afterward repeatedly under similar conditions in the expectorations of patients and in the normal human saliva. It was studied more closely by Gaffky and named "Micrococcus tetragenus."

They are rather large, perfectly-round cells, forming in culture dense heaps without any particular kind of arrangement. But they look very differently when developed in living tissue and taken from the animal body.

Four single cells are seen inclosed in a thick gelatinous capsule, transparent as glass, in which the bacteria are imbedded and from which they stand out like the spots on dice; for the capsule remains uncolored and appears as a transparent border. Sometimes only two or three cocci are thus cemented together, but this is only when one of the members exceeds the others in size and bulk, thus indicating that it is on the point of dividing and generating the missing cell. This striking kind of union at once resembles the sarcina. On examining unstained objects, however, a hanging blood-drop, etc., it will be seen that no division takes place here in the third direction of space.

This thickening of the membrane in the Micrococcus tetragenus is almost exclusively found when it has thriven in an animal organism. We, therefore, meet here with the features already observed in Fraenkel's and Friedländer's bacilli.

The M. tetragenus belongs to the aërobic bacteria and is immobile; it develops at ordinary and at incubator temperatures.

It is easily stained with any anilin color and is an especially favorable subject for Gram's double staining.

On the plate the colonies appear at first as small white dots in the depth of the gelatin, they advance to the surface rather rapidly, and rise above the medium like arched elevations of a china lustre. They do not, however, liquefy the gelatin or alter it in any other way. Microscopically, round or oval, dense and yellowish-brown discs are seen; they are of a slightly-granular structure; their borders are mostly smooth and sharp-edged.

In the test-tube there arise along the entire inoculation puncture, thick, globular, dense, white masses and a moderately large glistening coat on the free surface.

On agar-agar and blood serum, a white, damp, extensive film is formed.

On potatoes there develops a thick, slimy coating which can be taken off in long threads.

The Micrococcus tetragenus is pathogenic for white mice and Guinea-pigs, and house and field mice are usually susceptible, rabbits, etc., always. The white mice perish in three or four days after the subcutaneous application of the bacteria. Guinea-pigs endure larger quantities of the poison; it is best to inject directly into the abdominal cavity.

Post-mortem examination of mice shows a perceptible change: whitish, rather extensive foci in the spleen (more rarely also in the liver). Microscopically very large quantities of cocci are found in the blood and all organs, having been distributed over the body by the blood-current and, therefore, met with only in the vessels.

They regularly appear in characteristic groups of four, inclosed in a common capsule.

The latter is beautifully developed after injecting the bacteria into the abdominal cavity of Guinea-pigs. A very considerable purulent peritonitis is found on dissection; microscopic examination shows that the accumulated masses of viscid mucus coating the intestines with a dense, compact layer, consist almost entirely of cocci with their greatly swollen capsules.

Having thus far treated of the bacteria in general, ascertained the properties of these micro-organisms, and become aquainted with a number of the better-known non-pathogenic and pathogenic bacteria, we shall now proceed to discuss the final part of our task, viz.: the application of recent methods of investigation to the chief constitutents of our natural surroundings, the air, soil, and water.

CHAPTER VII.

INVESTIGATION OF AIR, SOIL, AND WATER.

1. AIR.

THE purpose of bacteriological investigations of the air is to ascertain the number and kind of micro-organisms inhabiting the strata of air surrounding us.

The entire method of culture on solid media now in use depends, as will be remembered, upon the fact that there are developed on the surface of slices of boiled and exposed potatoes a series of divers bacterial colonies owing their origin to germs dropped from the air. The soiling of plate cultures established the fact that the air contains vast quantities of micro-organisms.

The quaint ideas of the past in regard to the enormous distribution of bacteria in the atmosphere were disproved by a more accurate examination. It is well known that on admitting a ray of the sun into a dark space, the illumined strata of air teem with small particles of organic or inorganic origin, the so-called "sun-dust." Every one of these particles was supposed to be either a germ or a bearer of one; an obviously erroneous view, as will be seen by a simple consideration, even without direct proofs to the contrary.

It is not easy for bacteria to rise into the air. It has been seen, during the discussion of the causes of tuberculosis and cholera, that a voluntary rising of the bacteria is absolutely impossible and that they cannot be torn from a medium on which they have once become firmly rooted, even by the strongest draught of air. The medium on which the micro-organisms are found and have been thriving, must dry up completely and crumble into fine powder or dust before they become a plaything for air-currents to scatter broadcast. But as the majority of them cannot endure such a drying without harm, it will be obvious that the quantity of micro-organisms, proved to exist in the air by bacteriological investigation, does not by any means come up to the former exaggerated opinions.

It is a matter of course that these conditions cannot be ascertained by the microscope alone. The method of cultivation has, therefore, been resorted to and applied in various forms.

The most primitive procedure is that with common plates coated with gelatin, freely exposed for a certain time and then preserved and treated in the manner already familiar. The microorganisms that descended upon them will, in a few days, develop into colonies whose numbers and kind will enable us to draw conclusions as to their character.

Simple and convenient as this method may appear, it still lacks perfection. Its most essential defect consists in the circumstance that the quantity of air coming in contact with the nourishing medium can by no means be carefully estimated and that there is no certainty that all germs capable of development have actually been deposited. The celerity of motion of the air varies so greatly every moment, that comparable results cannot at all be reached in this way.

This defect was remedied to a certain extent by a procedure devised by Koch after the introduction of solid transparent media. The flat glass dish destined to receive the gelatin is 1 cm. high and has a diameter of 5 cm.; it is placed at the bottom of a cylindrical glass vessel 6 cm. in diameter and 18 cm. high.

This glass dish can be easily lifted from the cylindrical vessel by means of a narrow strip of sheet iron bent into a right angle, for the purpose of future microscopical examination of colonies. The glass is now closed with a solid, large plug of wadding and the whole apparatus is sterilized in the hot-air oven. The plug is then raised, the cup taken out, filled with gelatin and at once lowered again and closed anew by the plug.

The gelatin having become hard, the plug is removed at the spot where the air is to be examined and carefully preserved, while the apparatus remains open for a certain number of hours. The strata of air in the glass-vessel may now be considered as in repose, i.e., we may count on approximately equal quantities of air depositing their germs upon the gelatin within a given time. Whenever the germs have developed to colonies, the apparatus is reopened, the gelatin-glass is lifted to the surface and at once examined under the microscope.

An exact examination of quantities of air can, it is true, not be obtained in this way. The procedure was decidedly improved by the work of Hesse.

Hesse's method of investigating air is essentially as follows:

A glass tube, about 70 cm. long and 4 cm. wide, is provided at one end with a thick rubber stopper with a central perforation for the reception of a little glass tube, 1 cm. wide and 10 cm. long, stopped at each end by a dense plug of wadding. The other aperture of the large tube is closed by two stiff rubber caps, the inner having a central round opening, the outer being solid. The whole apparatus is then sterilized, for about an hour, in the steam-sterilizer. Then remove the rubber plug, pour in 50 c.cm. of sterile, liquid nutrient gelatin, close again, and distribute the mass on the walls of the tube, as in Esmarch's procedure. Put this under a stream of water and roll it on its horizontal axis as quickly as possible. When the gelatin begins to become viscid, the rotation should be stopped; the greater mass of the medium will then slowly sink to the most dependent part of the tube, and form a rather thick layer.

We must see that the tube remains constantly turned downward. Fasten the apparatus upon a movable tripod and begin investigation. Connect the small tube in the rubber plug with an aspirator, remove the outer, unperforated cap from the other aperture and let the suction begin. The water flowing over into the lower aspiration-flask is replaced in the upper by inflowing air. But this air, in order to reach the flask, must first pass through the long pipe and its germs will thus find an opportunity of depositing.

The result has shown that this happens satisfactorily in the anterior sections of the tube. Even if some micro-organism should be carried off by the current of air to the other end of the tube, it is compelled to settle at the cotton wadding closing the small inner tube. The wadding being likewise soaked with gelatin, these stray germs can develop.

The velocity of the current of air must, of course, not exceed a certain point. It is determined by regulating the quantity of the water passing between the two aspirator-flasks by cocks, glass tubes, etc.

One litre of water is generally allowed to run over in about two minutes, just as much air meanwhile passing through the tube. The upper flask having become empty, the vessels are changed by merely shifting their places. The apparatus works quite satisfactorily and has only a single, though very essential, drawback.

It is impossible to examine large quantities of air as to their bacterial contents in a short time. We dispose, at best, of fragmentary quantities only, supplied by small parts of the surrounding atmosphere, but incapable of generalization.

This defect is remedied by Petri's method, which satisfies all demands and can be designated as perfect. It requires rather considerable preparations and a larger amount of special apparatus.

Petri passes the air to be examined for bacteria through a small filter of previously-heated and, therefore, sterile sand. He thus catches all the micro-organisms present. He then transfers the sand into liquid nutrient gelatin, pours the latter into little saucers, and observes the colonies as they develop. The sand has a uniform grain of a diameter of $\frac{1}{4}$ to $\frac{1}{2}$ mm. It is brought (in the form of small plugs, 3 cm. long and $1\frac{1}{2}$ cm. thick, and supported on both sides by a fine wire screen) into a glass tube, 8 to 9 cm. long, in which two filters of the sizes just mentioned can be placed one behind the other. Everything being ready, the whole is once more sterilized in the drying oven; one aperture of the glass is closed by a rubber plug simply perforated, bearing a tight small glass tube, which is now connected with a strong aspirating apparatus (an oscillating air-pump being preferable), whose rotations can be exactly controlled and indicate the quantity of air sucked in through the sand (generally 10 litres in one to two minutes in Petri's experiments). Each of the two little filters is then mixed with gelatin and the work is continued. In properly-executed experiments the filter adjacent to the pump should contain no germs at all, as they should have been caught by the anterior filter.

The results reached by all these observations generally agree.

It has first been shown that the number of germs in the air is by no means very large and that evidently the quantity of these micro-organisms varies extremely as to place and time.

It would lead us too far to discuss these conditions more in detail. We will merely mention that the air of our dwelling-places contains, on an average, three to five germs in a litre; that the surrounding atmosphere is usually of the same composition; that there are, on the whole, fewer micro-organisms in winter than in summer; and, finally, that their proportion deviates considerably from the mean, only under special circumstances, such as strong motion, violent agitation, etc. The air of high regions is freer from bacteria than that of low grounds, and the atmosphere on the sea and the tops of mountains seems to contain no micro-organisms.

The kinds of germs developing into colonies in the culture-vessels are likewise subject to great variations. Saccharomyces, penicillia, and bacteria appear in confused masses, and among the last-named micrococci are found in great varieties. Pathogenic species, parasitic bacteria (except the Staphylococcus pyogenes aureus) have not yet been surely detected in the air by direct investigation.

2. Investigation of the Soil.

The method of bacteriological examination of the soil is much less complete than that of the air.

For the purpose of obtaining comparable results, weighed or measured quantities of earth are brought into more or less intimate contact with gelatin. The sample is distributed as uniformly as possible by means of a sterilized scalpel over the surface of a plate supplied with the medium. But all the germs are not developed nor can they produce independent colonies. The particles of earth lie but loosely on the medium and retain numerous micro-organisms in their interior, thus depriving the result of all its value.

Some try to mix the sample of soil directly with the gelatin by pouring it into the test-tube before the contents of the latter are poured on the plate. But a large part of the material cannot be prevented from remaining behind in the tube and thus being lost for examination. Even if this defect is remedied as far as possible, by preserving the emptied test-tube and taking into consideration the colonies developed in it, no certain results have been obtained.

The number of germs in the *upper* strata of the earth is usually extremely great.

Others have washed the sample of soil for hours with sterilized distilled water before placing measured quantities of it on gelatin; but this investigation is complicated and clumsy, and there is no absolute certainty of having actually loosened and washed out all the germs from their substratum.

Relatively the best, though by no means perfect procedure, is this: pour the sample of soil directly into the liquid gelatin in the test-tube, rub and stir it thoroughly by means of a strong platinum needle and then distribute it, according to Esmarch's method, on the walls of the tube. All germs approximately will thus be developed and comparable results will be obtained.

The most difficult, and at the same time most important part of the whole procedure, is to procure available, reliable material. Superficial strata of soil can be taken up without difficulty; but the more we penetrate into the earth, the more it will be felt that samples can hardly be obtained which are without any doubt derived from the locality needed.

For this procedure it will be safest if we are able to dig soil from suitable depths and to utilize portions on the sides of the excavation. Such a favorable opportunity is but rarely offered, however, and investigation will, as a rule, surely be discouraged under such circumstances. The use of common boring tools does not prevent particles of earth from sliding down from the upper layers and sinking into the bore. The whole procedure is thus rendered doubtful. It is, therefore, necessary to use a special instrument, a lock-fast bore, provided at its lower end with a section having a movable shell. This section remains covered during the rotations to the right as the shell shifts. The closed bore can be lowered, opened at any depth, filled with earth, closed again and drawn up for bacteriological examination.

This tool can be applied only within definite limits. Whenever we endeavor to go down about 4 or 5 m. the ground resists the bore so strongly, that soon very massive and unhandy inclosures must be used, and a further advance will soon have to be abandoned.

Unobjectionable results can only be expected, in any examination of the soil, when the samples obtained are transferred to the artificial medium as quickly as possible. Otherwise, a considerable increase of the originally existing germs in the particles of earth will likewise prevent the discovery of the natural conditions.

The bacteriological examination of the soil, is, therefore, as difficult as it is complicated, and the rarity with which it is undertaken need not be wondered at, the less so because all the observations hitherto made have led essentially to the same results, hardly requiring continued examination. It has been ascertained that the upper parts of the earth, almost everywhere, contain vast quantities of diverse bacteria, partly of pathogenic nature (such as œdema, tetanus, and anthrax bacilli), while the lower strata, even those belonging to the region of underground water (unless forcibly torn from their natural conditions by man), are wanting in or are even free from bacteria.

3. Investigation of Water.

The method of bacteriological examination of water is almost perfect. There is very little difficulty from the beginning and its manipulation is exceedingly easy. Water is a substance of which accurately-measured quantities can be taken and it is not difficult to mix it so intimately and uniformly with the nourishing gelatin that the germs are completely separated and without exception develop into colonies. The latter will, therefore, both in number and kind, correspond with the germs sown, and furnish perfectly clear results.

The objection that our nourishing gelatin does not supply the conditions of growth for all the kinds of bacteria existing in the water and that, therefore, the results must be unreliable, has already been rejected as untenable. We might rather find fault with investigators for using only liquid media and exposing them to incubator temperature, because all the micro-organisms that do not grow at all at higher temperatures (and they constitute quite a large number) are lost by such procedure.

Certain measures of precaution should be taken in examining water. One point in particular should be carefully considered, since the success of the procedure absolutely depends on it; the bacteriological examination of water must be made as early as possible, directly or, at the latest, *a few hours after having been collected,* because an incessant and extremely extensive increase of the germs will begin at a very early period.

The fact has already been stated, in the discussion of some kinds of bacteria usually found in water, that some of them are immeasurably reproduced even in the purest water imaginable. On being placed in surroundings differing from their natural conditions, and especially under the influence of the higher temperature nearly always found in our investigation-rooms, they make use of this capacity of reproduction. After finding, for instance, 200 germs in a cubic centimetre of water, there will be 5,000 on the second day, 20,000 on the third, and actually innumerable quantities on the succeeding days. This process usually reaches its climax after a short time; the nourishing substances are appropriated to a certain degree, and the number of living micro-organisms in the water slowly diminishes. This shows, at any rate, that investigation should be undertaken as soon as possible, and that the report of the examination of waters sent or transported from a distance should be given with great reserve.

The samples obtained should, of course, be at once placed in vessels free from germs and well closed (as in Erlenmeyer's flasks) and transferred to the liquid nutrient gelatin by thoroughly sterilized pipettes. Before doing so, the water must be vigorously shaken to obtain an accurate mixture of the germs, it having been noticed that the majority of micro-organisms will very soon sink to the bottom and easily escape observation.

The water having been added to the gelatin, the tube is slowly inclined up and down for a few times. The medium should then be poured directly upon the plate. This plate must not be too small; for, the greater the area, the more surely we shall succeed in separating the germs and obtaining well-defined colonies.

The details of the procedure to be applied in our examination of water will now be readily understood.

The water is placed in sterile Erlenmeyer's flasks, specially closed with a rubber cap above the cotton plug for greater safety.

An accurately-measured quantity is now taken from the flask, which has previously been well-shaken, and transferred into liquid gelatin. This is done by means of graduated sterile pipettes, a different one being used for each portion. The quantity of water to be tested is generally placed in two test-tubes, one having a capacity of 1 c.cm. and the other $\frac{1}{2}$ c.cm.

A double purpose is thus gained. Given a large quantity of micro-organisms in the water, the colonies on the first plate may develop in such dense masses that enumeration and examination is out of the question, while the plate with half the quantity may still give an available result.

Besides, the second plate will serve to control the first, as it were, for there will appear only half as many colonies; but even if this should not always prove to be absolutely correct, striking deviations from the rule will point to some mistake in the procedure.

The water having been poured into the gelatin, the tube is tipped to and fro and its contents at once poured out upon as large a plate as possible.

The germs will have grown into colonies after a few days and examination may now be commenced. Should the number be small, they may be counted with the naked eye. But the quantities are sometimes so great that this simple procedure must be abandoned and a special counting apparatus resorted to. A glass plate, divided by a diamond pencil into small quadrangles, is placed over the gelatin plate which rests upon a dark background of black glass.

The number of colonies developed within such a square is ascertained by a magnifying glass; repeat this observation in six or more squares, take the average number per square, and multiply it by the number of the quadrangles corresponding to the extent of the gelatin area.

The number of germs, of course, varies, according to the water investigated. River-water, especially if near thickly populated districts, contains sometimes so many micro-organisms that even one drop (=$\frac{1}{20}$ c.cm.) gives rise to several thousand colonies on the plate. Temporal conditions are also of some influence, the figures being higher in summer than in winter, etc.

The species of micro-organisms found in water belong mainly to the class of bacteria. Some of these have already been considered; most of them are insignificant. Even pathogenic bacteria have been ascertained in some cases by immediate investigation, as, for instance, cholera vibrios in an Indian tank and typhoid bacilli in the drinking-water of several small cities.

The bacteriological investigation of water is of great practical importance as compared with that of the soil and air. Soil and air are but rarely examined, owing to the many difficulties met with. The complicated apparatus required for it will not be employed very often because the results are, at best, not decisive and surely not in proportion to the means employed. It having been ascertained, in general, that but few micro-organisms exist in the air, very many in the superficial strata of the earth, and none at all in the deeper strata, we have, for the present, about reached the limits of investigation. Bacteriological examination cannot, at this date, furnish any valuable conclusions (as to whether this or that soil is objectionable or this or that air healthier, from a sanitary point of view) and investigation may never be able to accomplish it.

But not so with water. The question as to whether it satisfies hygienic requirements and can be used without hesitation is, in many cases and under definite conditions, decided only by the result of bacteriological examination. We repeat "under definite conditions" in order to oppose, from the beginning, the great mischief often done by bacteriological investigations and calculated to discredit them. The danger of overestimating our ability in this very sphere is not inconsiderable, and it may not be amiss to briefly state the principles of procedure.

Hygiene has for many years repeatedly declared and demanded that available drinking-water must, above all, be free from infectious matter. So long as the nature of these suspicious admixtures was unknown, their occurrence was surmised rather than really and positively ascertained. But it was discovered that these infectious substances were living organisms which should be classed among the bacteria. The more infectious substances water contained, the more replete it was with micro-organisms.

The number of bacteria found is of far less importance than their nature. Water containing 5,000 germs of the hay-bacillus or the fluorescent bacillus, etc., in 1 c.cm. is altogether harmless; but water with only ten germs, two of which are cholera vibrios and two typhoid bacilli, is exceedingly dangerous. It may be concluded from all this, that the single colonies must be carefully examined as to their nature before we can decide whether some drinking-water is to be considered as the cause of the outbreak of some typhoid epidemic. But such an investigation is very difficult. Only great experience, a skilled eye, and well-disciplined technique will be able to accomplish the end; and even then we may not succeed in separating a few typhus or cholera colonies from among countless other bacteria.

Let us take a case of typhoid fever. It has not been noticed, perhaps hardly diagnosticated. The next few days bring an accumulation of such cases. An epidemic has made its appearance. The examination of the drinking water is urgently demanded. But typhoid fever takes some time for incubation. From the moment of the reception of the fatal germ by the first individual to that moment when the bacteriologist completes his examination, days and even weeks and months will have elapsed. This accounts for the fact that investigations almost always come too late and that the results are usually negative.

It may now be asked how the bacteriological results can be utilized, if the number of bacteria is not decisive and the establishment of their nature is but rarely satisfactory.

Entire communities as well as individuals are frequently obliged to use bad and suspicious water, though previously improved and cleaned by proper measures, i.e., freed from infectious matter. This is generally done by sand-filtration on a large scale, and for individual purposes by domestic filters, boiling, etc. But bacteriological investigation here proves its great power, for it alone can decide whether water has been deprived of its injurious elements. If any method of purifying water is certain of removing infectious matter from it, it must be able to remove all the existing micro-organisms both pathogenic and non-pathogenic, because the former can be recognized as such only with difficulty and the latter may serve as objects of comparison, and because we can trust to the measures employed only when they also prove efficacious as to the harmless bacteria.

In other words, water having gone through such a purifying process must prove free from germs through bacteriological investigation. This demand is, indeed, imperative only in reference to individual cases, for instance, in the use of domestic filters. The water supplied must in this case be completely sterile under all circumstances—a claim, by the way, not yet satisfied by any of the small filters (which, indeed, deteriorate the water instead of improving it). But water free from germs proves to be unattainable in large quantities, be cause its control is more difficult and it is more liable to contamination.

It has been agreed to regard a certain quantity of micro-organisms as an unavoidable evil, due in part to defects of apparatus, and partly to subsequent pollution. One hundred and fifty to two hundred germs in the cubic centimeter is the conventional limit which is not to be exceeded in filtered aqueduct water.

Bacteriological investigation can, therefore, in every case surely determine whether a purified water can be admitted for use or not. It would be vain to dispute this commanding position of such an investigation in favor of an older competitor. The time was when the exclusion of infectious matter from water was unknown. It had been merely ascertained that water was injurious whenever it came from localities harboring garbage of human or animal origin and exhibiting processes of decomposition, and these processes were looked upon as the direct cause of the unwholesome nature of the water. When certain chemical bodies began to appear under the same conditions, they were made use of (as long as the infectious substances themselves could not be obtained) as indicators of the occurrence of those substances. There was no danger in the quantities of chlorine, ammonia, nitrites, or organic substance as such, which were declared inadmissible by chemical examination of water; but experience had shown that a transgression of the chemical limits still permitted (established by numerous comparative observations), frequently coincided with a hygienically defective quality of the water.

We need no longer content ourselves with such shifts and indirect conclusions. Quite a number of infectious substances formerly looked for in vain are now quite well known, and it is probable that such as may yet be unknown pertain to the lowest class of micro-organisms. We may, then, disregard chlorine, ammonia, etc., and deal with the infectious substances themselves or with their nearest relatives. Chemical investigation has, therefore, forfeited its claim as to the determination of the purity of water; it is retained and practised only on the ground of thoughtless tradition.

Chemical investigation can never, and bacteriological examination only exceptionally, ascertain infectious substances as such.

Common surface water, from creeks, rivers, and lakes, is regularly liable to be contaminated, especially by human refuse and excrements. It is, therefore, regarded, by recent hygiene, as absolutely suspicious of infection and excluded from use. It does not matter whether bacteriological or chemical examination results satisfactorily, and appearance, taste, etc., are ever so inviting, germs of typhoid and cholera may have found access to the water but a few hours before and have rendered it positively dangerous.

Surface water should, therefore, be always purified before use, unless it has just sprung from the earth and could not have received any contaminating substances. But such aqueducts do not contain surface, but underground water now appearing as a spring and formerly raised artificially by wells.

The underground water as such is, as a rule, free from germs. The filtering power of the upper strata of the earth is so great that all micro-organisms are prevented from penetrating deeply. If there should be a change in these conditions, such water must be regarded with suspicion, if not actually infected. It often happens that water, faultless *per se*, is contaminated subsequently by being received in improperly constructed reservoirs accessible to pollution. It is thus changed into surface-water and just as much in need of purification as river or lake water.

There remains only underground water unobjectionably obtained, raised by tube-wells or rising naturally. It may also (as before mentioned) become infected. The ground above that water may in the first place, be wanting in filtering power owing to physical properties, coarseness of grain, etc., or because the water is too near to the surface, or a source of pollution may flow at a depth.

Can chemical or bacteriological investigation discover this fact with certainty? This question should be answered with great caution. The bacteriological results can hardly be utilized, because germs (harmless water bacteria) are almost always reproduced within the conduit-pipes, etc., and because, therefore, the number of micro-organisms does not afford any actual proof of the original condition. The difficulty of judging the species found has already been discussed. We are justified in assuming a direct connection of water with some focus of putrefaction and decomposition only when very different bacteria appear side by side, or when germs are discovered which we know come from human intestines, as, for instance, Emmerich's bacillus.

But the result of chemical examination may here be relied upon more safely. If great quantities of organic substances, of chlorine, etc., are detected in a certain water, it must be regarded with positive distrust, and its surroundings should be carefully examined. The circumstances will be determined still more clearly and hardly admit of any doubt, wheneverbacteriological and chemicals investigations agree in condemning such water.

APPENDIX.

MOULD AND YEAST FUNGI.

We will, in conclusion, briefly consider some micro-organisms nearly related to the bacteria.

It has already been stated that all plants destitute of chlorophyl are designated as "fungi," while the bacteria, by reason of their mode of reproduction by division, were contrasted as "schizomycetes," "yeast" ("sprouting"), and "mould fungi."

These latter two evidently resemble the bacteria, but differ from them in many important properties.

The mould fungi belong to the flowerless plants, the cryptogams, and constitute the subdivision of Thallophytes which bear a simple foliage (thallus). This thallus is composed of cells destitute of chlorophyl, having, like the bacteria, a membrane and protoplasmic contents, but no nuclei. They are never reproduced by transverse division (splitting), but are developed, by progressive growth of the ends, into long filaments (hyphæ) which frequently become articulated without being disconnected. These hyphæ are, besides, characterized by an early ramification uniting the threads into a compact network, the mycelium.

At the period of sporulation, some hyphæ rise from the mycelium and assume other shapes and conditions of growth as "fruit-hyphæ" ("fruit-bearers"). On these the "fruit" (the spores, also called conidia) will arise. This occurs with the single mould fungi in such a peculiar manner that it has been utilized in their classification.

The number of the various mould fungi is very large and estimated by the thousand. We shall consider only such as are particularly important to us.

The generally undivided and unarticulated hyphæ of the mucorineæ rise perpendicularly from the delicate mycelium; a sporangium rises on their top, i.e., a globular mass rich in protoplasm, a mother spore-cell, is developed, the contents of which are divided into roundish spores by numerous partition walls. The sporangium is capped toward the end by an arched plate, called columella. The sporangium perishes when the spores are ripe and only the columella hangs frequently for some length of time like an empty, overturned cap over the hypha.

The end of the hypha (likewise one-celled) of the aspergillia swells in the shape of a club (like the head of asparagus) and is then covered by a great many sterigma, flask-shaped, small structures with spores arranged like chains. The straight, articulated hyphæ of the penicillia divide, by tree-shaped, forked division in their upper third, into compact tufts of short and erect pedicles, called basidia, on which the spores lie in long rows.

Nearly related to these genuine mould fungi are a number of the lowest plants (the best known species is the oidium) which are more simply organized, both in form and structure, and form, as it were, the transition to the yeast fungi. The hyphæ are but little developed, destitute of particular fruit-heads which are even sometimes absent, so that the conidia articulate directly with the mycelium.

The real "sprouting" or yeast fungi do not, as a rule, develop true mycelial filaments. They are rather single, oval cells without chlorophyl, with a thin membrane and a granular protoplasm interspersed with vacuoles. The spores are developed within the protoplasm; they are large, irregular, roundish bodies inclosed in a membrane and set free by the dissolution of the mother-cell membrane.

The "sprouting" (yeast) fungi multiply (as indicated by their name) by sprouting. At one or more points, on the surface of a cell, there arise small, bud or button-shaped protuberances gradually increasing in size and circumference, and finally separated from the mother cell by a constriction. But they frequently remain connected with the latter, and as the same process is usually repeated in every newly-formed member, long rows of these yeast cells are joined into extensive combinations of yeast fungi.

The yeast fungi (saccharomyces) reveal their affinity with the higher fungi by peculiar deviations from the common phenomena of growth. A distinct tendency to produce mycelium is sometimes noticed, especially on solid media; the members dwindle away to short, somewhat irregular hyphæ.

The ways and means of preparing these micro-organisms for investigation essentially agree with those used for bacteria; but a few differences must be noted.

The mould fungi generally resist the common staining substances, the species of aspergillus being the most accessible. But Löffler's methyl-blue will always bring to light the mycelium and hyphæ, and sometimes even the spores whose membrane does not seem to be as thick as that of the bacteria.

The observation of the moulds in unstained condition is simple and fully satisfactory. Since the fungi are not wet by water, other means must be resorted to in their preparation. We use, for this purpose, 50% alcohol and add to it a few drops of ammonia. Tear the objects into as minute pieces as possible by means of dissecting needles; endeavor to remove the omnipresent air-bubbles, and transfer the preparations to glycerin. If it be desired to preserve them, surround the edge of the cover glass with asphalt-lacquer.

The finer peculiarities of form in the mould fungi can be seen even with moderate magnifying powers.

The same is true of the oidium and saccharomyces.

The artificial cultivation of fungi is done exactly as with bacteria. Moulds thrive better on acid media, gelatin, etc. Sterilized bread-paste is especially favorable for their development.

Some of these lower vegetable organisms are of importance, as they possess, within certain limits, pathogenic properties. Grohé, in 1870, stated, as a result of a long series of examinations, that rabbits injected with spores of mould fungi (directly into the blood-vessels) perished in a short time in consequence of an extensive moulding of their internal organs. These observations were declared to be unfounded by many, until Grawitz confirmed them in the most essential points. He started with the view that the mould fungi rarely make use of their pathogenic capacity only because they first have to become accustomed to a parasitic mode of existence foreign to them. He then endeavored to cultivate them artificially.

He was apparently successful. By a gradual change of the conditions of nourishment, he prepared the mould-fungi step by step for their new position, and saw small numbers of spores of originally benign moulds killing the test animals. Far-reaching conclusions were drawn regarding both the hyphomycetes and the bacteria.

But the structure built on these conclusions was unsafe. Koch and Gaffky showed that Grawitz had fallen into a quite pardonable error and that his results did not correspond to facts. They ascertained that there are, among the mould fungi, species which are originally and constantly pathogenic.

This fact has been placed beyond doubt and confirmed by a great number of further investigations, especially those by Lichtheim.

We now know that definite species among the aspergilleæ and mucorineæ (Aspergillus flavescens and fumigatus, Mucor corymbifer and rhizopodiformis) can become pernicious to animals.

By soaking a large quantity of the spores of these fungi in sterile bouillon, straining the cloudy mixture through fine gauze (to keep out the coarser particles) and injecting it into the jugular or, easier still, the vein of the ear of a rabbit, its death will ensue after from twenty-four to seventy-two hours.

On post-mortem examination small and whitish nodules will be found spread over all the organs, especially the kidneys and liver, proving, on microscopical examination, to be densely felted mycelial layers of the respective species of mould. The vessels of larger calibre are, here and there, distorted by the confused network of the vigorously grown filaments which, however, never develop fructifying organs, hyphæ or conidia. This will be best seen by staining the sections with Löffler's blue or Ziehl's carbol-fuchsin. Proper media, especially bread-paste, will easily reproduce, at breeding temperature, luxuriant growths of fungi from the organs.

These facts led to a closer observation of more or less extensive mycoses in man under natural conditions. They had been caused by pathogenic kinds of aspergillus and mucor. The external auditory canal, the nasal cavities and cornea, also the internal organs, intestines, lungs and brain, were seen covered with mould filaments whose germs had found entrance in some way.

Among the vegetable organisms standing mid-way between the mould and yeast fungi, some are distinguished by pathogenic or, rather, parasitic properties, as the fungi of favus, herpes, and thrush. Injurious kinds among the real yeast fungi are unknown.

PENICILLIUM GLAUCUM.

Let us now turn to a species of the fungi.

The most widely diffused among the mould fungi is the Penicillium glaucum, whose green, dense films are found everywhere. Wherever a "moulding" of any substance occurs, a Penicillium glaucum will generally be found. Investigation of air proves that the germs of this fungus are present everywhere.

Penicillium does not thrive at breeding temperatures and is, therefore, destitute of pathogenic properties.

Its colonies appear on the plate as whitish flakes which rapidly increase in size and become covered from the centre with a superficial green, indicating sporulation. The gelatin is soon liquefied around the colonies.

The peculiar little brushes formed by the hyphæ can be seen even under low powers.

On bread-paste a low, fine-flaked film of a white color forms at the beginning, but soon turns decidedly green.

Among the aspergilli, the non-pathogenic albus and glaucus

should be mentioned, which thrive at ordinary room-temperature; also the niger, thriving at breeding heat.

ASPERGILLUS FLAVESCENS.

The pathogenic Aspergillus flavescens grows almost exclusively at high temperatures; it is distinguished by large, strong "fruit-heads" and the greenish-yellow color of its cultures. It exhibits on the plate (like all the aspergilli), even with slight magnification, the fructifying organs densely covered with sterigmata and their spores resembling little thorn-apples.

ASPERGILLUS FUMIGATUS.

Aspergillus fumigatus has exceedingly delicate and neat "fruit heads" and at breeding temperature forms a film, at first bluish-green and afterward ashy-gray. This film never extends far from the surface of the gelatin. Its germs are widely diffused and very frequently found in our common bread. In the incubator, non-sterilized bread-paste is almost regularly covered with a dense culture of this mould in a very few days.

MUCOR MUCEDO.

The Mucor mucedo is the best known among the mucorineæ; it is, next to the Penicillium glaucum, the most common mould. It only grows at ordinary temperature, and, on the gelatin plate, quickly forms dense, luxuriantly growing films, whose black poppy-seed-sized "fruit-heads" can easily be perceived with the naked eye; when magnified they appear as smooth, round structures. On bread-paste there is developed a dense yellowish-brown growth of filaments shooting upward like a miniature forest.

The Mucor stolonifer has a still more striking growth and usually develops mycelium growing to some distance from the surface of the culture medium.

The species Mucor corymbifer and M. rhizopodiformis, described by Lichtheim, are pathogenic. Grown on bread in the incubator, the former appears as a dense snow-white lawn or film, looking like plucked cotton. M. rhizopodiformis grows lower and has black fruit-heads; it is readily recognized by a peculiar ethereal or aromatic odor in bread cultures, presumably caused by the fermentation of this medium.

OIDIUM LACTIS.

The Oidium lactis belongs to the class of simple mycelial fungi destitute of germinative organs.

It is found in nearly all milk, especially when it begins to sour, also almost regularly in butter. It thrives both at ordinary and breeding temperatures and is easily stained by the anilin colors.

On the gelatin plate the colonies appear as neat, white stars, growing rather rapidly, advancing to the surface, and spreading there as flat, white, dry masses.

The medium is not liquefied. Under the microscope there are seen glassy, greatly ramified hyphæ, radiating from the middle of the colony in all directions.

In the test-tube their growth is found along the entire inoculation line, particularly on the surface of the gelatin.

Here too, the ramified network of the vigorously developed fungus-lawn is seen again. The oidium thrives in milk without causing striking transformations.

TRICHOPHYTON TONSURANS AND ACHORION SCHŒNLEINII.

The investigations of Grawitz, Quincke, and others have given detailed information regarding favus and herpes tonsurans.

The regular appearance of thread-shaped structures had been demonstrated long before in the scaly aggregations produced by these two skin diseases. The fungus of favus, the Achorion Schönleinii, and that of herpes, the Trichophyton tonsurans, were the first recognized vegetable parasites of man. Grawitz succeeded not only in cultivating them outside the body, with the aid of more recent methods, but also in demonstrating beyond doubt their causative significance by the successful reproduction of skin diseases on man from artificial cultures.

On microscopic examination both micro-organisms appear as pretty well ramified, flat, filamentous fungi with clearly-articulated hyphæ. These are frequently strangely twisted in favus and distinguished by completely rectangular ramifications. Both fungi are destitute of special generative organs. A degeneration of the mycelium into small, roundish members, arranged like "rolls," may sometimes be observed, under certain conditions, and best on blood-serum at 30° C.; they are characterized as conidia. The mycelium generally remains absolutely sterile on gelatin and agar.

The favus and herpes fungi thrive at ordinary temperature, best at about 30° C. The differences appearing between them in the course of development on gelatin or agar are, indeed, not very evident; but a direct comparison between cultures will suffice for differentiation.

On the plate they develop pretty rapidly, the fungus of herpes tonsurans more so than that of favus. The colonies are of a chalky-white color, star-shaped, and lumpy in the centre. They liquefy gelatin quickly and extensively.

In the test-tube the trichophyton forms on the surface of the medium a film several millimetres thick, arranged in crusty folds, white, as if dusted with flour, its lower surface being of a sulphur yellow color. Only a restricted growth takes place in the lower strata of the gelatin. Liquefaction is less speedy with the favus fungus and the film developed is not quite so thick; the lower surface is of a brighter yellow color. On agar-agar a white and dry film forms and adheres firmly to the medium.

THRUSH.

The micro-organism of thrush is the most perfect link between the thread and yeast fungi. Under certain conditions of nutrition, for instance, almost always on the gelatin plate, on very sweet media, etc., it appears in a yeast-like form, as a pronounced saccharomyces; but under other conditions likewise, for instance, at the bottom of the test-tube cultures, it develops long thread-shaped mycelia. The gelatin is not liquefied.

The thrush fungus is pathogenic for rabbits, as proved by Klemperer's investigations. The animals perish from within twenty-four to forty-eight hours after injection of a pure culture into the blood-vessels. The internal organs are traversed by the mycelium grown out into long threads.

SACCHAROMYCES.

The common beer-yeast, Saccharomyces cerevisiæ, is the most widely diffused of the yeast fungi and corresponds to the general description of these organisms.

Various kinds of yeast fungi have been found in the examination of air. They often appear on the plates as accidental contaminations and are noticable by the color of their colonies. The "red yeast" is the most common; it produces a pale red pigment. The "white yeast" forms snow-white, lustrous cultures. None of them liquefy gelatin; all thrive at ordinary temperatures. They do not seem of any special importance and are destitute of the ability of fermenting sugar solutions, as true yeast does.

ACTINOMYCES.

The actinomyces (ray fungus), finally, is a peculiar vegetable micro-organism whose classification is still in doubt.

Whitish, rather compact swellings are sometimes noticed on the jaws of cattle. They spring from the bone, grow rapidly, and finally rupture into the mouth or outside. Nodules of similar formation are generally found in the larynx and the lymphatic glands. Numerous abscess-like foci are seen on section; they surround yellow, nearly hemp-seed-sized, rough, compact bodies. By crushing such a particle between two cover-glasses, it will fall apart in many small pieces whose composition will be shown by proper staining.

Leave the preparations for twenty-four hours in an anilin-water gentian-violet solution, or for one-half hour in hot carbol-fuchsin; then for a few minutes—about a quarter of an hour—in iodide of potassium solution and thence into alcohol, etc. It will soon be seen that the globules just described, consist of a dense mass of hypha-like anastomosing threads. They radiate uniformly in all directions from a dense centre, gradually widen toward the end, and terminate in club-shaped, very characteristic swellings. The whole thus presents the appearance of a crystallized piece of ore or a filled aster.

The same structures have not only been observed in cattle and hogs (within the striated muscles), but recently and frequently also in man. They usually give rise to extensive suppurations, peritonitis, etc., generally terminating fatally.

A successful culture of the actinomyces has recently been reported from different sources. But these statements are still doubtful because transmission from artificial cultures to animals failed.

This gap has but very recently been filled by the combined experiments of M. Wolff and J. Israel. On agar-agar, and also within raw hen's eggs—according to Hueppe's method—yellowish-white vegetations of a dense and greatly-twisted mycelium were developed. The injection of such masses into the abdominal cavity of a rabbit produced changes in the peritoneum whose actinomycotic character has been established by microscopic investigation.

INDEX.

www.ingramcontent.com/pod-product-compliance
Lightning Source LLC
Chambersburg PA
CBHW030349230426
43664CB00007BB/580
* 9 7 8 3 7 4 4 6 8 9 4 6 5 *